Practice Problems *for the* Electrical and Computer Engineering PE Exam

A Companion to the *Electrical Engineering Reference Manual*

Sixth Edition

John A. Camara, PE

Professional Publications, Inc. • Belmont, CA

How to Locate Errata and Other Updates for This Book

At Professional Publications, we do our best to bring you error-free books. But when errors do occur, we want to make sure that you know about them so they cause as little confusion as possible.

A current list of known errata and other updates for this book is available on the PPI website at **www.ppi2pass.com**. From the website home page, click on "Errata." We update the errata page as often as necessary, so check in regularly. You will also find instructions for submitting suspected errata. We are grateful to every reader who takes the time to help us improve the quality of our books by pointing out an error.

PRACTICE PROBLEMS FOR THE ELECTRICAL AND COMPUTER ENGINEERING PE EXAM
Sixth Edition

Current printing of this edition: 3

Printing History

edition number	printing number	update
6	1	New edition.
6	2	Minor corrections.
6	3	Minor corrections.

Copyright © 2002 by Professional Publications, Inc. All rights reserved. No part of this publication may be reproduced, stored in a retrieval system, or transmitted, in any form or by any means, electronic, mechanical, photocopying, recording, or otherwise, without the prior written permission of the publisher.

Printed in the United States of America

Professional Publications, Inc.
1250 Fifth Avenue, Belmont, CA 94002
(650) 593-9119
www.ppi2pass.com

ISBN 1-888577-57-6

Topics

Where do I find help solving these Practice Problems?

Practice Problems for the Electrical and Computer Engineering PE Exam presents complete, step-by-step solutions for 445 problems covering topics on the Electrical and Computer PE exam. This book is a companion to the *Electrical Engineering Reference Manual*; the subjects covered in the two books correspond chapter for chapter, and the solutions in this book often cite the *Reference Manual*. You can find all the background information you need to solve these problems—including equations, figures, charts, and tables of data—in the *Electrical Engineering Reference Manual*.

The *Electrical Engineering Reference Manual* may be purchased from Professional Publications (800-426-1178 or **www.ppi2pass.com**) or from your favorite bookstore.

Table of Contents

The topics and chapters listed here correspond to those in the *Electrical Engineering Reference Manual*, sixth edition. Some chapters do not have practice problems.

Preface to the Sixth Edition

The sixth edition of *Practice Problems for the Electrical and Computer Engineering PE Exam* is completely new, yet it owes much to its predecessors and to the outstanding guidance of Professional Publications, Inc.

When you work the problems presented here, you'll be preparing for the NCEES Principles and Practice of Engineering (PE) examination in electrical and computer engineering in the most efficient way possible. This book explores the exam topics systematically, using a broad approach that will assist you in performing well on the *breadth* session of the exam. It also gives you a comprehensive review: the problems are written to suggest areas of further study, the incorrect answers themselves can enlighten, and the correct answers are thoroughly documented. When you take full advantage of the practice problems and solutions, you'll gain knowledge essential for peak performance on the *depth* session of the exam.

If you're a student of electrical engineering, you'll find this book to be an all-inclusive review of the fundamentals of electrical engineering—useful regardless of your current topic of interest. If you're a practicing engineer, you'll find that the logic followed in the solutions can help you as you attempt to resolve similar everyday engineering problems.

Reading the *Electrical Engineering Reference Manual for the Electrical and Computer PE Exam* in conjunction with this text will greatly enhance your study efforts. This *Practice Problems* book is written as a companion to the *Reference Manual*, and the subjects covered in the two books correspond chapter for chapter. The solutions given here often cite the *Reference Manual*, so it would be helpful to have a copy on hand for quick access to background information, including equations, figures, charts, and tables of data. The *Reference Manual* includes many essential appendices of mathematical, basic theoretical, and practical information you can refer to while solving problems. It also features a glossary of commonly used electrical terms and an in-depth index to aid in the search for specific information.

The sources used in assembling the content of this book include: (1) information on the electrical and computer PE exam made public by the National Council of Examiners for Engineering and Surveying (NCEES); (2) PE exam review course material published by the National Society of Professional Engineers; (3) electrical engineering curricula at leading colleges and universities; (4) current literature in the field of electrical engineering; (5) information on the numerous electrical engineering websites on the Internet; and (6) survey comments from those who have recently taken the PE exam and/or purchased copies of previous editions of this text.

Changes from the previous edition are numerous. All are meant to enhance the enduring usefulness of the book. Here are some of the major differences between the fifth and sixth editions.

- Mathematics chapters are added for those who need an extra review before moving on to the electrical engineering information.
- Theory and Fields topics are added, both to improve readers' understanding of principles and to prepare them for changes to the PE exam structure.
- The Power topic is expanded to include Generation Systems; Transmission and Distribution Systems; and Lightning Protection and Grounding.
- The Electronics topic is updated to reflect the latest advances.
- The Computers topic is added. The Digital Systems information is expanded to include the basics of interfaces, protocols, and standards.
- The Communications information is gathered into a single topic and expanded to reflect recent progress in the field.
- A chapter on Biomedical Engineering is incorporated.
- The National Electric Code is addressed as a separate topic.
- SI units are used, except where common practice dictates the use of customary U.S. units (as in the National Electric Code topic).

Should you find an error in this book, know that it is mine, and that I regret it. Beyond that, I hope two things happen. First, please let me know about it, either by using the "Errata" section on the PPI website at **www.ppi2pass.com** or by filling out the errata card found in this book. Second, I hope you learn something from the error—I know I will! I would appreciate constructive comments and suggestions for improvement, additional questions, and recommendations for expansion so that new editions or similar texts will more nearly meet the needs of future examinees.

John A. Camara, PE

Acknowledgments for the Sixth Edition

It is with enduring gratitude that I thank Michael Lindeburg for giving me the opportunity to realize my dream of authorship of a significant engineering text. It is my hope that this book will match the comprehensive and cohesive quality of the best of Professional Publications' family of engineering books, which I have admired for two decades.

The team at Professional Publications contributed substantially to this book, making it better than it ever could have been without them. Though I don't yet know them all by name, I thank them one and all. In particular, I owe this book to the dedication and professional competence of Aline Magee. Without her encouragement, guidance, and gentle chastisement, this book would never have come to fruition. My personal thanks to the very professional Jessica Whitney-Holden, who ably guided me through production, and to her daughter, Emily, who reminded me of the important things in life.

The National Society of Professional Engineers provided questions from previous professional engineering examination review courses for my study and guidance. Michael Lindeburg provided Chapters 1–14 on mathematics, as well as the chapters on economic analysis, law, ethics, and PE registration. This saved considerable time. Additionally, there was little, if anything, I could do to improve them, so I'm grateful for their use. The late Raymond Yarbrough's fifth edition provided the basis for expansion, and numerous figures, tables, and problems were taken, with minor modifications, directly from his book. Steven B. Edington reviewed my initial Communication topic outline and greatly improved its scope and timeliness. John Goularte, Jr. taught me first-hand the applications of biomedical engineering, and the real-world practicality of the questions in this topic is attributable to him. Becky Owens provided guidance on computer fundamentals and showed me practical applications, as well as typed the majority of the index. More importantly, she was patient and supportive through the entire process, even when I wasn't.

Gregg Wagener, PE, reviewed each of the chapters, providing valuable insight, correcting numerous errors and, in general, teaching me a great deal along the way.

A very special thanks to my son, Jac Camara, whose annotated periodic table became the basis for the Electrical Materials topic. Additionally, our discussions about quantum mechanics and a variety of other topics helped stir the intellectual curiosity I thought was fading. His knowledge will one day surpass mine, and I will be proud when that day comes to pass (though he probably believes that day has already come and gone).

Thanks also to my daughter, Cassiopeia, who has been a constant source of inspiration, challenge, and pride. Her knowledge of the healing arts does exceed mine. I am sure that veterinary or medical schools will see the treasure that awaits in her.

The knowledge expressed in this book represents years of training, instruction, and self-study in electrical, electronic, mechanical, nuclear, marine, space, and a variety of other types of engineering—all of which I find fascinating. I would not understand any of it were it not for the teachers, instructors, mentors, family, and friends who've spent countless hours with me and from whom I've learned so much. A few of them include: Mrs. Mary Avila, Capt. C.E. Ellis, Mr. Claude Estes, Capt. Karl Hasslinger, Mr. Jerome Herbeck, Mr. Jack Hunnicutt, Mr. Ralph Loya, Harry Lynch, Mr. Harold Mackey, Michael O'Neal, Mark Richwine, Mr. Charles Taylor, Mrs. Abigail Thyarks, and Mr. Jim Triguerio. There are others who have touched my life in special ways: Jim, Marla, Lora, Todd, Tom, and Rick. There are many others, though here they will remain nameless, to whom I am indebted. Thanks.

Finally, thanks to my mom, for without her steadfast love little in this world would be worthwhile.

John A. Camara, PE

Codes, Handbooks, and References

This edition of *Practice Problems for the Electrical and Computer Engineering PE Exam* is based on the following codes, standards, and references. The most current versions or editions available were used. However, the PE examination is not always based on the most currently available codes, as adoption by state and local controlling authorities often lags issuance by several years.

The PPI website (**www.ppi2pass.com**) provides the dates of the codes, standards, and regulations on which NCEES has announced the current exams are based. Use this information to decide which versions of these books should be part of your exam preparation.

The minimum recommended library for the electrical exam consists of this book, the current National Electric Code, a standard handbook of electrical engineering, and two textbooks that cover fundamental circuit theory (both electrical and electronic). Numerous textbooks covering basic electrical engineering topics are listed. These texts, or their equivalent (see the topics in brackets), should be used in preparation for the examination. Adequate preparation, not an extensive portable library, is the key to success.

Codes

Code of Federal Regulations, Title 47—Telecommunications, Ch. 73—"Radio Broadcast Rules." 47CFR73. U.S. Government.[1]

National Electric Code, NFPA 70. National Fire Protection Association.

National Electric Safety Code, NESC.

Standards Organizations of Interest:

ANSI: American National Standards Institute

EIA: Electronic Industries Alliance

FCC: Federal Communications Commission

IEEE: Institute of Electrical and Electronic Engineers

ISA: Instrument Society of America

ISO: International Organization for Standardization[2]

NEMA: National Electrical Manufacturers Association

[1]The chapters within this title of the Code of Federal Regulations are known as the *FCC Rules*.
[2]ISO is not the acronym for the organization. Rather, "iso" is from the Greek word for "equal."

Handbooks

American Electrician's Handbook. Terrell Croft and Wilford I. Summers. McGraw-Hill.

CRC Materials Science and Engineering Handbook. James F. Shackelford, William Alexander, and Jun S. Park, eds. CRC Press, Inc.

Electronic Engineer's Handbook. Donald Christiansen, ed. McGraw-Hill.

McGraw-Hill Internetworking Handbook. Ed Taylor. McGraw-Hill.

National Electric Code Handbook. Mark W. Earley, Joseph V. Sheehan, and John M. Caloggero. National Fire Protection Association.

Standard Handbook for Electrical Engineers. Donald G. Fink and H. Wayne Beaty. McGraw-Hill.

The Communications Handbook. Jerry D. Gibson, ed. CRC Press, Inc.

The Computer Science and Engineering Handbook. Allen B. Tucker, Jr., ed. CRC Press, Inc.

References

CRC Standard Mathematical Tables. William H. Beyer, ed. CRC Press, Inc.

McGraw-Hill Dictionary of Scientific and Technical Terms. Sybil P. Parker, ed. McGraw-Hill.

Schaum's Outline Series (Electronics and Electrical Engineering). McGraw-Hill.

The Internet for Scientists and Engineers, Brian J. Thomas. SPIE Press & IEEE Press.

Texts

An Introduction to Digital and Analog Integrated Circuits and Applications. Sanjit K. Mitra. Harper & Row, Publishers. [Digital Circuit Fundamentals]

Applied Electromagnetics. Martin A. Plonus. McGraw-Hill. [Electromagnetic Theory]

Introduction to Computer Engineering. Taylor L. Booth. John Wiley & Sons. [Computer Design Basics]

Electrical Power Technology. Theodore Wildi. John Wiley & Sons. [Power Theory and Application]

Linear Circuits. M.E. Van Valkenburg and B.K. Kinariwala. Prentice-Hall, Inc. [AC/DC Fundamentals]

Microelectronics. Jacob Millman. McGraw-Hill. [Electronic Fundamentals]

How to Use This Book

This book is meant as a companion to the *Electrical Engineering Reference Manual*—a compendium of the electrical engineering field. Admittedly, despite its size, the *Reference Manual* can not cover all the details of electrical engineering. Nevertheless, a thorough study of that volume and its equations, terms, diagrams, and tables, as well as of the problem statements and solutions given in this *Practice Problems* book, will expose you to many electrical engineering topics. This broad overview of the field can benefit you in a number of ways, not the least of which is by thoroughly preparing you for the electrical and computer engineering PE exam. Further, the range of subjects covered allows you to tailor your review toward the topics of your choice, and to study at a level that suits your needs.

Before you start studying, read about the format and content of the PE exam in the *Electrical Engineering Reference Manual* or on the Professional Publications website at **www.ppi2pass.com**. Then, begin your studies with the fundamentals and progressively expand your review to include more specific topics. Keep in mind that *experience in problem solving is the key to success on the exam.*

Depending on your specific study methods, you may wish to study the "easier" material first to build confidence. Alternately, you might delay reading the easy material until just prior to the exam. The choice is yours.

If you are using this text as a review of a particular area of study, go directly to the topic of interest and begin. Any weakness perceived while attempting to solve the problems should prompt you to review the applicable topic in the *Reference Manual.*

Allow yourself three to six months to prepare. You'll get the most benefit out of your exam preparation efforts if you make a plan and stick to it. Reserve time to review problems from topics you've read, up until the examination date, in order to maintain maximum proficiency.

I'm confident that with diligent study, you will find the electrical engineering information you seek within these pages or within the pages of the *Reference Manual.* At the very least, these companion books will allow you to conduct an intelligent search for information. I hope they serve you well. Enjoy the adventure of learning!

John A. Camara, PE

1 Systems of Units

PRACTICE PROBLEMS

1. Convert 250°F to degrees Celsius.

(A) 115°Cblah blah blah blah blah blah blah blah blah blah blah
(B) 121°C
(C) 124°C
(D) 420°C

2. Convert the Stefan-Boltzmann constant (0.1713×10^{-8} Btu/ft^2-hr-°R^4) from English to SI units.

(A) 5.14×10^{-10} W/m^2·K^4
(B) 0.95×10^{-8} W/m^2·K^4
(C) 5.67×10^{-8} W/m^2·K^4
(D) 7.33×10^{-6} W/m^2·K^4

3. How many U.S. tons (2000 lbm per ton) of coal with a heating value of 13,000 Btu/lbm must be burned to provide as much energy as a complete nuclear conversion of 1 g of its mass? (Hint: Use Einstein's equation: $E = mc^2$.)

(A) 1.7 tons
(B) 14 tons
(C) 779 tons
(D) 3300 tons

SOLUTIONS

1. The conversion to degrees Celsius is given by

$$\begin{aligned} ^\circ\text{C} &= \left(\tfrac{5}{9}\right)(^\circ\text{F} - 32^\circ\text{F}) \\ &= \left(\tfrac{5}{9}\right)(250^\circ\text{F} - 32^\circ\text{F}) \\ &= \left(\tfrac{5}{9}\right)(218^\circ\text{F}) \\ &= \boxed{121.1^\circ\text{C}\ (121^\circ\text{C})} \end{aligned}$$

The answer is (B).

2. In U.S. customary units, the Stefan-Boltzmann constant, σ, is 0.1713×10^{-8} Btu/hr-ft^2-°R^4.

Use the following conversion factors.

$$\begin{aligned} 1\ \text{Btu/hr} &= 0.2931\ \text{W} \\ 1\ \text{ft} &= 0.3048\ \text{m} \\ T_{^\circ\text{R}} &= \tfrac{5}{9}T_{\text{K}} \end{aligned}$$

Performing the conversion gives

$$\begin{aligned} \sigma &= \left(0.1713 \times 10^{-8} \frac{\text{Btu}}{\text{hr-ft}^2\text{-}^\circ\text{R}^4}\right)\left(0.2931 \frac{\text{W}}{\frac{\text{Btu}}{\text{hr}}}\right) \\ &\quad \times \left(\frac{1\ \text{ft}}{0.3048\ \text{m}}\right)^2 \left(\frac{1^\circ\text{R}}{\frac{5}{9}\text{K}}\right)^4 \\ &= \boxed{5.67 \times 10^{-8}\ \text{W/m}^2\cdot\text{K}^4} \end{aligned}$$

The answer is (C).

3. The energy produced from the nuclear conversion of any quantity of mass is given as

$$E = mc^2$$

The speed of light, c, is 3×10^8 m/s.

For a mass of 1 g (0.001 kg),

$$\begin{aligned} E &= mc^2 \\ &= (0.001\ \text{kg})\left(3 \times 10^8\ \frac{\text{m}}{\text{s}}\right)^2 \\ &= 9 \times 10^{13}\ \text{J} \end{aligned}$$

Convert to U.S. customary units with the conversion 1 Btu = 1055 J.

$$\begin{aligned} E &= (9 \times 10^{13}\ \text{J})\left(\frac{1\ \text{Btu}}{1055\ \text{J}}\right) \\ &= 8.53 \times 10^{10}\ \text{Btu} \end{aligned}$$

The number of tons of 13,000 Btu/lbm coal is

$$\frac{8.53 \times 10^{10}\ \text{Btu}}{\left(13{,}000\ \frac{\text{Btu}}{\text{lbm}}\right)\left(2000\ \frac{\text{lbm}}{\text{ton}}\right)} = \boxed{3281\ \text{tons}\ (3300\ \text{tons})}$$

The answer is (D).

2 Energy, Work, and Power

PRACTICE PROBLEMS

There are no practice problems corresponding with Ch. 2 of the *Electrical Engineering Reference Manual.*

3 Engineering Drawing Practice

PRACTICE PROBLEMS

There are no practice problems corresponding with Ch. 3 of the *Electrical Engineering Reference Manual.*

4 Algebra

PRACTICE PROBLEMS

Series

1. Calculate the following sum.

$$\sum_{j=1}^{5}[(j+1)^2-1]$$

(A) 15
(B) 24
(C) 35
(D) 85

Logarithms

2. If every 0.1 sec a quantity increases by 0.1% of its current value, calculate the doubling time.

(A) 14 sec
(B) 70 sec
(C) 690 sec
(D) 69,000 sec

SOLUTIONS

1. Let $S_n = (j+1)^2 - 1$.

For $j = 1$,

$$S_1 = (1+1)^2 - 1 = 3$$

For $j = 2$,

$$S_2 = (2+1)^2 - 1 = 8$$

For $j = 3$,

$$S_3 = (3+1)^2 - 1 = 15$$

For $j = 4$,

$$S_4 = (4+1)^2 - 1 = 24$$

For $j = 5$,

$$S_5 = (5+1)^2 - 1 = 35$$

Substituting the above expressions gives

$$\begin{aligned}\sum_{j=1}^{5}\left((j+1)^2-1\right) &= \sum_{j=1}^{5} S_j \\ &= S_1 + S_2 + S_3 + S_4 + S_5 \\ &= 3 + 8 + 15 + 24 + 35 \\ &= \boxed{85}\end{aligned}$$

The answer is (D).

2. Let n represent the number of elapsed periods of 0.1 sec, and let y_n represent the amount present after n periods.

y_0 represents the initial quantity.

$$y_1 = 1.001y_0$$

$$y_2 = 1.001y_1 = (1.001)\big((1.001)y_0\big) = (1.001)^2 y_0$$

Therefore, by deduction,

$$y_n = (1.001)^n y_0$$

The expression for a doubling of the original quantity is

$$2y_0 = y_n$$

Substitute for y_n.

$$2y_0 = (1.001)^n y_0$$
$$2 = (1.001)^n$$

Take the logarithm of both sides.

$$\begin{aligned}\log(2) &= \log(1.001)^n \\ &= n\log(1.001)\end{aligned}$$

Solve for n.

$$n = \frac{\log(2)}{\log(1.001)} = 693.5$$

Since each period is 0.1 sec, the time is given by

$$\begin{aligned}t &= n(0.1 \text{ sec}) \\ &= (693.5)(0.1 \text{ sec}) \\ &= \boxed{69.35 \text{ sec}}\end{aligned}$$

The answer is (B).

5 Linear Algebra

PRACTICE PROBLEMS

Determinants

1. What is the determinant of matrix **A**?

$$\mathbf{A} = \begin{bmatrix} 8 & 2 & 0 & 0 \\ 2 & 8 & 2 & 0 \\ 0 & 2 & 8 & 2 \\ 0 & 0 & 2 & 4 \end{bmatrix}$$

(A) 459
(B) 832
(C) 1552
(D) 1776

Simultaneous Linear Equations

2. Use Cramer's rule to solve for the values of $x, y,$ and z that simultaneously satisfy the following equations.

$$x + y = -4$$
$$x + z - 1 = 0$$
$$2z - y + 3x = 4$$

(A) $(x, y, z) = (3, 2, 1)$
(B) $(x, y, z) = (-3, -1, 2)$
(C) $(x, y, z) = (3, -1, -3)$
(D) $(x, y, z) = (-1, -3, 2)$

SOLUTIONS

1. Expand by cofactors of the first row since there are two zeros in that row.

$$D = 8\begin{vmatrix} 8 & 2 & 0 \\ 2 & 8 & 2 \\ 0 & 2 & 4 \end{vmatrix} - 2\begin{vmatrix} 2 & 0 & 0 \\ 2 & 8 & 2 \\ 0 & 2 & 4 \end{vmatrix} + 0 - 0$$

by first row:

$$\begin{vmatrix} 8 & 2 & 0 \\ 2 & 8 & 2 \\ 0 & 2 & 4 \end{vmatrix} = (8)[(8)(4) - (2)(2)] - (2)[(2)(4) - (2)(0)] = (8)(28) - (2)(8) = 208$$

by first row:

$$\begin{vmatrix} 2 & 0 & 0 \\ 2 & 8 & 2 \\ 0 & 2 & 4 \end{vmatrix} = (2)[(8)(4) - (2)(2)] = 56$$

$$D = (8)(208) - (2)(56) = \boxed{1552}$$

The answer is (C).

2. Rearrange the equations.

$$\begin{aligned} x + y \quad\quad &= -4 \\ x \quad + z &= 1 \\ 3x - y + 2z &= 4 \end{aligned}$$

Write the set of equations in matrix form: $\mathbf{AX} = \mathbf{B}$.

$$\begin{bmatrix} 1 & 1 & 0 \\ 1 & 0 & 1 \\ 3 & -1 & 2 \end{bmatrix}\begin{bmatrix} x \\ y \\ z \end{bmatrix} = \begin{bmatrix} -4 \\ 1 \\ 4 \end{bmatrix}$$

Find the determinant of the matrix **A**.

$$|\mathbf{A}| = \begin{vmatrix} 1 & 1 & 0 \\ 1 & 0 & 1 \\ 3 & -1 & 2 \end{vmatrix}$$

$$\begin{aligned} &= 1\begin{vmatrix} 0 & 1 \\ -1 & 2 \end{vmatrix} - 1\begin{vmatrix} 1 & 0 \\ -1 & 2 \end{vmatrix} + 3\begin{vmatrix} 1 & 0 \\ 0 & 1 \end{vmatrix} \\ &= (1)\big((0)(2) - (1)(-1)\big) \\ &\quad - (1)\big((1)(2) - (-1)(0)\big) \\ &\quad + (3)\big((1)(1) - (0)(0)\big) \\ &= (1)(1) - (1)(2) + (3)(1) \\ &= 1 - 2 + 3 \\ &= 2 \end{aligned}$$

Find the determinant of the substitutional matrix $\mathbf{A}_1$.

$$|\mathbf{A}_1| = \begin{vmatrix} -4 & 1 & 0 \\ 1 & 0 & 1 \\ 4 & -1 & 2 \end{vmatrix}$$

$$\begin{aligned} &= -4\begin{vmatrix} 0 & 1 \\ -1 & 2 \end{vmatrix} - 1\begin{vmatrix} 1 & 0 \\ -1 & 2 \end{vmatrix} + 4\begin{vmatrix} 1 & 0 \\ 0 & 1 \end{vmatrix} \\ &= (-4)\big((0)(2) - (1)(-1)\big) \\ &\quad - (1)\big((1)(2) - (-1)(0)\big) \\ &\quad + (4)\big((1)(1) - (0)(0)\big) \\ &= (-4)(1) - (1)(2) + (4)(1) \\ &= -4 - 2 + 4 \\ &= -2 \end{aligned}$$

Find the determinant of the substitutional matrix $\mathbf{A}_2$.

$$|\mathbf{A}_2| = \begin{vmatrix} 1 & -4 & 0 \\ 1 & 1 & 1 \\ 3 & 4 & 2 \end{vmatrix}$$

$$\begin{aligned} &= 1\begin{vmatrix} 1 & 1 \\ 4 & 2 \end{vmatrix} - 1\begin{vmatrix} -4 & 0 \\ 4 & 2 \end{vmatrix} + 3\begin{vmatrix} -4 & 0 \\ 1 & 1 \end{vmatrix} \\ &= (1)\big((1)(2) - (4)(1)\big) \\ &\quad - (1)\big((-4)(2) - (4)(0)\big) \\ &\quad + (3)\big((-4)(1) - (1)(0)\big) \\ &= (1)(-2) - (1)(-8) + (3)(-4) \\ &= -2 + 8 - 12 \\ &= -6 \end{aligned}$$

Find the determinant of the substitutional matrix $\mathbf{A}_3$.

$$|\mathbf{A}_3| = \begin{vmatrix} 1 & 1 & -4 \\ 1 & 0 & 1 \\ 3 & -1 & 4 \end{vmatrix}$$

$$\begin{aligned} &= 1\begin{vmatrix} 0 & 1 \\ -1 & 4 \end{vmatrix} - 1\begin{vmatrix} 1 & -4 \\ -1 & 4 \end{vmatrix} + 3\begin{vmatrix} 1 & -4 \\ 0 & 0 \end{vmatrix} \\ &= (1)\big((0)(4) - (-1)(1)\big) \\ &\quad - (1)\big((1)(4) - (-1)(-4)\big) \\ &\quad + (3)\big((1)(1) - (0)(-4)\big) \\ &= (1)(1) - (1)(0) + (3)(1) \\ &= 1 - 0 + 3 \\ &= 4 \end{aligned}$$

Use Cramer's rule.

$$x = \frac{|\mathbf{A}_1|}{|\mathbf{A}|} = \frac{-2}{2} = \boxed{-1}$$

$$y = \frac{|\mathbf{A}_2|}{|\mathbf{A}|} = \frac{-6}{2} = \boxed{-3}$$

$$z = \frac{|\mathbf{A}_3|}{|\mathbf{A}|} = \frac{4}{2} = \boxed{2}$$

The answer is (D).

6 Vectors

PRACTICE PROBLEMS

There are no practice problems corresponding with Ch. 6 of the *Electrical Engineering Reference Manual.*

7 Trigonometry

PRACTICE PROBLEMS

1. A 5 lbm (5 kg) block sits on a 20° incline without slipping. (a) Draw the freebody with respect to axes parallel and perpendicular to the surface of the incline. (b) Determine the magnitude of the frictional force on the block.

(A) 1.71 lbf (16.8 N)
(B) 3.35 lbf (32.9 N)
(C) 4.70 lbf (46.1 N)
(D) 5.00 lbf (49.1 N)

SOLUTIONS

1. (a)

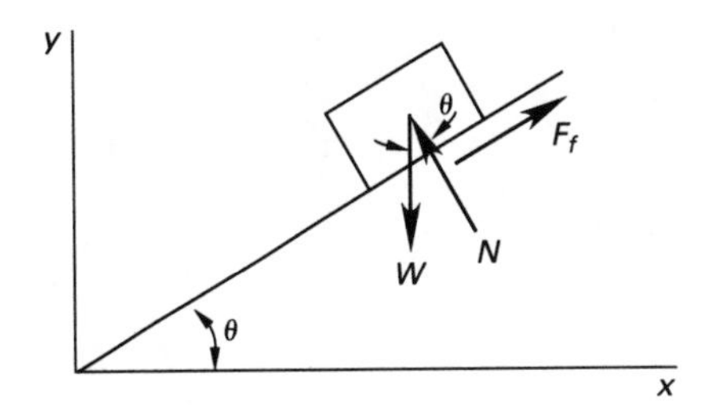

Customary U.S. Solution

(b) The mass of the block is $m = 5$ lbm.

The angle of inclination is $\theta = 20°$. The weight is

$$W = \frac{mg}{g_c} = \frac{(5\ \text{lbm})\left(32.2\ \frac{\text{ft}}{\text{sec}^2}\right)}{32.2\ \frac{\text{lbm-ft}}{\text{lbf-sec}^2}} = 5\ \text{lbf}$$

The frictional force is

$$F_f = W\sin\theta = (5\ \text{lbf})(\sin 20°) = \boxed{1.71\ \text{lbf}}$$

The answer is (A).

SI Solution

(b) The mass of the block is $m = 5$ kg.

The angle of inclination is $\theta = 20°$. The gravitational force is

$$W = mg = (5\ \text{kg})\left(9.81\ \frac{\text{m}}{\text{s}^2}\right) = 49.1\ \text{N}$$

The frictional force is

$$F_f = W\sin\theta = (49.1\ \text{N})(\sin 20°) = \boxed{16.8\ \text{N}}$$

The answer is (A).

8 Analytic Geometry

PRACTICE PROBLEMS

1. The diameter of a sphere and the base of a cone are equal. What percentage of that diameter must the cone's height be so that both volumes are equal?

(A) 133%
(B) 150%
(C) 166%
(D) 200%

SOLUTIONS

1. Let d be the diameter of the sphere and the base of the cone.

The volume of the sphere is given by

$$\begin{aligned} V_{\text{sphere}} &= \left(\tfrac{4}{3}\right)\pi r^3 = \left(\tfrac{4}{3}\right)\pi\left(\frac{d}{2}\right)^3 \\ &= \left(\frac{\pi}{6}\right)d^3 \end{aligned}$$

The volume of the circular cone is given by

$$\begin{aligned} V_{\text{cone}} &= \left(\tfrac{1}{3}\right)\pi r^2 h = \left(\tfrac{1}{3}\right)\pi\left(\frac{d}{2}\right)^2 h \\ &= \left(\frac{\pi}{12}\right)d^2 h \end{aligned}$$

Since the volume of the sphere and cone are equal,

$$\begin{aligned} V_{\text{cone}} &= V_{\text{sphere}} \\ \left(\frac{\pi}{12}\right)d^2 h &= \left(\frac{\pi}{6}\right)d^3 \\ h &= 2d \end{aligned}$$

The height of the cone must be 200% of the diameter.

The answer is (D).

9 Differential Calculus

PRACTICE PROBLEMS

1. What are the values of a, b, and c in the following expression such that $n(\infty) = 100$, $n(0) = 10$, and $dn(0)/dt = 0.5$?

$$n(t) = \frac{a}{1 + be^{ct}}$$

(A) $a = 10,\ b = 9,\ c = 1$
(B) $a = 100,\ b = 10,\ c = 1.5$
(C) $a = 100,\ b = 9,\ c = -0.056$
(D) $a = 1000,\ b = 10,\ c = 0.056$

2. Find all minima, maxima, and inflection points for

$$y = x^3 - 9x^2 - 3$$

(A) maximum at $x = 0$
inflection at $x = 3$
minimum at $x = 6$
(B) maximum at $x = 0$
inflection at $x = -3$
minimum at $x = -6$
(C) minimum at $x = 0$
inflection at $x = 3$
maximum at $x = 6$
(D) minimum at $x = 3$
inflection at $x = 0$
maximum at $x = -3$

SOLUTIONS

1. If c is positive, then $n(\infty) = 0$, which is contrary to the data given. Therefore, $c \leq 0$. If $c = 0$, then $n(\infty) = a/(1 + b) = 100$, which is possible depending on a and b. However, $n(0) = a/(1+b)$ would also equal 100, which is contrary to the given data. Therefore, $c \neq 0$.

c must be less than 0.

Since $c < 0$, then $n(\infty) = a$, so $\boxed{a = 100.}$

Applying the condition $t = 0$ gives

$$n(0) = \frac{a}{1+b} = 10$$

Since $a = 100$,

$$n(0) = \frac{100}{1+b}$$
$$100 = (10)(1+b)$$
$$10 = 1+b$$
$$\boxed{b = 9}$$

Substitute the results for a and b into the expression.

$$n(t) = \frac{100}{1 + 9e^{ct}}$$

Take the first derivative.

$$\frac{d}{dt}n(t) = \left(\frac{100}{(1 + 9e^{ct})^2}\right)(-9ce^{ct})$$

Apply the initial condition.

$$\frac{d}{dt}n(0) = \left(\frac{100}{(1 + 9e^{c(0)})^2}\right)\left(-9ce^{c(0)}\right) = 0.5$$
$$\left(\frac{100}{(1+9)^2}\right)(-9c) = 0.5$$
$$(1)(-9c) = 0.5$$
$$c = \frac{-0.5}{9}$$
$$= \boxed{-0.0556}$$

The answer is (C).

Substitute the terms a, b, and c into the expression.

$$n(t) = \frac{100}{1 + 9e^{-0.0556t}}$$

2. Determine the critical points by taking the first derivative of the function and setting it equal to zero.

$$\begin{aligned} \frac{dy}{dx} &= 3x^2 - 18x = 3x(x-6) \\ 3x(x-6) &= 0 \\ x(x-6) &= 0 \end{aligned}$$

The critical points are located at $x = 0$ and $x = 6$.

Determine the inflection points by setting the second derivative equal to zero. Take the second derivative.

$$\begin{aligned} \frac{d^2y}{dx^2} &= \left(\frac{d}{dx}\right)\left(\frac{dy}{dx}\right) = \frac{d}{dx}(3x^2 - 18x) \\ &= 6x - 18 \end{aligned}$$

Set the second derivative equal to zero.

$$\begin{aligned} \frac{d^2y}{dx^2} &= 0 = 6x - 18 = (6)(x-3) \\ (6)(x-3) &= 0 \\ x - 3 &= 0 \\ x &= 3 \end{aligned}$$

This inflection point is at $x = 3$.

Determine the local maximum and minimum by substituting the critical points into the expression for the second derivative.

At the critical point $x = 0$,

$$\begin{aligned} \left.\frac{d^2y}{dx^2}\right|_{x=0} &= (6)(x-3) = (6)(0-3) \\ &= -18 \end{aligned}$$

Since $-18 < 0$, $x = 0$ is a local maximum.

At the critical point $x = 6$,

$$\begin{aligned} \left.\frac{d^2y}{dx^2}\right|_{x=6} &= (6)(x-3) = (6)(6-3) \\ &= 18 \end{aligned}$$

Since $18 > 0$, $x = 6$ is a local minimum.

The answer is (A).

10 Integral Calculus

PRACTICE PROBLEMS

There are no practice problems corresponding with Ch. 10 of the *Electrical Engineering Reference Manual.*

11 Differential Equations

PRACTICE PROBLEMS

1. Solve the following differential equation for y.

$$y'' - 4y' - 12y = 0$$

(A) $A_1e^{6x} + A_2e^{-2x}$
(B) $A_1e^{-6x} + A_2e^{2x}$
(C) $A_1e^{6x} + A_2e^{2x}$
(D) $A_1e^{-6x} + A_2e^{-2x}$

2. Solve the following differential equation for y.

$$y' - y = 2xe^{2x} \qquad y(0) = 1$$

(A) $y = 2e^{-2x}(x-1) + 3e^{-x}$
(B) $y = 2e^{2x}(x-1) + 3e^{x}$
(C) $y = -2e^{-2x}(x-1) + 3e^{-x}$
(D) $y = 2e^{2x}(x-1) + 3e^{-x}$

3. The oscillation exhibited by the top story of a certain building in free motion is given by the following differential equation.

$$x'' + 2x' + 2x = 0 \qquad x(0) = 0 \qquad x'(0) = 1$$

(a) What is x as a function of time?
(A) $e^{-2t}\sin t$
(B) $e^{t}\sin t$
(C) $e^{-t}\sin t$
(D) $e^{-t}\sin t + e^{-t}\cos t$

(b) What is the building's fundamental natural frequency of vibration?
(A) $1/2$
(B) 1
(C) $\sqrt{2}$
(D) 2

(c) What is the amplitude of oscillation?
(A) 0.32
(B) 0.54
(C) 1.7
(D) 6.6

(d) What is x as a function of time if a lateral wind load is applied with a form of $\sin t$?
(A) $\frac{6}{5}e^{-t}\sin t + \frac{2}{5}e^{-t}\cos t$
(B) $\frac{6}{5}e^{t}\sin t + \frac{2}{5}e^{t}\cos t$
(C) $\frac{2}{5}e^{-t}\sin t + \frac{6}{5}e^{-t}\cos t + \frac{2}{5}\sin t - \frac{1}{5}\cos t$
(D) $\frac{6}{5}e^{-t}\sin t + \frac{2}{5}e^{-t}\cos t + \frac{1}{5}\sin t - \frac{2}{5}\cos t$

4. (*Time limit: one hour*) A 90 lbm (40 kg) bag of a chemical is accidentally dropped in an aerating lagoon. The chemical is water soluble and nonreacting. The lagoon is 120 ft (35 m) in diameter and filled to a depth of 10 ft (3 m). The aerators circulate and distribute the chemical evenly throughout the lagoon.

Water enters the lagoon at a rate of 30 gal/min (115 L/min). Fully mixed water is pumped into a reservoir at a rate of 30 gal/min (115 L/min).

The established safe concentration of this chemical is 1 ppb (part per billion). How many days will it take for the concentration of the discharge water to reach this level?
(A) 25 days
(B) 50 days
(C) 100 days
(D) 200 days

5. A tank contains 100 gal (100 L) of brine made by dissolving 60 lbm (60 kg) of salt in pure water. Salt water with a concentration of 1 lbm/gal (1 kg/L) enters the tank at a rate of 2 gal/min (2 L/min). A well-stirred mixture is drawn from the tank at a rate of 3 gal/min (3 L/min). Find the mass of salt in the tank after 1 hr.
(A) 13 lbm (13 kg)
(B) 37 lbm (37 kg)
(C) 43 lbm (43 kg)
(D) 51 lbm (51 kg)

SOLUTIONS

1. Obtain the characteristic equation by replacing each derivative with a polynomial term of equal degree.

$$r^2 - 4r - 12 = 0$$

Factor the characteristic equation.

$$(r-6)(r+2) = 0$$

The roots are $r_1 = 6$ and $r_2 = -2$.

Since the roots are real and distinct, the solution is

$$\begin{aligned} y &= A_1 e^{r_1 x} + A_2 e^{r_2 x} \\ &= \boxed{A_1 e^{6x} + A_2 e^{-2x}} \end{aligned}$$

The answer is (A).

2. The equation is a first-order linear differential equation of the form

$$\begin{aligned} y' + p(x)y &= g(x) \\ p(x) &= -1 \\ g(x) &= 2xe^{2x} \end{aligned}$$

The integration factor $u(x)$ is given by

$$\begin{aligned} u(x) &= \exp\left(\int p(x)dx\right) \\ &= \exp\left(\int (-1)dx\right) \\ &= e^{-x} \end{aligned}$$

The closed form of the solution is given by

$$\begin{aligned} y &= \left(\frac{1}{u(x)}\right)\left(\int u(x)g(x)dx + C\right) \\ &= \left(\frac{1}{e^{-x}}\right)\left(\int (e^{-x})(2xe^{2x})dx + C\right) \\ &= e^x\left(2(xe^x - e^x) + C\right) \\ &= e^x\left(2e^x(x-1) + C\right) \\ &= 2e^{2x}(x-1) + Ce^x \end{aligned}$$

Apply the initial condition $y(0) = 1$ to obtain the integration constant C.

$$\begin{aligned} y(0) = 2e^{(2)(0)}(0-1) + Ce^0 &= 1 \\ (2)(1)(-1) + C(1) &= 1 \\ -2 + C &= 1 \\ C &= 3 \end{aligned}$$

Substituting in the value for the integration constant C, the solution is

$$y = \boxed{2e^{2x}(x-1) + 3e^x}$$

The answer is (B).

3. (a) The differential equation is a homogeneous second-order linear differential equation with constant coefficients. Write the characteristic equation.

$$r^2 + 2r + 2 = 0$$

This is a quadratic equation of the form $ar^2 + br + c = 0$ where $a = 1$, $b = 2$, and $c = 2$.

Solve for r.

$$\begin{aligned} r &= \frac{-b \pm \sqrt{b^2 - 4ac}}{2a} \\ &= \frac{-2 \pm \sqrt{(2)^2 - (4)(1)(2)}}{(2)(1)} \\ &= \frac{-2 \pm \sqrt{4-8}}{2} \\ &= \frac{-2 \pm \sqrt{-4}}{2} \\ &= \frac{-2 \pm 2\sqrt{-1}}{2} \\ &= -1 \pm \sqrt{-1} \\ &= -1 \pm i \end{aligned}$$

$$r_1 = -1 + i \text{ and } r_2 = -1 - i$$

Since the roots are imaginary and of the form $\alpha + i\omega$ and $\alpha - i\omega$ where $\alpha = -1$ and $\omega = 1$, the general form of the solution is given by

$$\begin{aligned} x(t) &= A_1 e^{\alpha t}\cos\omega t + A_2 e^{\alpha t}\sin\omega t \\ &= A_1 e^{-1t}\cos(1t) + A_2 e^{-1t}\sin(1t) \\ &= A_1 e^{-t}\cos t + A_2 e^{-t}\sin t \end{aligned}$$

Apply the initial conditions $x(0) = 0$ and $x'(0) = 1$ to solve for A_1 and A_2.

First, apply the initial condition $x(0) = 0$.

$$\begin{aligned} x(t) = A_1 e^0 \cos 0 + A_2 e^0 \sin 0 &= 0 \\ A_1(1)(1) + A_2(1)(0) &= 0 \\ A_1 &= 0 \end{aligned}$$

Substituting, the solution of the differential equation becomes

$$x(t) = A_2 e^{-t}\sin t$$

To apply the second initial condition, take the first derivative.

$$\begin{aligned} x'(t) &= \frac{d}{dt}(A_2 e^{-t} \sin t) \\ &= A_2 \frac{d}{dt}(e^{-t} \sin t) \\ &= A_2 \left(\sin t \frac{d}{dt}(e^{-t}) + e^{-t} \frac{d}{dt} \sin t \right) \\ &= A_2 \Big(\sin t(-e^{-t}) + e^{-t}(\cos t) \Big) \\ &= A_2(e^{-t})(-\sin t + \cos t) \end{aligned}$$

Apply the initial condition, $x'(0) = 1$.

$$\begin{aligned} x(0) = A_2(e^0)(-\sin 0 + \cos 0) &= 1 \\ A_2(1)(0+1) &= 1 \\ A_2 &= 1 \end{aligned}$$

The solution is

$$\begin{aligned} x(t) &= A_2 e^{-t} \sin t \\ &= (1)e^{-t} \sin t \\ &= \boxed{e^{-t} \sin t} \end{aligned}$$

The answer is (C).

(b) To determine the natural frequency, set the damping term to zero. The equation has the form

$$x'' + 2x = 0$$

This equation has a general solution of the form

$$x(t) = x_0 \cos \omega t + \left(\frac{v_0}{\omega}\right) \sin \omega t$$

ω is the natural frequency.

Given the equation $x'' + 2x = 0$, the characteristic equation is

$$\begin{aligned} r^2 + 2 &= 0 \\ r &= \sqrt{-2} \\ &= \pm\sqrt{2}i \end{aligned}$$

Since the roots are imaginary and of the form $\alpha + i\omega$ and $\alpha - i\omega$ where $\alpha = 0$ and $\omega = \sqrt{2}$, the general form of the solution is given by

$$\begin{aligned} x(t) &= A_1 e^{\alpha t} \cos \omega t + A_2 e^{\alpha t} \sin \omega t \\ &= A_1 e^{0t} \cos \sqrt{2}t + A_2 e^{0t} \sin \sqrt{2}t \\ &= A_1(1) \cos \sqrt{2}t + A_2(1) \sin \sqrt{2}t \\ &= A_1 \cos \sqrt{2}t + A_2 \sin \sqrt{2}t \end{aligned}$$

Apply the initial conditions, $x(0) = 0$ and $x'(0) = 1$ to solve for A_1 and A_2. Applying the initial condition $x(0) = 0$ gives

$$\begin{aligned} x(0) = A_1 \cos\big((\sqrt{2})(0)\big) + A_2 \sin(\sqrt{2})(0) &= 0 \\ A_1 \cos 0 + A_2 \sin 0 &= 0 \\ A_1(1) + A_2(0) &= 0 \\ A_1 &= 0 \end{aligned}$$

Substituting, the solution of the different equation becomes

$$x(t) = A_2 \sin \sqrt{2}t$$

To apply the second initial condition, take the first derivative.

$$\begin{aligned} x'(t) &= \frac{d}{dt}(A_2 \sin \sqrt{2}t) \\ &= A_2\sqrt{2} \cos \sqrt{2}t \end{aligned}$$

Apply the second initial condition, $x'(0) = 1$.

$$\begin{aligned} x'(0) = A_2\sqrt{2}\cos(\sqrt{2})(0) &= 1 \\ A_2\sqrt{2}\cos(0) &= 1 \\ A_2(\sqrt{2})(1) &= 1 \\ A_2\sqrt{2} &= 1 \\ A_2 = \frac{1}{\sqrt{2}} &= \frac{\sqrt{2}}{2} \end{aligned}$$

Substituting, the undamped solution becomes

$$x(t) = \left(\frac{\sqrt{2}}{2}\right) \sin \sqrt{2}t$$

Therefore, the undamped natural frequency is

$$\boxed{\omega = \sqrt{2}}$$

The answer is (C).

(c) The amplitude of the oscillation is the maximum displacement.

Take the derivative of the solution, $x(t) = e^{-t} \sin t$.

$$\begin{aligned} x'(t) &= \frac{d}{dt}(e^{-t} \sin t) \\ &= \sin t \frac{d}{dt}(e^{-t}) + e^{-t} \frac{d}{dt} \sin t \\ &= \sin(t)(-e^{-t}) + e^{-t} \cos t \\ &= e^{-t}(\cos t - \sin t) \end{aligned}$$

The maximum displacement occurs at $x'(t) = 0$.

Since $e^{-t} \neq 0$ except as t approaches infinity,

$$\begin{aligned}\cos t - \sin t &= 0\\ \tan t &= 1\\ t &= \tan^{-1}(1)\\ &= 0.785 \text{ rad}\end{aligned}$$

At t= 0.785 rad, the displacement is maximum. Substitute into the orginal solution to obtain a value for the maximum displacement.

$$\begin{aligned}x(0.785) &= e^{-0.785}\sin(0.785)\\ &= 0.322\end{aligned}$$

The amplitude is $\boxed{0.322.}$

The answer is (A).

(d) (An alternative solution using Laplace transforms follows this solution.) The application of a lateral wind load with the form $\sin t$ revises the differential equation to the form

$$x'' + 2x' + 2x = \sin t$$

Express the solution as the sum of the complementary x_c and particular x_p solutions.

$$x(t) = x_c(t) + x_p(t)$$

From part (a),

$$x_c(t) = A_1e^{-t}\cos t + A_2e^{-t}\sin t$$

The general form of the particular solution is given by

$$x_p(t) = x^s(A_3\cos t + A_4\sin t)$$

Determine the value of s; check to see if the terms of the particular solution solve the homogeneous equation.

Examine the term $A_3\cos(t)$.

Take the first derivative.

$$\frac{d}{dt}(A_3\cos t) = -A_3\sin t$$

Take the second derivative.

$$\begin{aligned}\frac{d}{dt}\left(\frac{d}{dt}(A_3\cos t)\right) &= \frac{d}{dt}(-A_3\sin t)\\ &= -A_3\cos t\end{aligned}$$

Substitute the terms into the homogeneous equation.

$$\begin{aligned}x'' + 2x' + 2x &= -A_3\cos t + (2)(-A_3\sin t)\\ &\quad + (2)(-A_3\cos t)\\ &= A_3\cos t - 2A_3\sin t\\ &\neq 0\end{aligned}$$

Except for the trival solution $A_3 = 0$, the term $A_3\cos t$ does not solve the homogeneous equation.

Examine the second term $A_4\sin t$.

Take the first derivative.

$$\frac{d}{dt}(A_4\sin t) = A_4\cos t$$

Take the second derivative.

$$\begin{aligned}\frac{d}{dt}\left(\frac{d}{dt}(A_4\sin t)\right) &= \frac{d}{dt}(A_4\cos t)\\ &= -A_4\sin t\end{aligned}$$

Substitute the terms into the homogeneous equation.

$$\begin{aligned}x'' + 2x' + 2x &= -A_4\sin t + (2)(A_4\cos t)\\ &\quad + (2)(A_4\sin t)\\ &= A_4\sin t + 2A_4\cos t\\ &\neq 0\end{aligned}$$

Except for the trival solution $A_4 = 0$, the term $A_4\sin t$ does not solve the homogeneous equation.

Neither of the terms satisfies the homogeneous equation $s = 0$; therefore, the particular solution is of the form

$$x_p(t) = A_3\cos t + A_4\sin t$$

Use the method of undetermined coefficients to solve for A_3 and A_4. Take the first derivative.

$$\begin{aligned}x_p'(t) &= \frac{d}{dt}(A_3\cos t + A_4\sin t)\\ &= -A_3\sin t + A_4\cos t\end{aligned}$$

Take the second derivative.

$$\begin{aligned}x_p''(t) &= \frac{d}{dt}\left(\frac{d}{dt}(A_3\cos t + A_4\sin t)\right)\\ &= \frac{d}{dt}(-A_3\sin t + A_4\cos t)\\ &= -A_3\cos t - A_4\sin t\end{aligned}$$

Substitute the expressions for the derivatives into the differential equation.

$$\begin{aligned}x'' + 2x' + 2x &= (-A_3\cos t - A_4\sin t)\\ &\quad + (2)(-A_3\sin t + A_4\cos t)\\ &\quad + (2)(A_3\cos t + A_4\sin t)\\ &= \sin t\end{aligned}$$

Rearranging terms gives

$$(-A_3 + 2A_4 + 2A_3)\cos t + (-A_4 - 2A_3 + 2A_4)\sin t = \sin t$$
$$(A_3 + 2A_4)\cos t + (-2A_3 + A_4)\sin t = \sin t$$

Equating coefficients gives

$$A_3 + 2A_4 = 0$$
$$-2A_3 + A_4 = 1$$

Multiplying the first equation by 2 and adding equations gives

$$\begin{aligned} A_3 + 2A_4 &= 0 \\ +(-2A_3 + A_4) &= 1 \\ \hline 5A_4 = 1 \text{ or } A_4 &= \tfrac{1}{5} \end{aligned}$$

From the first equation for $A_4 = {}^1\!/_5$, $A_3 + (2)({}^1\!/_5) = 0$ and $A_3 = -{}^2\!/_5$.

Substituting for the coefficients, the particular solution becomes

$$x_p(t) = -\tfrac{2}{5}\cos t + \tfrac{1}{5}\sin t$$

Combining the complementary and particular solutions gives

$$\begin{aligned} x(t) &= x_c(t) + x_p(t) \\ &= A_1 e^{-t}\cos t + A_2 e^{-t}\sin t - \tfrac{2}{5}\cos t + \tfrac{1}{5}\sin t \end{aligned}$$

Apply the initial conditions to solve for the coefficients A_1 and A_2; then apply the first initial condition, $x(0) = 0$.

$$\begin{aligned} x(t) = A_1 e^0 \cos 0 + A_2 e^0 \sin 0 \\ - \tfrac{2}{5}\cos 0 + \tfrac{1}{5}\sin 0 &= 0 \\ A_1(1)(1) + A_2(1)(0) + \left(-\tfrac{2}{5}\right)(1) + \left(\tfrac{1}{5}\right)(0) &= 0 \\ A_1 - \tfrac{2}{5} &= 0 \\ A_1 &= \tfrac{2}{5} \end{aligned}$$

Substituting for A_1, the solution becomes

$$x(t) = \tfrac{2}{5}e^{-t}\cos t + A_2 e^{-t}\sin t - \tfrac{2}{5}\cos t + \tfrac{1}{5}\sin t$$

Take the first derivative.

$$\begin{aligned} x'(t) &= \frac{d}{dt}\left(\tfrac{2}{5}e^{-t}\cos t + A_2 e^{-t}\sin t\right) \\ &\quad + \left(\left(-\tfrac{2}{5}\right)\cos t + \tfrac{1}{5}\sin t\right) \\ &= \left(\tfrac{2}{5}\right)\left(-e^{-t}\cos t - e^{-t}\sin t\right) \\ &\quad + A_2\left(-e^{-t}\sin t + e^{-t}\cos t\right) \\ &\quad + \left(-\tfrac{2}{5}\right)(-\sin t) + \tfrac{1}{5}\cos t \end{aligned}$$

Apply the second initial condition, $x'(0) = 1$.

$$\begin{aligned} x'(0) &= \left(\tfrac{2}{5}\right)\left(-e^0\cos 0 - e^0 \sin 0\right) \\ &\quad + A_2\left(-e^0 \sin 0 + e^0\cos 0\right) \\ &\quad + \left(-\tfrac{2}{5}\right)(-\sin 0) + \tfrac{1}{5}\cos 0 \\ &= 1 \end{aligned}$$

$$\begin{aligned} \left(\tfrac{2}{5}\right)\left(-(1)(1) - (1)(0)\right) + A_2\big(-(1)(0) \\ + (1)(1)\big) + \left(-\tfrac{2}{5}\right)(0) + \left(\tfrac{1}{5}\right)(1) &= 1 \\ \left(\tfrac{2}{5}\right)(-1) + A_2(1) + \left(\tfrac{1}{5}\right) &= 1 \\ A_2 &= \tfrac{6}{5} \end{aligned}$$

Substituting for A_2, the solution becomes

$$x(t) = \boxed{\begin{aligned} &\tfrac{2}{5}e^{-t}\cos t + \tfrac{6}{5}e^{-t}\sin t \\ &- \tfrac{2}{5}\cos t + \tfrac{1}{5}\sin t \end{aligned}}$$

The answer is (D).

(d) *Alternate solution:*

Use the Laplace transform method.

$$x'' + 2x' + 2x = \sin t$$

$$\mathcal{L}(x'') + 2\mathcal{L}(x') + 2\mathcal{L}(x) = \mathcal{L}(\sin t)$$

$$s^2\mathcal{L}(x) - 1 + 2s\mathcal{L}(x) + 2\mathcal{L}(x) = \frac{1}{s^2 + 1}$$

$$\mathcal{L}(x)(s^2 + 2s + 2) - 1 = \frac{1}{s^2 + 1}$$

$$\begin{aligned} \mathcal{L}(x) &= \frac{1}{s^2 + 2s + 2} + \frac{1}{(s^2 + 1)(s^2 + 2s + 2)} \\ &= \frac{1}{(s + 1)^2 + 1} + \frac{1}{(s^2 + 1)(s^2 + 2s + 2)} \end{aligned}$$

Use partial fractions to expand the second term.

$$\frac{1}{(s^2 + 1)(s^2 + 2s + 2)} = \frac{A_1 + B_1 s}{s^2 + 1} + \frac{A_2 + B_2 s}{s^2 + 2s + 2}$$

Cross multiply.

$$= \frac{\begin{aligned} &A_1 s^2 + 2A_1 s + 2A_1 + B_1 s^3 + 2B_1 s^2 \\ &+ 2B_1 s + A_2 s^2 + A_2 + B_2 s^3 + B_2 s \end{aligned}}{(s^2 + 1)(s^2 + 2s + 2)}$$

$$= \frac{\begin{aligned} &s^3(B_1 + B_2) + s^2(A_1 + A_2 + 2B_1) \\ &+ s(2A_1 + 2B_1 + B_2) + 2A_1 + A_2 \end{aligned}}{(s^2 + 1)(s^2 + 2s + 2)}$$

Compare numerators to obtain the following four simultaneous equations.

$$\begin{aligned} B_1 + B_2 &= 0 \\ A_1 + A_2 + 2B_1 &= 0 \\ 2A_1 + 2B_1 + B_2 &= 0 \\ 2A_1 + A_2 &= 1 \end{aligned}$$

Use Cramer's rule to find A_1.

$$A_1 = \frac{\begin{vmatrix} 0 & 0 & 1 & 1 \\ 0 & 1 & 2 & 0 \\ 0 & 0 & 2 & 1 \\ 1 & 1 & 0 & 0 \end{vmatrix}}{\begin{vmatrix} 0 & 0 & 1 & 1 \\ 1 & 1 & 2 & 0 \\ 2 & 0 & 2 & 1 \\ 2 & 1 & 0 & 0 \end{vmatrix}} = \frac{-1}{-5} = \frac{1}{5}$$

The rest of the coefficients are found similarly.

$$\begin{aligned} A_1 &= \tfrac{1}{5} \\ A_2 &= \tfrac{3}{5} \\ B_1 &= -\tfrac{2}{5} \\ B_2 &= \tfrac{2}{5} \end{aligned}$$

Then,

$$\begin{aligned} \mathcal{L}(x) &= \frac{1}{(s+1)^2+1} + \frac{\frac{1}{5}}{s^2+1} + \frac{-\frac{2}{5}s}{s^2+1} \\ &= + \frac{\frac{3}{5}}{s^2+2s+2} + \frac{\frac{2}{5}s}{s^2+2s+2} \end{aligned}$$

Take the inverse transform.

$$\begin{aligned} x(t) &= \mathcal{L}^{-1}\{\mathcal{L}(x)\} \\ &= e^{-t}\sin t + \tfrac{1}{5}\sin t - \tfrac{2}{5}\cos t + \tfrac{3}{5}e^{-t}\sin t \\ &= + \tfrac{2}{5}(e^{-t}\cos t - e^{-t}\sin t) \\ &= \boxed{\tfrac{6}{5}e^{-t}\sin t + \tfrac{2}{5}e^{-t}\cos t + \tfrac{1}{5}\sin t - \tfrac{2}{5}\cos t} \end{aligned}$$

The answer is (D).

4. *Customary U.S. Solution*

The differential equation is given as

$$m'(t) = a(t) - \frac{m(t)o(t)}{V(t)}$$

$$\begin{aligned} a(t) &= \text{rate of addition of chemical} \\ m(t) &= \text{mass of chemical at time } t \\ o(t) &= \text{volumetric flow out of the lagoon} \\ &= (= 30 \text{ gal/min}) \\ V(t) &= \text{volume in the lagoon at time } t \end{aligned}$$

Water flows into the lagoon at a rate of 30 gal/min, and a water-chemical mix flows out of the lagoon at rate of 30 gal/min. Therefore, the volume of the lagoon at time t is equal to the initial volume.

$$\begin{aligned} V(t) &= \left(\frac{\pi}{4}\right)(\text{diameter of lagoon})^2(\text{depth of lagoon}) \\ &= \left(\frac{\pi}{4}\right)(120 \text{ ft})^2(10 \text{ ft}) \\ &= 113{,}097 \text{ ft}^3 \end{aligned}$$

Use a conversion factor of 7.48 gal/ft^3.

$$\begin{aligned} o(t) &= \frac{30 \frac{\text{gal}}{\text{min}}}{7.48 \frac{\text{gal}}{\text{ft}^3}} \\ &= 4.01 \text{ ft}^3/\text{min} \end{aligned}$$

Substituting into the general form of the differential equation gives

$$\begin{aligned} m'(t) &= a(t) - \frac{m(t)o(t)}{V(t)} \\ &= (0) - m(t)\left(\frac{4.01 \frac{\text{ft}^3}{\text{min}}}{113{,}097 \text{ ft}^3}\right) \\ &= -\left(\frac{3.55\times10^{-5}}{\text{min}}\right)m(t) \end{aligned}$$

$$m'(t) + \left(\frac{3.55\times10^{-5}}{\text{min}}\right)m(t) = 0$$

The differential equation of the problem has a characteristic equation.

$$r + \frac{3.55\times10^{-5}}{\text{min}} = 0$$

$$r = -3.55\times10^{-5}/\text{min}$$

The general form of the solution is given by

$$m(t) = Ae^{rt}$$

Substituting for the root, r, gives

$$m(t) = Ae^{\left(\frac{-3.55\times10^{-5}}{\text{min}}\right)t}$$

Apply the initial condition $m(0)$= 90 lbm at time $t = 0$.

$$\begin{aligned} m(0) = Ae^{\left(\frac{-3.55\times10^{-5}}{\text{min}}\right)(0)} &= 90 \text{ lbm} \\ Ae^0 &= 90 \text{ lbm} \\ A &= 90 \text{ lbm} \end{aligned}$$

Therefore,

$$m(t) = (90 \text{ lbm})\, e^{\left(\frac{-3.55\times 10^{-5}}{\text{min}}\right)t}$$

Solve for t.

$$\frac{m(t)}{90 \text{ lbm}} = e^{\left(\frac{-3.55\times 10^{-5}}{\text{min}}\right)t}$$

$$\ln\left(\frac{m(t)}{90 \text{ lbm}}\right) = \ln\left(e^{\left(\frac{-3.55\times 10^{-5}}{\text{min}}\right)t}\right)$$

$$= \left(\frac{-3.55 \times 10^{-5}}{\text{min}}\right)t$$

$$t = \frac{\ln\left(\frac{m(t)}{90 \text{ lbm}}\right)}{\frac{-3.55 \times 10^{-5}}{\text{min}}}$$

The initial mass of the water in the lagoon is given by

$$m_i = V\rho$$
$$= (113{,}097 \text{ ft}^3)\left(62.4 \,\frac{\text{lbm}}{\text{ft}^3}\right)$$
$$= 7.05 \times 10^6 \text{ lbm}$$

The final mass of chemicals is achieved at a concentration of 1 ppb or

$$m_f = \frac{7.06 \times 10^6 \text{ lbm}}{1 \times 10^9}$$
$$= 7.06 \times 10^{-3} \text{ lbm}$$

Find the time required to achieve a mass of 7.06×10^{-3} lbm.

$$t = \left(\frac{\ln\left(\frac{m(t)}{90 \text{ lbm}}\right)}{\frac{-3.55 \times 10^{-5}}{\text{min}}}\right)\left(\frac{1 \text{ hr}}{60 \text{ min}}\right)\left(\frac{1 \text{ day}}{24 \text{ hr}}\right)$$

$$= \left(\frac{\ln\left(\frac{7.06 \times 10^{-3} \text{ lbm}}{90 \text{ lbm}}\right)}{\frac{-3.55 \times 10^{-5}}{\text{min}}}\right)\left(\frac{1 \text{ hr}}{60 \text{ min}}\right)\left(\frac{1 \text{ day}}{24 \text{ hr}}\right)$$

$$= \boxed{185 \text{ days}}$$

The answer is (D).

SI Solution

The differential equation is given as

$$m'(t) = a(t) - \frac{m(t)o(t)}{V(t)}$$

$a(t)$ = rate of addition of chemical

$m(t)$ = mass of chemical at time t

$o(t)$ = volumetric flow out of the lagoon = (= 115 L/min)

$V(t)$ = volume in the lagoon at time t

Water flows into the lagoon at a rate of 115 L/min, and a water-chemical mix flows out of the lagoon at a rate of 115 L/min. Therefore, the volume of the lagoon at time t is equal to the initial volume.

$$V(t) = \left(\frac{\pi}{4}\right)(\text{diameter of lagoon})^2(\text{depth of lagoon})$$
$$= \left(\frac{\pi}{4}\right)(35 \text{ m})^2(3 \text{ m})$$
$$= 2886 \text{ m}^3$$

Using a conversion factor of 1 m^3/1000 L gives

$$o(t) = \left(115 \,\frac{\text{L}}{\text{min}}\right)\left(\frac{1 \text{ m}^3}{1000 \text{ L}}\right)$$
$$= 0.115 \text{ m}^3/\text{min}$$

Substitute into the general form of the differential equation.

$$m'(t) = a(t) - \frac{m(t)o(t)}{V(t)}$$
$$= 0 - m(t)\left(\frac{0.115 \,\frac{\text{m}^3}{\text{min}}}{2886 \text{ m}^3}\right)$$
$$= -\left(\frac{3.985\times 10^{-5}}{\text{min}}\right)m(t)$$

$$m'(t) + \left(\frac{3.985\times 10^{-5}}{\text{min}}\right)m(t) = 0$$

The differential equation of the problem has the following characteristic equation.

$$r + \frac{3.985 \times 10^{-5}}{\text{min}} = 0$$
$$r = -3.985 \times 10^{-5}/\text{min}$$

The general form of the solution is given by

$$m(t) = Ae^{rt}$$

Substituting in for the root, r, gives

$$m(t) = Ae^{\left(\frac{-3.985\times10^{-5}}{\text{min}}\right)t}$$

Apply the initial condition $m(0) = 40$ kg at time $t = 0$.

$$m(0) = Ae^{\left(\frac{-3.985\times10^{-5}}{\text{min}}\right)0} = 40 \text{ kg}$$

$$Ae^0 = 40 \text{ kg}$$

$$A = 40 \text{ kg}$$

Therefore,

$$m(t) = (40 \text{ kg})e^{\left(\frac{-3.985\times10^{-5}}{\text{min}}\right)t}$$

Solve for t.

$$\frac{m(t)}{40 \text{ kg}} = e^{\left(\frac{-3.985\times10^{-5}}{\text{min}}\right)t}$$

$$\ln\left(\frac{m(t)}{40 \text{ kg}}\right) = \ln\left(e^{\left(\frac{-3.985\times10^{-5}}{\text{min}}\right)t}\right)$$

$$= \left(\frac{-3.985\times10^{-5}}{\text{min}}\right)t$$

$$t = \ln\left(\frac{\frac{m(t)}{40 \text{ kg}}}{\frac{-3.985\times10^{-5}}{\text{min}}}\right)$$

The initial mass of water in the lagoon is given by

$$m_i = V\rho$$

$$= (2886 \text{ m}^3)\left(1000 \frac{\text{kg}}{\text{m}^3}\right)$$

$$= 2.886\times10^6 \text{ kg}$$

The final mass of chemicals is achieved at a concentration of 1 ppb or

$$m_f = \frac{2.886\times10^6 \text{ kg}}{1\times10^9}$$

$$= 2.886\times10^{-3} \text{ kg}$$

Find the time required to achieve a mass of 2.886×10^{-3} kg.

$$t = \left(\frac{\ln\frac{m(t)}{40 \text{ kg}}}{\frac{-3.985\times10^{-5}}{\text{min}}}\right)\left(\frac{1 \text{ h}}{60 \text{ min}}\right)\left(\frac{1 \text{ day}}{24 \text{ h}}\right)$$

$$= \left(\frac{\ln\left(\frac{2.886\times10^{-3} \text{ kg}}{40 \text{ kg}}\right)}{\frac{-3.985\times10^{-5}}{\text{min}}}\right)\left(\frac{1 \text{ h}}{60 \text{ min}}\right)\left(\frac{1 \text{ day}}{24 \text{ h}}\right)$$

$$= \boxed{166 \text{ days}}$$

The answer is (D).

5. Let

$$m(t) = \text{mass of salt in tank at time } t$$

$$m_0 = 60 \text{ mass units}$$

$$m'(t) = \text{rate at which salt content is changing}$$

Two mass units of salt enter each minute, and three volumes leave each minute. The amount of salt leaving each minute is

$$\left(3 \frac{\text{vol}}{\text{min}}\right)\left(\text{concentration in } \frac{\text{mass}}{\text{vol}}\right)$$

$$= \left(3 \frac{\text{vol}}{\text{min}}\right)\left(\frac{\text{salt content}}{\text{volume}}\right)$$

$$= \left(3 \frac{\text{vol}}{\text{min}}\right)\left(\frac{m(t)}{100-t}\right)$$

$$m'(t) = 2 - (3)\left(\frac{m(t)}{100-t}\right) \text{ or } m'(t) + \frac{3m(t)}{100-t}$$

$$= 2 \text{ mass/min}$$

This is a first-order linear differential equation. The integrating factor is

$$m = \exp\left[3\int \frac{dt}{100-t}\right]$$

$$= \exp\left[(3)(-\ln(100-t))\right]$$

$$= (100-t)^{-3}$$

$$m(t) = (100-t)^3\left[2\int \frac{dt}{(100-t)^3} + k\right]$$

$$= 100 - t + k(100-t)^3$$

But $m = 60$ mass units at $t = 0$, so $k = -0.00004$.

$$m(t) = 100 - t - (0.00004)(100-t)^3$$

At $t = 60$ min,

$$m = 100 - 60 \text{ min} - (0.00004)(100 - 60 \text{ min})^3$$

$$= \boxed{37.44 \text{ mass units}}$$

The answer is (B).

12 Probability and Statistical Analysis of Data

PRACTICE PROBLEMS

Probability

1. Four military recruits whose respective shoe sizes are 7, 8, 9, and 10 report to the supply clerk to be issued boots. The supply clerk selects one pair of boots in each of the four required sizes and hands them at random to the recruits.

(a) What is the probability that all recruits will receive boots of an incorrect size?

(A) 0.25
(B) 0.38
(C) 0.45
(D) 0.61

(b) What is the probability that exactly three recruits will receive boots of the correct size?

(A) 0
(B) 0.063
(C) 0.17
(D) 0.25

Probability Distributions

2. The time taken by a toll taker to collect the toll from vehicles crossing a bridge is an exponential distribution with a mean of 23 sec. What is the probability that a random vehicle will be processed in 25 sec or more (i.e., will take longer than 25 sec)?

(A) 0.17
(B) 0.25
(C) 0.34
(D) 0.52

3. The number of cars entering a toll plaza on a bridge during the hour after midnight follows a Poisson distribution with a mean of 20.

(a) What is the probability that 17 cars will pass through the toll plaza during that hour on any given night?

(A) 0.076
(B) 0.12
(C) 0.16
(D) 0.23

(b) What is the probability that three or fewer cars will pass through the toll plaza at that hour on any given night?

(A) 0.0000032
(B) 0.0019
(C) 0.079
(D) 0.11

4. A mechanical component exhibits a negative exponential failure distribution with a mean time to failure of 1000 hr. What is the maximum operating time such that the reliability remains above 99%?

(A) 3.3 hr
(B) 5.6 hr
(C) 8.1 hr
(D) 10 hr

5. A survey field crew measures one leg of a traverse four times. The following results are obtained.

repetition	measurement	direction
1	1249.529	forward
2	1249.494	backward
3	1249.384	forward
4	1249.348	backward

The crew chief is under orders to obtain readings with confidence limits of 90%.

(a) Which readings are acceptable?

(A) No readings are acceptable.
(B) Two readings are acceptable.
(C) Three readings are acceptable.
(D) All four readings are acceptable.

(b) Which readings are not acceptable?

(A) No readings are unacceptable.
(B) One reading is unacceptable.
(C) Two readings are unacceptable.
(D) All four readings are unacceptable.

(c) Explain how to determine which readings are not acceptable.

(A) Readings inside the 90% confidence limits are unacceptable.
(B) Readings outside the 90% confidence limits are unacceptable.
(C) Readings outside the upper 90% confidence limit are unacceptable.
(D) Readings outside the lower 90% confidence limit are unacceptable.

(d) What is the most probable value of the distance?

(A) 1249.399
(B) 1249.410
(C) 1249.439
(D) 1249.452

(e) What is the error in the most probable value (at 90% confidence)?

(A) 0.08
(B) 0.11
(C) 0.14
(D) 0.19

(f) If the distance is one side of a square traverse whose sides are all equal, what is the most probable closure error?

(A) 0.14
(B) 0.20
(C) 0.28
(D) 0.35

(g) What is the probable error of part (f) expressed as a fraction?

(A) 1:17,600
(B) 1:14,200
(C) 1:12,500
(D) 1:10,900

(h) What is the order of accuracy of the closure?

(A) first order
(B) second order
(C) third order
(D) fourth order

(i) Define accuracy and distinguish it from precision.

(A) If an experiment can be repeated with identical results, the results are considered accurate.
(B) If an experiment has a small bias, the results are considered precise.
(C) If an experiment is precise, it cannot also be accurate.
(D) If an experiment is unaffected by experimental error, the results are accurate.

(j) Give an example of systematic error.

(A) measuring river depth as a motorized ski boat passes by
(B) using a steel tape that is too short to measure consecutive distances
(C) locating magnetic north near a large iron ore deposit along an overland route
(D) determining local wastewater BOD after a toxic spill

Statistical Analysis

6. California law requires a statistical analysis of the average speed driven by motorists on a road prior to the use of radar speed control. The following speeds (all in mi/hr) were observed in a random sample of 40 cars.

44, 48, 26, 25, 20, 43, 40, 42, 29, 39, 23, 26, 24, 47, 45, 28, 29, 41, 38, 36, 27, 44, 42, 43, 29, 37, 34, 31, 33, 30, 42, 43, 28, 41, 29, 36, 35, 30, 32, 31

(a) Tabulate the frequency distribution of the data.

(b) Draw the frequency histogram.

(c) Draw the frequency polygon.

(d) Tabulate the cumulative frequency distribution.

(e) Draw the cumulative frequency graph.

(f) What is the upper quartile speed?

(A) 30 mph
(B) 35 mph
(C) 40 mph
(D) 45 mph

(g) What is the median speed?

(A) 31 mph
(B) 33 mph
(C) 35 mph
(D) 37 mph

(h) What is the standard deviation of the sample data?

(A) 2.1 mph
(B) 6.1 mph
(C) 6.8 mph
(D) 7.4 mph

(i) What is the sample standard deviation?

(A) 7.5 mph
(B) 18 mph
(C) 35 mph
(D) 56 mph

(j) What is the sample variance?

(A) 56 mi^2/hr^2
(B) 324 mi^2/hr^2
(C) 1225 mi^2/hr^2
(D) 3136 mi^2/hr^2

7. A spot speed study is conducted for a stretch of roadway. During a normal day, the speeds were found to be normally distributed with a mean of 46 and a standard deviation of 3.

(a) What is the 50th percentile speed?

(A) 39
(B) 43
(C) 46
(D) 49

(b) What is the 85th percentile speed?

(A) 47.1
(B) 48.3
(C) 49.1
(D) 52.7

(c) What is the upper two standard deviation speed?

(A) 47.2
(B) 49.3
(C) 51.1
(D) 52.0

(d) The daily average speeds for the same stretch of roadway on consecutive normal days were determined by sampling 25 vehicles each day. What is the upper two-standard deviation average speed?

(A) 46.6
(B) 47.2
(C) 52.0
(D) 54.7

8. The diameters of bolt holes drilled in structural steel members are normally distributed with a mean of 0.502 in and a standard deviation of 0.005 in. Holes are out of specification if their diameters are less than 0.497 in or more than 0.507 in.

(a) What is the probability that a hole chosen at random will be out of specification?

(A) 0.16
(B) 0.22
(C) 0.32
(D) 0.68

(b) What is the probability that 2 holes out of a sample of 15 will be out of specification?

(A) 0.074
(B) 0.12
(C) 0.15
(D) 0.32

Hypothesis Testing

9. 100 bearings were tested to failure. The average life was 1520 hr, and the standard deviation was 120 hr. The manufacturer claims a 1600 hr life. Evaluate using confidence limits of 95% and 99%.

(A) The claim is accurate at both 95% and 99% confidence.
(B) The claim is inaccurate only at 95%.
(C) The claim is inaccurate only at 99%.
(D) The claim is inaccurate at both 95% and 99% confidence.

Curve Fitting

10. (a) Find the best equation for a line passing through the points given. (b) Find the correlation coefficient.

x	y
400	370
800	780
1250	1210
1600	1560
2000	1980
2500	2450
4000	3950

11. Find the best equation for a line passing through the points given.

s	t
20	43
18	141
16	385
14	1099

12. The number of vehicles lining up behind a flashing railroad crossing has been observed for five trains of different lengths, as given. What is the mathematical formula that relates the two variables?

no. of cars in train	no. of vehicles
2	14.8
5	18.0
8	20.4
12	23.0
27	29.9

13. The following yield data are obtained from five identical treatment plants. (a) Develop a mathematical equation to correlate the yield and average temperature. (b) What is the correlation coefficient?

treatment plant	average temperature (T)	average yield (Y)
1	207.1	92.30
2	210.3	92.58
3	200.4	91.56
4	201.1	91.63
5	203.4	91.83

14. The following data are obtained from a soil compaction test. What is the mathematical formula that relates the two variables?

x	y
−1	0
0	1
1	1.4
2	1.7
3	2
4	2.2
5	2.4
6	2.6
7	2.8
8	3

15. Two resistances, the meter resistance and a shunt resistor, are connected in parallel in an ammeter. Most of the current passing through the meter goes through the shunt resistor. In order to determine the accuracy of the resistance of shunt resistors being manufactured for a line of ammeters, a manufacturer tests a sample of 100 shunt resistors. The numbers of shunt resistors with the resistance indicated (to the nearest hundredth of an ohm) are as follows.

0.200 Ω, 1; 0.210 Ω, 3; 0.220 Ω, 5; 0.230 Ω, 10; 0.240 Ω, 17; 0.250 Ω, 40; 0.260 Ω, 13; 0.270 Ω, 6; 0.280 Ω, 3; 0.290 Ω, 2

(a) What is the mean resistance?

(A) 0.235 Ω
(B) 0.247 Ω
(C) 0.251 Ω
(D) 0.259 Ω

(b) What is the sample standard deviation?

(A) 0.0003
(B) 0.010
(C) 0.016
(D) 0.24

(c) What is the median resistance?

(A) 0.221 Ω
(B) 0.244 Ω
(C) 0.252 Ω
(D) 0.259 Ω

(d) What is the sample variance?

(A) 0.00027
(B) 0.0083
(C) 0.0114
(D) 0.0163

SOLUTIONS

1. (a) There are 4! = 24 different possible outcomes. By enumeration, there are 9 completely wrong combinations.

$$p\{\text{all wrong}\} = \frac{9}{24} = \boxed{0.375}$$

The answer is (B).

(b) If three recruits get the correct size, the fourth recruit will also since there will be only one pair remaining.

$$p\{\text{exactly 3}\} = \boxed{0}$$

The answer is (A).

2. For an exponential distribution function, the mean is given as

$$\mu = \frac{1}{\lambda}$$

For a mean of 23,

$$\mu = 23 = \frac{1}{\lambda}$$
$$\lambda = 0.0435$$

For an exponential distribution function,

$$\begin{aligned} p\{X < x\} &= F(x) = 1 - e^{-\lambda x} \\ p\{x > X\} &= 1 - p\{X < x\} \\ &= 1 - F(x) \\ &= 1 - \left(1 - e^{-\lambda x}\right) \\ &= e^{-\lambda x} \end{aligned}$$

The probability of a random vehicle being processed in 25 sec or more is given by

$$\begin{aligned} p\{x > 25\} &= e^{-(0.0435)(25)} \\ &= e^{-1.0875} \\ &= \boxed{0.337} \end{aligned}$$

The answer is (C).

3. (a) The distribution is a Poisson distribution with an average of $\lambda = 20$. The probability for a Poisson distribution is given by

$$p\{x\} = f(x) = \frac{e^{-\lambda}\lambda^x}{x!}$$

The probability of 17 cars is

$$\begin{aligned} p\{x = 17\} = f(17) &= \frac{e^{-20} \times 20^{17}}{17!} \\ &= \boxed{0.076 \enspace (7.6\%)} \end{aligned}$$

The answer is (A).

(b) The probability of three or fewer cars is given by

$$\begin{aligned} p\{x \leq 3\} &= p\{x=0\} + p\{x=1\} + p\{x=2\} \\ &\quad + p\{x=3\} \\ &= f(0) + f(1) + f(2) + f(3) \\ &= \frac{e^{-20} \times 20^0}{0!} + \frac{e^{-20} \times 20^1}{1!} \\ &\quad + \frac{e^{-20} \times 20^2}{2!} + \frac{e^{-20} \times 20^3}{3!} \\ &= 2 \times 10^{-9} + 4.1 \times 10^{-8} \\ &\quad + 4.12 \times 10^{-7} + 2.75 \times 10^{-6} \\ &= 3.2 \times 10^{-6} \quad (3.2 \times 10^{-4}\%) \\ &= \boxed{0.0000032} \end{aligned}$$

The answer is (A).

4.

$$\lambda = \frac{1}{\text{MTTF}}$$

$$\frac{1}{1000} = 0.001$$

The reliability function is

$$R\{t\} = e^{-\lambda t} = e^{-0.001t}$$

Since the reliability is greater than 99%,

$$\begin{aligned} e^{-0.001t} &> 0.99 \\ \ln(e^{-0.001t}) &> \ln 0.99 \\ -0.001t &> \ln 0.99 \\ t &< -1000 \ln 0.99 \\ &\boxed{t < 10.05} \end{aligned}$$

The maximum operating time such that the reliability remains above 99% is 10.05 hr.

The answer is (D).

5. Find the average.

$$\begin{aligned} \overline{x} &= \frac{\sum x_i}{n} \\ &= \frac{1249.529 + 1249.494 + 1249.384 + 1249.348}{4} \\ &= 1249.439 \end{aligned}$$

Since the sample population is small, use the sample standard deviation.

$$\begin{aligned} s &= \sqrt{\frac{\sum (x_i - \overline{x})^2}{n-1}} \\ &= \sqrt{\frac{\begin{array}{l}(1249.529 - 1249.439)^2 + (1249.494 - 1249.439)^2 \\ + (1249.384 - 1249.439)^2 + (1249.348 - 1249.439)^2\end{array}}{4-1}} \\ &= 0.08647 \end{aligned}$$

From the standard deviation table, a 90% confidence limit falls within $1.645s$ of $\overline{x}$.

$$1249.439 \pm (1.645)(0.08647) = 1249.439 \pm 0.142$$

Therefore, (1249.297, 1249.581) is the 90% confidence range.

(a) By observation, all the readings fall within the 90% confidence range.

The answer is (D).

(b) No readings are unacceptable.

The answer is (A).

(c) Readings outside the 90% confidence limits are unacceptable.

The answer is (B).

(d) The unbiased estimate of the most probable distance is 1249.439.

The answer is (C).

(e) The error for the 90% confidence range is 0.142.

The answer is (C).

(f) If the surveying crew places a marker, measures a distance x, places a second marker, and then measures the same distance x back to the original marker, the ending point should coincide with the original marker. If, due to measurement errors, the ending and starting points do not coincide, the difference is the closure error.

In this example, the survey crew moves around the four sides of a square, so there are two measurements in the x-direction and two measurements in the y-direction. If the errors E_1 and E_2 are known for two measurements, x_1 and x_2, the error associated with the sum or difference $x_1 \pm x_2$ is

$$E\{x_1 \pm x_2\} = \sqrt{E_1^2 + E_2^2}$$

In this case, the error in the x-direction is

$$\begin{aligned} E_x &= \sqrt{(0.1422)^2 + (0.1422)^2} \\ &= 0.2011 \end{aligned}$$

The error in the y-direction is calculated the same way and is also 0.2011. E_x and E_y are combined by the Pythagorean theorem to yield

$$E_{\text{closure}} = \sqrt{(0.2011)^2 + (0.2011)^2} = \boxed{0.2844}$$

The answer is (C).

(g) In surveying, error may be expressed as a fraction of one or more legs of the traverse. Assume that the total of all four legs is to be used as the basis.

$$\frac{0.2844}{(4)(1249)} = \boxed{\frac{1}{17{,}567}}$$

The answer is (A).

(h) In surveying, a class 1 third-order error is smaller than 1/10,000. The error of 1/17,567 is smaller than the third-order error; therefore, the error is within the third-order accuracy.

The answer is (C).

(i) An experiment is accurate if it is unchanged by experimental error. Precision is concerned with the repeatability of the experimental results. If an experiment is repeated with identical results, the experiment is said to be precise. However, it is possible to have a highly precise experiment with a large bias.

The answer is (D).

(j) A systematic error is one that is always present and is unchanged from sample to sample. For example, a steel tape that is 0.02 ft short introduces a systematic error.

The answer is (B).

6. (a) and (d) Tabulate the frequency distribution data.

(Note that the lowest speed is 20 mi/hr and the highest speed is 48 mi/hr; therefore, the range is 28 mi/hr. Choose 10 cells with a width of 3 mi/hr.)

midpoint	interval (mi/hr)	frequency	cumulative frequency	cumulative percent
21	20–22	1	1	3
24	23–25	3	4	10
27	26–28	5	9	23
30	29–31	8	17	43
33	32–34	3	20	50
36	35–37	4	24	60
39	38–40	3	27	68
42	41–43	8	35	88
45	44–46	3	38	95
48	47–49	2	40	100

(b)

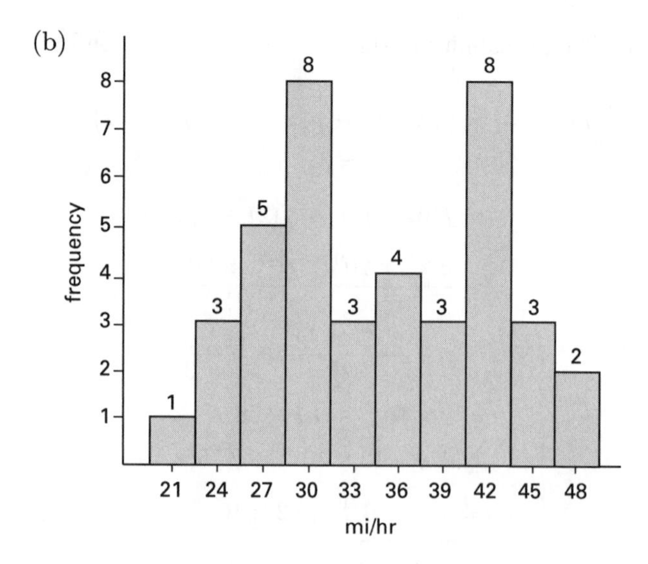

(c)

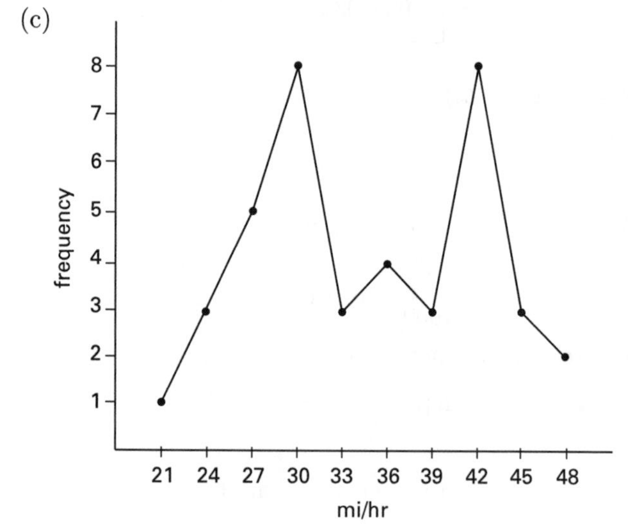

(e)

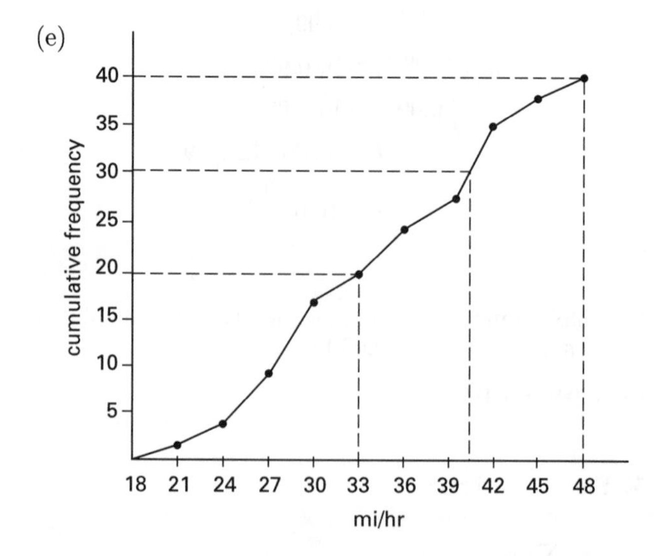

(f) From the cumulative frequency graph in part (e), the upper quartile speed occurs at 30 cars or 75%, which corresponds to approximately 40 mi/hr.

The answer is (C).

(g) The mode occurs at two speeds (the frequency at each of the two speeds is 8), which are 30 mi/hr and 42 mi/hr.

From the cumulative frequency chart, the median occurs at 50% or 20 cars and corresponds to 33 mi/hr.

$$\sum x_i = 1390 \text{ mi/hr}$$
$$n = 40$$

The mean is computed as

$$\overline{x} = \frac{\sum x_i}{n} = \frac{1390 \ \frac{\text{mi}}{\text{hr}}}{40} = \boxed{34.75 \text{ mi/hr}}$$

The answer is (B).

(h) The standard deviation of the sample data is given as

$$\sigma = \sqrt{\frac{\sum x^2}{n} - \mu^2}$$
$$\sum x^2 = 50{,}496 \text{ mi}^2/\text{hr}^2$$

Use the sample mean as an unbiased estimator of the population mean, μ.

$$\sigma = \sqrt{\frac{\sum x^2}{n} - \mu^2} = \sqrt{\frac{50{,}496 \ \frac{\text{mi}^2}{\text{hr}^2}}{40} - \left(34.75 \ \frac{\text{mi}}{\text{hr}}\right)^2} = \boxed{7.405 \text{ mi/hr}}$$

The answer is (D).

(i) The sample standard deviation is given by

$$s = \sqrt{\frac{\sum x^2 - \frac{\left(\sum x\right)^2}{n}}{n-1}} = \sqrt{\frac{50{,}496 \ \frac{\text{mi}^2}{\text{hr}^2} - \frac{\left(1390 \ \frac{\text{mi}}{\text{hr}}\right)^2}{40}}{40-1}} = \boxed{7.500 \text{ mi/hr}}$$

The answer is (A).

(j) The sample variance is given by the square of the sample standard deviation.

$$s^2 = \left(7.500 \ \frac{\text{mi}}{\text{hr}}\right)^2 = \boxed{56.25 \text{ mi}^2/\text{hr}^2}$$

The answer is (A).

7. (a) The 50th percentile speed is the mean speed,

$$\boxed{46}$$

The answer is (C).

(b) The 85th percentile speed is the speed that is exceeded by only 15% of the measurements. Since this is a normal distribution, App. 12.A can be used. 15% in the upper tail corresponds to 35% between the mean and the 85th percentile. This occurs at approximately $= 1.04\sigma$. The 85th percentile speed is

$$x_{85\%} = \mu + 1.04\sigma = 46 + (1.04)(3) = \boxed{49.12}$$

The answer is (C).

(c) The upper 2σ speed is

$$x_{2\sigma} = \mu + 2\sigma = 46 + (2)(3) = \boxed{52}$$

The answer is (D).

(d) According to the central limit theorem, the mean of the average speeds is the same as the distribution mean, and the standard deviation of sample means is

$$\sigma_{\overline{x}} = \frac{\sigma_x}{\sqrt{K}} = \frac{3}{\sqrt{25}} = 0.6$$
$$\overline{x}_{2\sigma} = \mu + 2\sigma_{\overline{x}} = 46 + (2)(0.6) = \boxed{47.2}$$

The answer is (B).

8. (a) From Eq. 12.43,

$$z_{\text{upper}} = \frac{0.507 \text{ in} - 0.502 \text{ in}}{0.005 \text{ in}} = +1$$

From App. 12.A, the area outside $z = +1$ is

$$0.5 - 0.3413 = 0.1587$$

Since these are symmetrical limits, $z_{\text{lower}} = -1$.

$$\text{total fraction defective} = (2)(0.1587) = \boxed{0.3174}$$

The answer is (C).

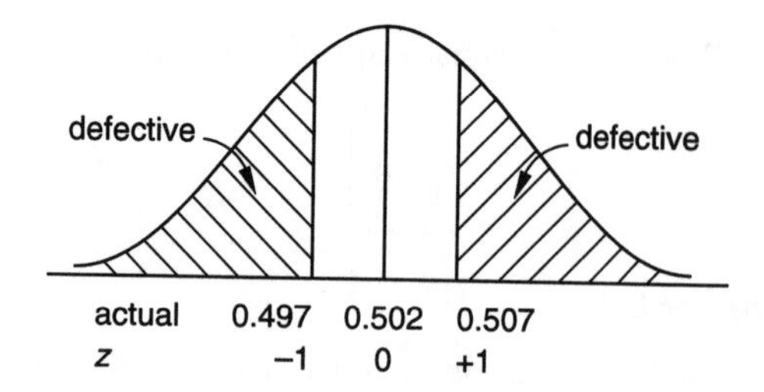

(b) This is a binomial problem.

$$p = p\{\text{defective}\} = 0.3174$$
$$q = 1 - p = 0.6826$$

From Eq. 12.28,

$$f(2) = \binom{15}{2}(0.3174)^2(0.6826)^{13}$$
$$= \left(\frac{15!}{13!2!}\right)(0.3174)^2(0.6826)^{13} = \boxed{0.0739}$$

The answer is (A).

9. This is a typical hypothesis test of two sample population means. The two populations are the original population the manufacturer used to determine the 1600 hr average life value and the new population the sample was taken from. The mean ($\overline{x} = 1520$ hr) of the sample and its standard deviation ($s = 120$ hr) are known, but the mean and standard deviation of a population of average lifetimes are unknown.

Assume that the average lifetime population mean and the sample mean are identical.

$$\overline{x} = \mu = 1520 \text{ hr}$$

The standard deviation of the average lifetime population is

$$\sigma_{\overline{x}} = \frac{s}{\sqrt{n}} = \frac{120 \text{ hr}}{\sqrt{100}} = 12 \text{ hr}$$

The manufacturer can be reasonably sure that the claim of a 1600 hr average life is justified if the average test life is near 1600 hr. "Reasonably sure" must be evaluated based on acceptable probability of being incorrect.

If the manufacturer is willing to be wrong with a 5% probability, then a 95% confidence level is required.

Since the direction of bias is known, a one-tailed test is required. To determine if the mean has shifted downward, test the hypothesis that 1600 hr is within the 95% limit of a distribution with a mean of 1520 hr and a standard deviation of 12 hr. From a standard normal table, 5% of a standard normal distribution is outside of $z = 1.645$. Therefore, the 95% confidence limit is

$$1520 \text{ hr} + (1.645)(12 \text{ hr}) = 1540 \text{ hr}$$

The manufacturer can be 95% certain that the average lifetime of the bearings is less than 1600 hr.

If the manufacturer is willing to be wrong with a probability of only 1%, then a 99% confidence limit is required. From the normal table, $z = 2.33$ and the 99% confidence limit is

$$1520 \text{ hr} + (2.33)(12 \text{ hr}) = 1548 \text{ hr}$$

The manufacturer can be 99% certain that the average bearing life is less than 1600 hr.

The answer is (D).

10. (a) Plot the data points to determine if the relationship is linear.

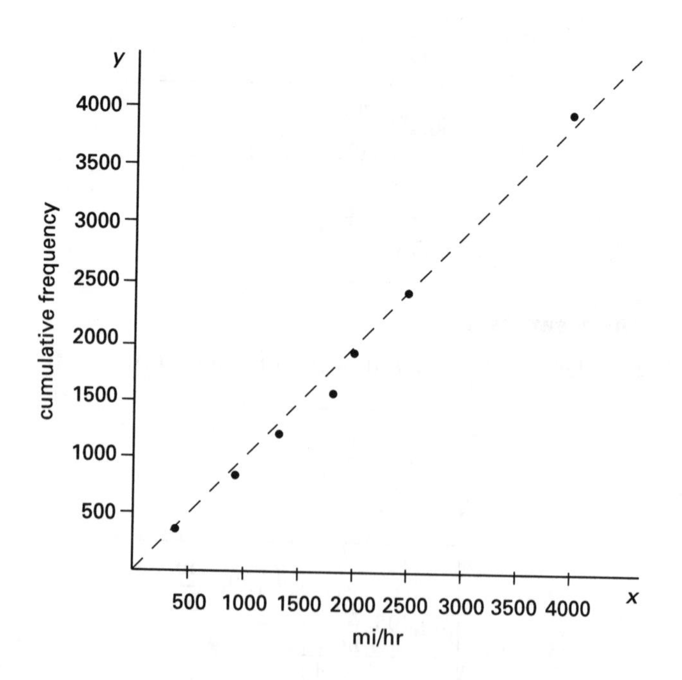

The data appear to be essentially linear. The slope, m, and the y-intercept, b, can be determined using linear regression.

The individual terms are

$$n = 7$$

$$\begin{aligned}\sum x_i &= 400 + 800 + 1250 + 1600 + 2000 + 2500 \\ &\quad + 4000 \\ &= 12{,}550\end{aligned}$$

$$\begin{aligned}\left(\sum x_i\right)^2 &= (12{,}550)^2 \\ &= 1.575 \times 10^8\end{aligned}$$

$$\begin{aligned}\overline{x} &= \frac{\sum x_i}{n} \\ &= \frac{12{,}550}{7} \\ &= 1792.9\end{aligned}$$

$$\begin{aligned}\sum x_i^2 &= (400)^2 + (800)^2 + (1250)^2 + (1600)^2 \\ &\quad + (2000)^2 + (2500)^2 + (4000)^2 \\ &= 3.117 \times 10^7\end{aligned}$$

Similarly,

$$\begin{aligned}\sum y_i &= 370 + 780 + 1210 + 1560 + 1980 \\ &= + 2450 + 3950 \\ &= 12{,}300\end{aligned}$$

$$\left(\sum y_i\right)^2 = (12{,}300)^2 = 1.513 \times 10^8$$

$$\overline{y} = \frac{\sum y_i}{n} = \frac{12{,}300}{7} = 1757.1$$

$$\begin{aligned}\sum y_i^2 &= (370)^2 + (780)^2 + (1210)^2 + (1560)^2 \\ &= + (1980)^2 + (2450)^2 + (3950)^2 \\ &= 3.017 \times 10^7\end{aligned}$$

Also,

$$\begin{aligned}\sum x_i y_i &= (400)(370) + (800)(780) + (1250)(1210) \\ &= + (1600)(1560) + (2000)(1980) \\ &= + (2500)(2450) + (4000)(3950) \\ &= 3.067 \times 10^7\end{aligned}$$

The slope is

$$\begin{aligned}m &= \frac{n \sum x_i y_i - \sum x_i \sum y_i}{n \sum x_i^2 - \left(\sum x_i\right)^2} \\ &= \frac{(7)(3.067 \times 10^7) - (12{,}550)(12{,}300)}{(7)(3.117 \times 10^7) - (12{,}550)^2} \\ &= 0.994\end{aligned}$$

The y-intercept is

$$\begin{aligned}b &= \overline{y} - m\overline{x} \\ &= 1757.1 - (0.994)(1792.9) \\ &= -25.0\end{aligned}$$

The least squares equation of the line is

$$\begin{aligned}y &= mx + b \\ &= \boxed{0.994x - 25.0}\end{aligned}$$

(b) The correlation coefficient is

$$\begin{aligned}r &= \frac{n \sum (x_i y_i) - \left(\sum x_i\right)\left(\sum y_i\right)}{\sqrt{\left[n \sum x_i^2 - \left(\sum x_i\right)^2\right]\left[n \sum y_i^2 - \left(\sum y_i\right)^2\right]}} \\ &= \frac{(7)(3.067 \times 10^7) - (12{,}500)(12{,}300)}{\sqrt{\begin{array}{l}[(7)(3.117 \times 10^7) - (12{,}500)^2] \\ \quad \times [(7)(3.017 \times 10^7) - (12{,}300)^2]\end{array}}} \\ &\approx \boxed{1.00}\end{aligned}$$

11. Plotting the data shows that the relationship is nonlinear.

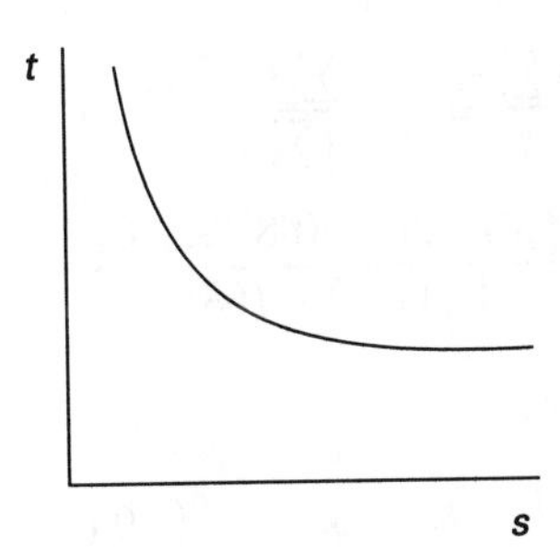

This appears to be an exponential with the form

$$t = ae^{bs}$$

Take the natural log of both sides.

$$\begin{aligned}\ln t &= \ln(ae^{bs}) \\ &= \ln a + \ln(e^{bs}) \\ &= \ln a + bs\end{aligned}$$

But, $\ln a$ is just a constant, c.

$$\ln t = c + bs$$

Make the transformation $R = \ln t$.

$$R = c + bs$$

s	R
20	3.76
18	4.95
16	5.95
14	7.00

This is linear.

$$\begin{aligned}
n &= 4 \\
\textstyle\sum s_i &= 20 + 18 + 16 + 14 = 68 \\
\overline{s} &= \frac{\sum s}{n} = \frac{68}{4} = 17 \\
\textstyle\sum s_i^2 &= (20)^2 + (18)^2 + (16)^2 + (14)^2 = 1176 \\
\left(\textstyle\sum s_i\right)^2 &= (68)^2 = 4624 \\
\textstyle\sum R_i &= 3.76 + 4.95 + 5.95 + 7.00 = 21.66 \\
\overline{R} &= \frac{\sum R_i}{n} = \frac{21.66}{4} = 5.415 \\
\textstyle\sum R_i^2 &= (3.76)^2 + (4.95)^2 + (5.95)^2 + (7.00)^2 \\
&= 123.04 \\
\left(\textstyle\sum R_i\right)^2 &= (21.66)^2 = 469.16 \\
\textstyle\sum s_i R_i &= (20)(3.76) + (18)(4.95) + (16)(5.95) \\
&= +\,(14)(7.00) \\
&= 357.5
\end{aligned}$$

The slope, b, of the transformed line is

$$\begin{aligned}
b &= \frac{n\sum s_i R_i - \sum s_i \sum R_i}{n\sum s_i^2 - (\sum s_i)^2} \\
&= \frac{(4)(357.5) - (68)(21.66)}{(4)(1176) - (68)^2} = -0.536
\end{aligned}$$

The intercept is

$$\begin{aligned}
c &= \overline{R} - b\overline{s} = 5.415 - (-0.536)(17) \\
&= 14.527
\end{aligned}$$

The transformed equation is

$$\begin{aligned}
R &= c + bs \\
&= 14.527 - 0.536s
\end{aligned}$$

$$\boxed{\ln t = 14.527 - 0.536s}$$

12. The first step is to graph the data.

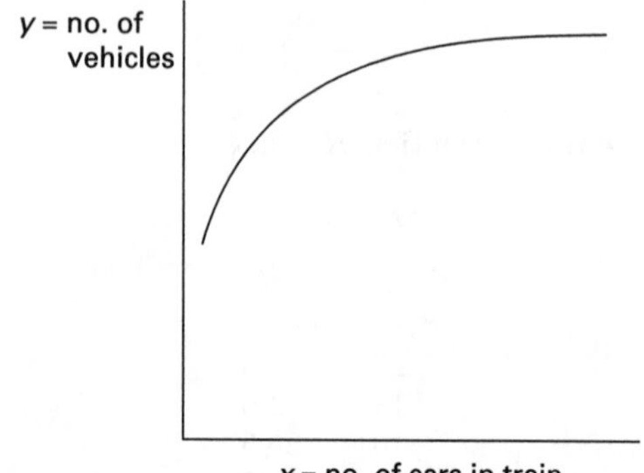

It is assumed that the relationship between the variables has the form $y = a + b \log x$. Therefore, the variable change $z = \log x$ is made, resulting in the following set of data.

z	y
0.301	14.8
0.699	18.0
0.903	20.4
1.079	23.0
1.431	29.9

$$\begin{aligned}
\textstyle\sum z_i &= 4.413 \\
\textstyle\sum y_i &= 106.1 \\
\textstyle\sum z_i^2 &= 4.6082 \\
\textstyle\sum y_i^2 &= 2382.2 \\
\left(\textstyle\sum z_i\right)^2 &= 19.475 \\
\left(\textstyle\sum y_i\right)^2 &= 11{,}257.2 \\
\overline{z} &= 0.8826 \\
\overline{y} &= 21.22 \\
\textstyle\sum z_i y_i &= 103.06 \\
n &= 5
\end{aligned}$$

The slope is

$$\begin{aligned}
m &= \frac{n\sum z_i y_i - \sum z_i \sum y_i}{n\sum z_i^2 - (\sum z_i)^2} \\
&= \frac{(5)(103.06) - (4.413)(106.1)}{(5)(4.6082) - 19.475} \\
&= 13.20
\end{aligned}$$

The y-intercept is

$$\begin{aligned}
b &= \overline{y} - m\overline{z} \\
&= 21.22 - (13.20)(0.8826) \\
&= 9.570
\end{aligned}$$

The resulting equation is

$$y = 9.570 + 13.20z$$

The relationship between x and y is approximately

$$\boxed{y = 9.570 + 13.20 \log x}$$

This is not an optimal correlation, as better correlation coefficients can be obtained if other assumptions about the form of the equation are made. For example, $y = 9.1 + 4\sqrt{x}$ has a better correlation coefficient.

13. (a) Plot the data to verify that they are linear.

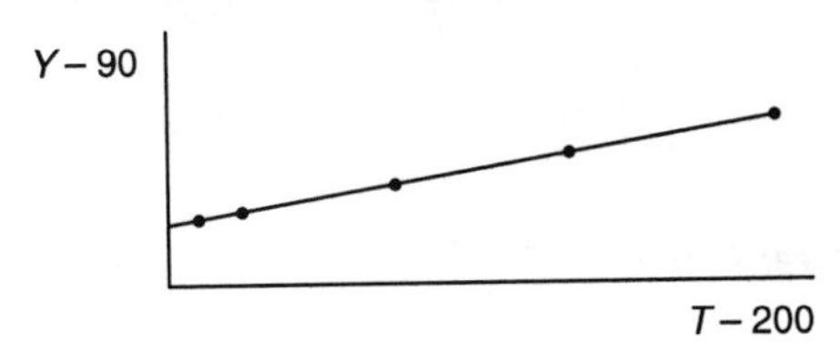

x $T-200$	y $Y-90$
7.1	2.30
10.3	2.58
0.4	1.56
1.1	1.63
3.4	1.83

step 1:

$$\begin{aligned} \sum x_i &= 22.3 & \sum y_i &= 9.9 \\ \sum x_i^2 &= 169.43 & \sum y_i^2 &= 20.39 \\ (\sum x_i)^2 &= 497.29 & (\sum y_i)^2 &= 98.01 \\ \overline{x} &= \frac{22.3}{5} = 4.46 & \overline{y} &= 1.98 \end{aligned}$$

$$\sum x_i y_i = 51.54$$

step 2: From Eq. 12.70, the slope is

$$m = \frac{(5)(51.54) - (22.3)(9.9)}{(5)(169.43) - 497.29} = 0.1055$$

step 3: From Eq. 12.71, the y-intercept is

$$b = 1.98 - (0.1055)(4.46) = 1.509$$

The equation of the line is

$$\begin{aligned} y &= 0.1055x + 1.509 \\ Y - 90 &= (0.1055)(T - 200) + 1.509 \\ Y &= \boxed{0.1055T + 70.409} \end{aligned}$$

(b) *step 4:* Use Eq. 12.72 to get the correlation coefficient.

$$r = \frac{(5)(51.54) - (22.3)(9.9)}{\sqrt{[(5)(169.43) - 497.29][(5)(20.39) - 98.01]}}$$

$$= \boxed{0.995}$$

14. Plot the data to see if they are linear.

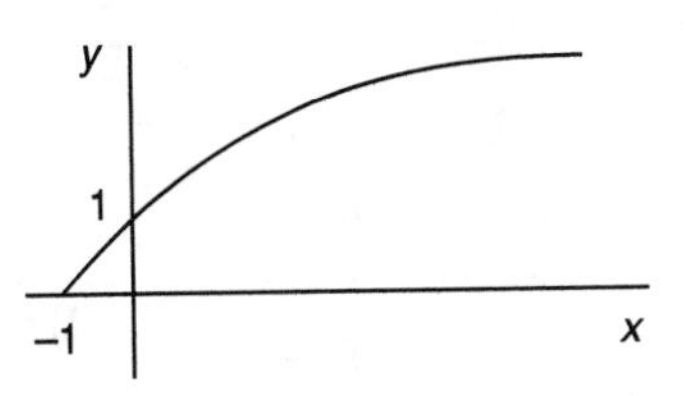

This looks like it could be of the form

$$y = a + b\sqrt{x}$$

However, when x is negative (as in the first point), the function is imaginary. Try shifting the curve to the right, replacing x with $x+1$.

$$y = a + bz$$

$$z = \sqrt{x+1}$$

z	y
0	0
1	1
1.414	1.4
1.732	1.7
2	2
2.236	2.2
2.45	2.4
2.65	2.6
2.83	2.8
3	3

Since $y \approx z$, the relationship is

$$\boxed{y = \sqrt{x+1}}$$

In this problem, the answer was found accidentally. Usually, regression would be necessary.

15. (a)

R	f	fR	fR^2
0.200	1	0.200	0.0400
0.210	3	0.630	0.1323
0.220	5	1.100	0.2420
0.230	10	2.300	0.5290
0.240	17	4.080	0.9792
0.250	40	10.000	2.5000
0.260	13	3.380	0.8788
0.270	6	1.620	0.4374
0.280	3	0.840	0.2352
0.290	2	0.580	0.1682
	100	24.730	6.1421

$$\overline{R} = \frac{\sum fR}{\sum f} = \frac{24.730\ \Omega}{100} = \boxed{0.2473\ \Omega}$$

The answer is (B).

(b) The sample standard deviation is given by Eq. 12.61.

$$s = \sqrt{\frac{\sum fR^2 - \frac{(\sum fR)^2}{n}}{n-1}}$$

$$= \sqrt{\frac{6.1421\ \Omega - \frac{(24.73\ \Omega)^2}{100}}{99}}$$

$$= \boxed{0.0163\ \Omega}$$

The answer is (C).

(c) 50% of the observations are below the median. 36 observations are below 0.240. 14 more are needed.

$$0.240\ \Omega + \left(\frac{14}{40}\right)(0.250\ \Omega - 0.240\ \Omega) = \boxed{0.2435\ \Omega}$$

The answer is (B).

(d) $$s^2 = (0.0163\ \Omega)^2 = \boxed{0.0002656\ \Omega^2}$$

The answer is (A).

13 Computer Mathematics

PRACTICE PROBLEMS

There are no practice problems corresponding with Ch. 13 of the *Electrical Engineering Reference Manual.*

14 Numerical Analysis

PRACTICE PROBLEMS

1. A function is given as $y = 3x^{0.93} + 4.2$. What is the percent error if the value of y at $x = 2.7$ is found by using straight-line interpolation between $x = 2$ and $x = 3$?

(A) 0.06%
(B) 0.18%
(C) 2.5%
(D) 5.4%

2. Given the following data points, find y by straight-line interpolation for $x = 2.75$.

x	y
1	4
2	6
3	2
4	−14

(A) 2.1
(B) 2.4
(C) 2.7
(D) 3.0

3. Using the bisection method, find all of the roots of $f(x) = 0$ to the nearest 0.000005.

$$f(x) = x^3 + 2x^2 + 8x - 2$$

SOLUTIONS

1. The actual value at $x = 2.7$ is given by

$$y(x) = 3x^{0.93} + 4.2$$
$$y(2.7) = (3)(2.7)^{0.93} + 4.2 = 11.756$$

At $x = 3$,

$$y(3) = (3)(3)^{0.93} + 4.2 = 12.534$$

At $x = 2$,

$$y(2) = (3)(2)^{0.93} + 4.2 = 9.916$$

Use straight-line interpolation.

$$\frac{x_2 - x}{x_2 - x_1} = \frac{y_2 - y}{y_2 - y_1}$$
$$\frac{3 - 2.7}{3 - 2} = \frac{12.534 - y}{12.534 - 9.916}$$
$$y = 11.749$$

The relative error is given by

$$\frac{\text{actual value} - \text{predicted value}}{\text{actual value}} = \frac{11.756 - 11.749}{11.756} = \boxed{0.0006 \;\; (0.06\%)}$$

The answer is (A).

2. Let $x_1 = 2$; therefore, from the table of data points, $y_1 = 6$. Let $x_2 = 3$; therefore, from the table of data points, $y_2 = 2$.

Let $x = 2.75$. By straight-line interpolation,

$$\frac{x_2 - x}{x_2 - x_1} = \frac{y_2 - y}{y_2 - y_1}$$
$$\frac{3 - 2.75}{3 - 2} = \frac{2 - y}{2 - 6}$$
$$\boxed{y = 3}$$

The answer is (D).

3. $f(x) = x^3 + 2x^2 + 8x - 2$

Try to find an interval in which there is a root.

x	$f(x)$
0	−2
1	9

A root exists in the interval [0,1].

Try $x = \left(\frac{1}{2}\right)(0+1) = 0.5$.

$$f(0.5) = (0.5)^3 + (2)(0.5)^2 + (8)(0.5) - 2 = 2.625$$

A root exists in [0,0.5].

Try $x = 0.25$.

$$f(0.25) = (0.25)^3 + (2)(0.25)^2 + (8)(0.25) - 2 = 0.1406$$

A root exists in [0,0.25].

Try $x = 0.125$.

$$\begin{aligned} f(0.125) &= (0.125)^3 + (2)(0.125)^2 + (8)(0.125) - 2 \\ &= -0.967 \end{aligned}$$

A root exists in [0.125,0.25].

Try $x = \left(\frac{1}{2}\right)(0.125 + 0.25) = 0.1875$.

Continuing,

$$\begin{aligned} f(0.1875) &= -0.42 \quad [0.1875,0.25] \\ f(0.21875) &= -0.144 \quad [0.21875,0.25] \\ f(0.234375) &= -0.002 \quad [\text{This is close enough.}] \end{aligned}$$

One root is $x_1 \approx \boxed{0.234375.}$

Try to find the other two roots. Use long division to factor the polynomial.

$$\begin{array}{r} x^2 + 2.234375x + 8.52368 \\ x - 0.234375 \overline{\big) \; x^3 + 2x^2 + 8x - 2} \\ -(x^3 - 0.234375x^2) \\ \hline 2.234375x^2 + 8x \\ -(2.234375x^2 - 0.52368x) \\ \hline 8.52368x - 2 \\ -(8.52368x - 1.9977) \\ \hline \approx 0 \end{array}$$

Use the quadratic equation to find the roots of $x^2 + 2.234375x + 8.52368$.

$$\begin{aligned} x_2, x_3 &= \frac{-2.234375 \pm \sqrt{(2.234375)^2 - (4)(1)(8.52368)}}{(2)(1)} \\ &= \boxed{-1.117189 \pm j2.697327} \quad [\text{both imaginary}] \end{aligned}$$

15 Advanced Engineering Mathematics

PRACTICE PROBLEMS

There are no practice problems corresponding with Ch. 15 of the *Electrical Engineering Reference Manual.*

16 Electromagnetic Theory

PRACTICE PROBLEMS

1. What is the magnitude of the electric force between a proton and electron 0.046 nm apart in free space?

(A) 9.651×10^{-19} N
(B) -5.018×10^{-18} N
(C) 5.018×10^{-18} N
(D) 1.090×10^{-7} N

2. What is the magnitude of the electric field strength for a proton in free space at 0.046 nm?

(A) 3.047×10^{-30} N/C
(B) 1.440×10^{-9} N/C
(C) 6.805×10^{11} N/C
(D) 31.3 N/C

3. What is the magnitude of the electric field intensity for a parallel plate capacitor with a potential of 12 V applied and a distance of 0.3 cm between plates?

(A) 0.12 V/m
(B) 0.40 V/m
(C) 40 V/m
(D) 4000 V/m

4. During design discussion, quartz is picked as the dielectric for certain capacitors. What is the expected permittivity?

(A) 1.8×10^{-12} $C^2/N{\cdot}m^2$
(B) 8.8×10^{-12} $C^2/N{\cdot}m^2$
(C) 4.4×10^{-11} $C^2/N{\cdot}m^2$
(D) 8.8×10^{-8} $C^2/N{\cdot}m^2$

5. What is the magnitude of the electric flux density 0.5 m from a wire with a uniform line charge of 30×10^{-5} C/m?

(A) 9.5×10^{-5} C/m^2
(B) 1.9×10^{-4} C/m^2
(C) 6.0×10^{-4} C/m^2
(D) 1.7×10^{-3} C/m^2

SOLUTIONS

1. The magnitude of the electric force is given by Coulomb's law. From Eq. 16.3,

$$F = \frac{|Q||Q|}{4\pi\epsilon_0 r^2}$$

$$= \frac{(1.6022 \times 10^{-19}\ \text{C})(1.6022 \times 10^{-19}\ \text{C})}{4\pi\left(8.854 \times 10^{-12}\,\frac{\text{C}^2}{\text{N·m}}\right)(0.046 \times 10^{-9}\ \text{m})^2}$$

$$= \boxed{1.090 \times 10^{-7}\ \text{N}}$$

The answer is (D).

Note that the distance 0.046 nm is the approximate distance between the proton and electron in the hydrogen atom.

2. From Eq. 16.6, the magnitude of the electric field strength, E, is

$$E = \frac{Q}{4\pi\epsilon_0 r^2}$$

$$= \frac{1.6022 \times 10^{-19}\ \text{C}}{4\pi\left(8.854 \times 10^{-12}\,\frac{\text{C}^2}{\text{N·m}^2}\right)(0.046 \times 10^{-9}\ \text{m})^2}$$

$$= \boxed{6.805 \times 10^{11}\ \text{N/C}}$$

The answer is (C).

3. From Table 16.6, the electric field intensity for a parallel plate capacitor (ignoring the effects of fringing) is

$$E = \frac{V}{r}$$

$$= \frac{12\ \text{V}}{(0.3\ \text{cm})\left(\frac{1\ \text{m}}{100\ \text{cm}}\right)}$$

$$= \boxed{4000\ \text{V/M}}$$

The answer is (D).

4. From Table 16.7,

$$\epsilon_r = 5$$

From Eq. 16.11, the permittivity is

$$\begin{aligned}\epsilon &= \epsilon_r \epsilon_0 \\ &= (5)\left(8.854 \times 10^{-12}\ \frac{\text{C}^2}{\text{N·m}^2}\right) \\ &= \boxed{4.4 \times 10^{-11}\ \frac{\text{C}^2}{\text{N·m}^2}}\end{aligned}$$

The answer is (C).

5. From Eq. 16.21, the electric flux density for a uniform line charge is

$$\begin{aligned}D &= \frac{\rho_l}{2\pi r} \\ &= \frac{30 \times 10^{-5}\ \frac{\text{C}}{\text{m}}}{2\pi(0.5\ \text{m})} \\ &= \boxed{9.5 \times 10^{-5}\ \text{C/m}^2}\end{aligned}$$

The answer is (A).

Note that 30×10^{-5} C/m correlates approximately with 30 A using an electron speed of 10^5 m/s, which is the thermal limit for electrons in silicon at room temperature.

17 Electronic Theory

PRACTICE PROBLEMS

1. Assuming no kinetic energy will be imparted to the electron, what frequency of incoming radiation is required to free an electron from the surface of silicon that has a work function of 4.8 V?

(A) 5.6×10^4 Hz
(B) 1.2×10^{15} Hz
(C) 3.5×10^{23} Hz
(D) 7.2×10^{33} Hz

2. Using the properties of germanium at room temperature, what is the conductivity of the holes?

(A) 7.6×10^{-7} S/m
(B) 7.6×10^{-2} S/m
(C) 4.8×10^{14} S/m
(D) 9.5×10^{14} S/m

3. Silicon and germanium are widely used in semiconductors. How many valence electrons exist for a given atom of either?

(A) 2
(B) 4
(C) 6
(D) 10

4. In an n-type silicon crystal, the number of electrons is

(A) equal to the number of holes
(B) greater than the number of holes
(C) less than the number of holes
(D) equal to the square of the number of holes

5. The n_i of gallium arsenide is 1.1×10^7 cm^{-3}. If the hole concentration is 1.5×10^2 cm^{-3}, what is the electron concentration?

(A) 1.5×10^2 cm^{-3}
(B) 7.3×10^4 cm^{-3}
(C) 1.1×10^7 cm^{-3}
(D) 8.1×10^{11} cm^{-3}

SOLUTIONS

1. From Eq. 17.6,

$$h\nu = \varphi + \tfrac{1}{2}m\mathrm{v}^2$$

From App. 17.A, the work function for silicon is 4.8 V. Substituting into the equation gives

$$\frac{\varphi}{h} = \frac{(4.8\ \mathrm{V})\left(1\,\frac{\mathrm{J}}{\mathrm{C{\cdot}V}}\right)\left(1.6022 \times 10^{-19}\,\frac{\mathrm{C}}{\mathrm{electron}}\right)}{6.6256 \times 10^{-34}\ \mathrm{J{\cdot}s}}$$

$$= \boxed{1.2 \times 10^{15}\ \mathrm{cycles/s\ (Hz)}}$$

The answer is (B).

2. From Eq. 17.8,

$$\sigma_h = \rho_h \mu_h$$

From App. 17.A

$$n_i = 2.5 \times 10^{12}\ \mathrm{cm^{-3}} = n_{\mathrm{holes}}$$

$$q_h = 1.6022 \times 10^{-19}\ \mathrm{C}$$

Since in an intrinsic semiconductor there are equal numbers of electrons and holes,

$$\begin{aligned}\rho_h &= n_i q_h \\ &= (2.5 \times 10^{12}\ \mathrm{cm^{-3}})(1.6022 \times 10^{-19}\ \mathrm{C}) \\ &= 4.0 \times 10^{-7}\ \frac{\mathrm{C}}{\mathrm{cm^3}}\end{aligned}$$

From App. 17.A,

$$\mu_h = 1900\ \frac{\mathrm{cm^2}}{\mathrm{V{\cdot}s}}$$

$$\begin{aligned}\sigma &= \rho_h \mu_h \\ &= \left(4.0 \times 10^{-7}\ \frac{\mathrm{C}}{\mathrm{cm^3}}\right)\left(1900\ \frac{\mathrm{cm^2}}{\mathrm{V{\cdot}s}}\right)\left(\frac{100\ \mathrm{cm}}{\mathrm{m}}\right) \\ &= \boxed{7.6 \times 10^{-2}\ \mathrm{S/m}}\end{aligned}$$

The answer is (B).

3. Silicon has an electronic structure of

$$1s^2 2s^2 2p^6 3s^2 3p^2$$

Germanium has a structure of

$$1s^2 2s^2 2p^6 3s^2 3p^6 3d^{10} 4s^2 4p^2$$

Both have four valence electrons in the outer shell.

The answer is (B).

4. An n-type crystal is doped with donor impurities, which increases the charge carrier density of electrons and thereby reduces the number of holes since by Eq. 17.13,

$$np = n_i^2 = \text{constant}$$

The answer is (B).

5. Using the law of mass action, Eq. 17.13,

$$np = n_i^2$$

$$n = \frac{n_i^2}{p} = \frac{\left(1.1\times 10^7\ \text{cm}^{-3}\right)^2}{1.5\times 10^2\ \text{cm}^{-3}}$$

$$= \boxed{8.1\times 10^{11}\ \text{cm}^{-3}}$$

The answer is (D).

18 Communication Theory

PRACTICE PROBLEMS

1. A 60 Hz signal is to be transformed into a digital signal. Determine the ideal sampling rate.

2. Determine the Nyquist interval for a 400 Hz signal.

3. A certain signal has an equivalent rectangular bandwidth of 6 MHz. Determine the response time of the system.

4. A 32 bit computer system has equal probabilities for each of its information elements, xi, or bits. Determine this probability.

5. Describe entropy as used in information theory.

6. The signal power is doubled at a given transmitter. What is the decibel increase?

(A) 0.3 dB
(B) 2 dB
(C) 3 dB
(D) 20 dB

7. A certain communication signal uses a bandwidth of 2 kHz at 101 MHz. What is the classification of this signal?

(A) narrowband, VLF
(B) voice-grade, HF
(C) voice-grade, VHF
(D) wideband, UHF

8. A certain communication system uses the left-most bit as an even parity bit. Which of the following messages is in error?

(A) 11101
(B) 00101
(C) 10001
(D) 01011

9. What is the maximum thermal noise voltage generated in a 5 kΩ resistor operating in a 6 MHz bandwidth and designed for a temperature service range of −10°C to 60°C?

(A) 9.71×10^{-6} V
(B) 1.14×10^{-5} V
(C) 2.35×10^{-5} V
(D) 2.46×10^{-5} V

10. What is the shot noise current for a microphone with a 16 kHz bandwidth drawing 500 mA?

(A) 2.60×10^{-12} A
(B) 2.56×10^{-8} A
(C) 2.53×10^{-8} A
(D) 5.06×10^{-8} A

SOLUTIONS

1. The ideal sampling rate is given by the Nyquist rate.

From Eq. 18.1,

$$\begin{aligned} f_S &= 2f_I \\ &= (2)(60\,\text{Hz}) \\ &= \boxed{120\,\text{Hz}} \end{aligned}$$

Note that, in practice, the rate would have to be greater than this to avoid a zero output, which would occur if the initial sampling occurred at a value of zero for the analog signal.

2. From Eq. 18.2, the Nyquist interval, T_S, is

$$T_S = \frac{1}{2f_I}$$

The information signal, or message signal, frequency is given as 400 Hz. therefore,

$$\begin{aligned} T_S &= \frac{1}{(2)(400\,\text{Hz})\left(\frac{\text{s}^{-1}}{1\,\text{Hz}}\right)} \\ &= \boxed{1.25\ \text{ms}} \end{aligned}$$

3. From Eq. 18.4, the bandwidth and duration, or response time, are related by the following uncertainty relationship.

$$\Delta T \Delta \Omega = 2\pi$$

Rearranging and substituting,

$$\begin{aligned} \Delta T &= \frac{2\pi}{\Delta\Omega} = \frac{2\pi}{6\times 10^6\ \text{s}^{-1}} \\ &= \boxed{1.047\times 10^{-6}\ \text{s}} \end{aligned}$$

4. The information content of a message is given as 32 bits. The information content is related to the probability as follows.

From Eq. 18.6,

$$I(x_i) = \log_2\left(\frac{1}{p(x_i)}\right)$$

Rearranging and substituting,

$$\begin{aligned} p(x_i) &= \frac{1}{\text{antilog}_2 I(x_i)} \\ &= \frac{1}{\text{antilog}_2(32)} \\ &= \boxed{\frac{1}{4.295\times 10^9}} \end{aligned}$$

5. Entropy is the mean or average value of a message. It can be considered a measure of the disorder of a system, or of uncertainty in a message. For a given message of alphabet size A, the entropy is determined by Eq. 18.7.

$$H(X) = \sum_{i=1}^{A} p(x_i)I(x_i)$$

It is normally given in base 2, in units of nats.

6. The decibel is defined by the following ratio of power, P_{dB}.

$$P_{\text{dB}} = 10\log_{10}\left(\frac{P_2}{P_1}\right)$$

Since power doubled, P_2/P_1, equals 2,

$$\begin{aligned} P_{\text{dB}} &= 10\log_{10}(2) \\ &= \boxed{3\ \text{dB}} \end{aligned}$$

The answer is (C).

7. Classification can mean a number of things. Given the possible choices, this signal is a voice-grade channel (BW 300 Hz to 4 kHz) in the VHF range (30-300 MHz).

The answer is (C).

8. The message is contained in the bits to the right of the parity bit. The parity represents an even or odd total of ones, with an even parity bit zero indicating even and one indicating odd. The following table is helpful.

	message	parity	signal
(A)	1101	1	11101
(B)	0101	0	00101
(C)	0001	1	10001
(D)	1011	1	11011

By comparison, message (D) is in error.

The answer is (D).

9. From Eq. 18.10, the applicable thermal noise voltage formula is

$$\overline{v^2} = 4kTR\,(\text{BW})$$

$$v = \sqrt{4kTR\,(\text{BW})}$$

The maximum voltage occurs at the maximum temperature (60°C or 333K), thus

$$v = \sqrt{\begin{array}{c}(4)\left(1.38 \times 10^{-23}\ \frac{\text{J}}{\text{K}}\right)(333\text{K}) \\ \times (5 \times 10^{3}\Omega)(6 \times 10^{6}\ \text{Hz})\end{array}}$$

$$= \boxed{2.35 \times 10^{-5}\ \text{V}}$$

The answer is (C).

10. From Eq. 18.13, the shot noise current is determined from

$$\overline{i_n^2} = 2qI\,(\text{BW})$$

$$i_n = \sqrt{2qI\,(\text{BW})}$$

$$= \sqrt{\begin{array}{c}(2)(1.6022 \times 10^{-19}\ \text{C}) \\ (500 \times 10^{-3}\ \text{A})(16 \times 10^{3}\ \text{Hz})\end{array}}$$

$$= \boxed{5.06 \times 10^{-8}\ \text{A}}$$

The answer is (D).

Theory

19 Acoustic and Transducer Theory

PRACTICE PROBLEMS

1. Explain two differences between sound waves and electromagnetic waves.

2. Define the term "transducer."

3. What is the sound pressure level in air for a signal at the pain sensation threshold of 1 W/m^2?

(A) −120 dB
(B) 3 dB
(C) 60 dB
(D) 120 dB

4. Explain the difference between photovoltaic and photoconductive transduction.

5. The reference pressure level, p_0, in air of 20 μPa correlates with the threshold audible frequency of 2000 Hz. What is the wavelength of this sound wave?

(A) 0.17 m
(B) 1.0 m
(C) 6.1 m
(D) 20 m

SOLUTIONS

1. Sound is a longitudinal or compression wave. Electromagnetic waves are transverse. That is, the electric and magnetic fields are at right angles to the direction of propagation. Additionally, sound waves require a medium in which to travel, while electromagnetic waves do not.

2. A transducer is a communication interface that provides a usable output in response to a specified measurand (i.e., a physical quantity, property, or condition).

3. From Ex. 19.2, the pressure level corresponding to 1 W/m^2 is 29.3 Pa. The reference pressure is defined as 20 μPa. Substituting from Eq. 19.8 gives

$$\begin{aligned} L_p &= 20 \log\left(\frac{p}{p_0}\right) \\ &= 20 \log\left(\frac{29.3\ \text{Pa}}{20 \times 10^{-6}\ \text{Pa}}\right) \\ &= \boxed{123\ \text{dB}\ (120\ \text{dB})} \end{aligned}$$

The answer is (D).

Note that the "pain" threshold is nominally 120 dB in most references.

4. The primary difference is that a photovoltaic device uses dissimilar metals and a photoconductive device uses a semiconductor *pn* junction.

5. From Table 19.1, the speed of sound in air is approximately 330 m/s. Substituting this and the given frequency into Eq. 19.1 gives

$$\text{v}_s = f\lambda$$

$$\begin{aligned} \lambda &= \frac{\text{v}_s}{f} = \frac{330\ \frac{\text{m}}{\text{s}}}{2000\ \text{Hz}} \\ &= \boxed{0.165\ \text{m}\quad (0.17\ \text{m})} \end{aligned}$$

The answer is (A).

20 Electrostatics

PRACTICE PROBLEMS

1. Determine the magnitude of the electric field necessary to place a 1 N force on an electron.

2. A point charge of 4.8×10^{-19} C is located at a point (3,1,0). What is the electric field strength at (3,0,1)?

(A) 1.5×10^{-9} V/m $\left(\frac{-\mathbf{y}}{\sqrt{2}} + \frac{\mathbf{z}}{\sqrt{2}}\right)$

(B) 2.2×10^{-9} V/m $\left(\frac{-\mathbf{y}}{\sqrt{2}} + \frac{\mathbf{z}}{\sqrt{2}}\right)$

(C) 3.1×10^{-9} V/m $(-\mathbf{y} + \mathbf{z})$

(D) 4.3×10^{-9} V/m $(-\mathbf{y} + \mathbf{z})$

3. Water molecules in vapor form have a fractional dipole charge of approximately 5.26×10^{-20} C. If the dipole bond length is 1.17 Å, what is the magnitude of the dipole moment?

(A) 6.2×10^{-32} C·m

(B) 6.2×10^{-30} C·m

(C) 5.3×10^{-20} C·m

(D) 6.1×10^{-20} C·m

4. The potential of a point charge, in spherical coordinates and relative to infinity, is

$$V = \frac{Q}{4\pi\varepsilon_0 r}$$

Determine the expression for the electric field strength, **E**.

5. The electric field in an n-type material outside the space charge region is given by

$$\mathbf{E} = 2\mathbf{x} + 2\mathbf{y} + 3\mathbf{z}$$

Determine the charge density, ρ, and explain the result.

SOLUTIONS

1. The magnitude of the force is given by

$$F = QE$$

Rearranging gives

$$\begin{aligned} E &= \frac{F}{Q} \\ &= \frac{1 \text{ N}}{1.602 \times 10^{-19} \text{ C}} \\ &= 6.2 \times 10^{18} \ \frac{\text{N}}{\text{C}} = 6.2 \times 10^{18} \text{ V/m} \end{aligned}$$

For determining the magnitude of the force and not the direction, the minus sign is not required on the charge.

2.

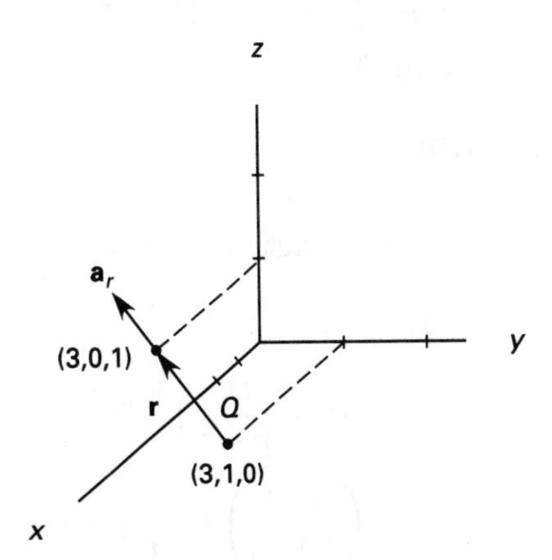

The electric field is given by

$$\mathbf{E} = \left(\frac{Q}{4\pi\epsilon r^2}\right)\mathbf{a}_r$$

First, determine the value of the unit vector, $\mathbf{a}_r$.

$$a_r = \sqrt{(3-3)^2 + (0-1)^2 + (1-0)^2} = \sqrt{2}$$

$$\mathbf{r} = 0\mathbf{x} - 1\mathbf{y} + 1\mathbf{z}$$

Thus,

$$\mathbf{a}_r = \frac{\mathbf{r}}{a_r} = \frac{1}{\sqrt{2}}(-\mathbf{y}+\mathbf{z})$$

Substituting this and the given information (free space is assumed) gives

$$\begin{aligned}\mathbf{E} &= \left(\frac{Q}{4\pi\epsilon r^2}\right)\mathbf{a}_r \\ &= \left(\frac{4.8\times10^{-19}\ \text{C}}{4\pi\left(8.854\times10^{-12}\ \frac{\text{C}^2}{\text{N}\cdot\text{m}^2}\right)\left(\sqrt{2}\ \text{m}\right)^2}\right) \\ &\quad\times\left(\left(\frac{1}{\sqrt{2}}\right)(-\mathbf{y}+\mathbf{z})\right) \\ &= \boxed{2.16\times10^{-9}\ \text{N/C}\left(-\frac{\mathbf{y}}{\sqrt{2}}+\frac{\mathbf{z}}{\sqrt{2}}\right)}\end{aligned}$$

The answer is (B).

3. The dipole moment is given by

$$\mathbf{p} = q\mathbf{d}$$

An angstrom, Å, equals 10^{-10} m. Substituting gives

$$\begin{aligned}p = qd &= \left(5.26\times10^{-20}\ \text{C}\right)\left(1.17\times10^{-10}\ \text{m}\right) \\ &= 6.18\times10^{-30}\ \text{C}\cdot\text{m}\end{aligned}$$

The answer is (B).

4. The electric field strength is given by

$$\begin{aligned}\mathbf{E} &= -\nabla V \\ &= -\nabla\left(\frac{Q}{4\pi\epsilon_0 r}\right) \\ &= \left(-\frac{Q}{4\pi\epsilon_0}\right)\left(\nabla\frac{1}{r}\right)\end{aligned}$$

In spherical coordinates, the del operator can be represented as

$$\nabla = \left(\frac{\delta}{\delta r}\right)\mathbf{r} + \left(\frac{1}{r}\right)\left(\frac{\delta}{\delta\theta}\right)\boldsymbol{\theta} + \left(\frac{1}{r\sin\theta}\right)\left(\frac{\delta}{\delta\phi}\right)\boldsymbol{\phi}$$

For a point charge, the potential and the electric field are constant with respect to θ and ϕ. Thus,

$$\begin{aligned}\mathbf{E} &= \left(-\frac{Q}{4\pi\epsilon_0}\right)\left(\frac{\delta\frac{1}{r}}{\delta r}\right)\mathbf{r} \\ &= \left(\frac{Q}{4\pi\epsilon_0 r^2}\right)\mathbf{r} = \left(\frac{Q}{4\pi\epsilon_0 r^2}\right)\mathbf{a}_r\end{aligned}$$

The last equation changes the symbology from $\mathbf{r}$ to $\mathbf{a}_r$ only. Many variations exist.

5. The charge density can be determined from Gauss's law.

$$\begin{aligned}\text{div}\ \mathbf{E} &= \nabla\cdot\mathbf{E} = \frac{\rho}{\epsilon} \\ \nabla\cdot\mathbf{E} &= \nabla\cdot(2\mathbf{x}+2\mathbf{y}+3\mathbf{z}) \\ &= \left(\frac{\delta 2}{\delta x}\right)\mathbf{x} + \left(\frac{\delta 2}{\delta y}\right)\mathbf{y} + \left(\frac{\delta 3}{\delta z}\right)\mathbf{z} = 0\end{aligned}$$

When the electric field is constant, the divergence is zero. This indicates there is no *net charge* in the region, as is the case for the *pn* junction not under the influence of an external electric field.

21 Electrostatic Fields

PRACTICE PROBLEMS

1. Determine the electric flux density magnitude if the polarization is 2.0×10^{-6} C/m^2 and the susceptibility is 2.2.

(A) 1.8×10^{-17} C/m^2
(B) 1.8×10^{-6} C/m^2
(C) 2.9×10^{-6} C/m^2
(D) 4.0×10^{-6} C/m^2

2. Given $\mathbf{D} = 2 \times 10^{-6}\ \mathbf{a}_r$ C/m^2 and $\varepsilon_r = 3.0$, what is the electric field strength?

(A) $6.7 \times 10^{-7}\ \mathbf{a}_r$ N/C
(B) $2.0 \times 10^{-4}\ \mathbf{a}_r$ N/C
(C) $7.5 \times 10^{4}\ \mathbf{a}_r$ N/C
(D) $2.3 \times 10^{5}\ \mathbf{a}_r$ N/C

3. The following values for relative permittivity are taken from a standard reference: paper, 2.0; quartz, 5.0; and marble, 8.3.

The electric field strength is 2.0×10^5 V/m. $\mathbf{D}$ must be greater than or equal to 7×10^{-6} C/m^2. Determine which of the given dielectrics is/are suitable.

4. Moist soil has a conductivity of approximately 10^{-3} S/m and an ε_r of 3.0. A certain electric field strength is given by

$$E = 3.5 \times 10^{-6} \sin\left(2.5 \times 10^{3}\right) t \ \frac{\text{V}}{\text{m}}$$

Determine the displacement current density.

5. In what circuit element is displacement current largest: conductor, inductor, or capacitor? Explain.

SOLUTIONS

1. The polarization is

$$P = \chi_e \varepsilon_0 E$$

This assumes that $\mathbf{E}$ and $\mathbf{P}$ are in the same direction, that is, the material is isotropic and linear. Given that, the electic flux density is

$$D = \varepsilon_0 E + P$$

Substituting P/χ_e for $\varepsilon_0 E$,

$$\begin{aligned} D &= \frac{P}{\chi_e} + P \\ &= \frac{2 \times 10^{-6}}{2.2}\ \frac{\text{C}}{\text{m}^2} + 2 \times 10^{-6}\ \frac{\text{C}}{\text{m}^2} \\ &= \boxed{2.9 \times 10^{-6}\ \text{C/m}^2} \end{aligned}$$

The answer is (C).

2. $\mathbf{D}$ is given by

$$\mathbf{D} = \varepsilon_0 \varepsilon_r \mathbf{E}$$

Rearranging gives

$$\begin{aligned} \mathbf{E} &= \frac{\mathbf{D}}{\varepsilon_0 \varepsilon_r} = \frac{2 \times 10^{-6}\ \mathbf{a}_r\ \frac{\text{C}}{\text{m}^2}}{\left(8.854 \times 10^{-12}\ \frac{\text{C}^2}{\text{N}\cdot\text{m}^2}\right)(3.0)} \\ &= \boxed{7.5 \times 10^{4}\ \mathbf{a}_r\ \text{N/C}} \end{aligned}$$

The answer is (C).

3. Determine the minimum value of ε_r required.

$$\mathbf{D} = \varepsilon_0 \varepsilon_r \mathbf{E}$$

Assuming the dielectrics to be isotropic and linear,

$$D = \varepsilon_0 \varepsilon_r \text{E}$$

$$\begin{aligned} \varepsilon_r &= \frac{D}{E\varepsilon_0} = \frac{7 \times 10^{-6}\ \frac{\text{C}}{\text{m}^2}}{\left(2.0 \times 10^{5}\ \frac{\text{V}}{\text{m}}\right)\left(8.854 \times 10^{-12}\ \frac{\text{C}^2}{\text{N}\cdot\text{m}^2}\right)} \\ &= 3.9 \end{aligned}$$

Thus, the quartz or marble is an acceptable dielectric.

4. Determine the value of D.

$$\begin{aligned} D &= \varepsilon_0 \varepsilon_r \mathbf{E} \\ &= \left(8.854 \times 10^{-12}\ \frac{\text{C}^2}{\text{N}\cdot\text{m}^2}\right)(3.0) \\ &\quad \times \left(3.5 \times 10^{-6} \sin\left(2.5 \times 10^3\right) t\ \frac{\text{V}}{\text{m}}\right) \\ &= 9.30 \times 10^{-17} \sin\left(2.5 \times 10^3\right) t\ \text{A/m}^2 \end{aligned}$$

The displacement current is

$$\begin{aligned} \mathbf{J}_d &= \frac{\delta \mathbf{D}}{\delta t} \\ &= \frac{\delta}{\delta t}\left(9.30 \times 10^{-17} \sin\left(2.5 \times 10^3\right) t\ \frac{\text{A}}{\text{m}^2}\right) \\ &= \left(9.30 \times 10^{-17}\right)\left(2.5 \times 10^3 \cos\left(2.5 \times 10^3\right) t\ \frac{\text{A}}{\text{m}^2}\right) \\ &= \boxed{2.32 \times 10^{-13} \cos\left(2.5 \times 10^3\right) t\ \text{A/m}^2} \end{aligned}$$

5. The displacement current is largest in a capacitor. The charge stored as a capacitor is

$$Q = CV$$

The capacitance is proportional to the area.

$$C \propto A$$

Capacitors, in general, have large areas and store high concentrations of charges. Thus, the electric field strength is large and consequently so is $\mathbf{D}$. At high frequencies this results in a significant displacement current.

22 Magnetostatics

PRACTICE PROBLEMS

1. The flux density a distance r from a current-carrying conductor of length $d\mathbf{l}$ varies as

(A) $\ln r$
(B) $1/r$
(C) $1/r^2$
(D) constant

2. Determine the magnitude of the core flux density for one turn of wire wrapped around a cast-iron core with a relative permeability, μ_r, of 400. The wire is rated at 15 A. The core measures 0.10 m × 0.15 m.

3. In a straight, infinitely long conductor, how much current does it take to produce a magnetic field 1 m away that is equivalent in strength to the Earth's, that is, 5×10^{-5} T?

4. Consider the following figure.

B ⊗ ⊗ ⊗ ← e^-
⊗ ⊗ ⊗
⊗ ⊗ ⊗

The magnetic field, **B**, is uniform and directed into the paper. An electron is injected into the field from the right with some initial velocity. In what direction does the electron initially deflect?

(A) up
(B) down
(C) left
(D) right

5. A galvanometer moving-coil, used in instrumentation, is placed between two permanent magnets with a **B**-field strength of 4.0×10^{-3} T in the directions shown. The radius of the coil is 0.5 cm. The coil contains 10 turns, or loops, and has a current rating of 1 mA. Determine the maximum torque that the meter can generate against the restoring spring.

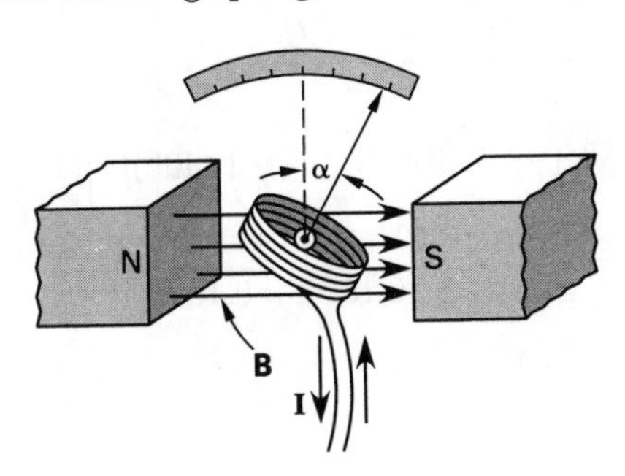

SOLUTIONS

1. The Biot-Savart law gives the magnitude of flux density in differential form as

$$dB = \left(\frac{\mu_0 I dl}{4\pi r^2}\right)\sin\theta$$

The answer is (C).

2. The flux is

$$\Phi = \mu N I l = BA$$

Rearranging to solve for B gives

$$B = \frac{\mu N I l}{A} = \frac{\mu_0 \mu_r N I l}{A}$$

$$p = \frac{\left(1.2566\times 10^{-6}\ \frac{\text{N}}{\text{A}^2}\right)(400)(1)(15\ \text{A})\times(0.10+0.15+0.10+0.15\ \text{m})}{0.10\ \text{m}\times 0.15\ \text{m}}$$

$$= 0.25\ \text{T}$$

3. The magnetic field, B, of a straight, infinitely long conductor is

$$B = \frac{\mu I}{2\pi r}$$

Since no specific condition was given, free space is assumed. Rearranging and substituting gives

$$I = \frac{2\pi r B}{\mu_0}$$

$$= \frac{2\pi\,(1\ \text{m})\,(5\times 10^{-5}\ \text{T})}{1.2566\times 10^{-6}\ \frac{\text{H}}{\text{m}}} = 250\ \text{A}$$

4. The force on a moving charged particle is given by

$$\mathbf{F} = q\mathbf{V}\times\mathbf{B}$$

Using the right-hand rule to determine the direction of the cross-product, the initial deflection is down toward the bottom of the paper.

The answer is (B).

5. The torque is

$$\mathbf{T} = \mathbf{m}\times\mathbf{B}$$

The strength of the **B**-field is given. The magnetic moment, m, can be determined from

$$m = IA$$

Since there are N turns of wire, each carrying a current I, the equation is

$$m = NIA = NI(\pi r^2)$$

$$= (10)\,(0.001\ \text{A})\,(\pi)\,(0.005\ \text{m})^2$$

$$= 7.85\times 10^{-7}\ \text{A}\cdot\text{m}^2$$

The maximum magnetic torque can now be determined by substitution.

$$\mathbf{T} = \mathbf{m}\times\mathbf{B}$$

$$T_{\text{max}} = mB\sin\left(\frac{\pi}{2}\right) = mB$$

$$= \left(7.85\times 10^{-7}\ \text{A}\cdot\text{m}^2\right)\left(4.0\times 10^{-3}\ \text{T}\right)$$

$$= 3.1\times 10^{-9}\ \text{N}\cdot\text{m}$$

23 Magnetostatic Fields

PRACTICE PROBLEMS

1. Define magnetization and state its mathematical formula with appropriate units.

2. Write the formula for the magnetic flux density that is always valid regardless of the linearity of the material. Use the SI form of the equation.

3. By what factor does the flux density increase if high-purity iron is substituted for commercial iron as the core of a transformer?

(A) 4
(B) 30
(C) 6000
(D) 2×10^5

4. The magnetic field strength of Earth is approximately 40 A/m. Determine the current required in a single turn of wire located at the equator to generate the same magnetic field at distances far from the Earth's surface.

5. What type of magnetic material generates an internal magnetic field that opposes the applied field?

(A) diamagnetic
(B) paramagnetic
(C) ferromagnetic
(D) antiferromagnetic

SOLUTIONS

1. Magnetization, **M**, is the magnetic moment per unit volume, and is given by

$$\mathbf{M} = \frac{\mathbf{m}}{v}$$

The units are A/m.

2. The formula is

$$\mathbf{B} = \mu_0 \mathbf{H} + \mu_0 \mathbf{M}$$

3. The relative permeability of commercial-grade iron is approximately 6000, while that of high-purity iron is 2×10^5. The flux density magnitude is

$$B = \mu_0 \mu_r H$$

The only change from one material to the next is μ_r. Thus, the factor, f, is

$$\begin{aligned} f &= \frac{\mu_r \text{ (high purity)}}{\mu_r \text{ (commercial)}} \\ &= \frac{2 \times 10^5}{6000} = 33 \quad (30) \end{aligned}$$

The answer is (B).

4. At distances far from Earth's surface, the applicable equation is

$$H = \frac{NI}{l}$$

Rearranging gives

$$I = \frac{Hl}{N}$$

The radius of the earth is approximately 6.4×10^6 m.

The path length is

$$\begin{aligned} l &= 2\pi r = 2\pi(6.4 \times 10^6 \text{ m}) \\ &= 4.0 \times 10^7 \text{ m} \end{aligned}$$

Fields

Substituting the known and calculated information gives

$$I = \frac{Hl}{N} = \frac{\left(40\ \frac{\text{A}}{\text{m}}\right)(4.0 \times 10^7\ \text{m})}{1}$$
$$= 1.6 \times 10^9\ \text{A}$$

5. Diamagnetic materials tend to reduce the flux by generating an opposing internal magnetic field.

The answer is (A).

24 Electrodynamics

PRACTICE PROBLEMS

1. The following information is taken from a certain portable generator: v = 1450 m/s, B = 1.2×10^{-2} T, loop length = 0.4 m, and output voltage = 120 V. Estimate the number of turns required to induce the indicated voltage.

(A) 1
(B) 15
(C) 20
(D) 120

The answer is (C).

2. Which law describes induced electromotance?

(A) Faraday's law
(B) Ampere's law
(C) Coulomb's law
(D) Gauss's law

The answer is (A).

3. Which of the following terms represents magnetomotive force, or mmf?

(A) $\oint \mathbf{E} \cdot d\mathbf{l}$
(B) $\oint \varepsilon \mathbf{D} \cdot d\mathbf{l}$
(C) $\mu_0 \mathbf{H}$
(D) $\oint \mathbf{H} \cdot d\mathbf{l}$

The answer is (D).

4. Consider the following figure.

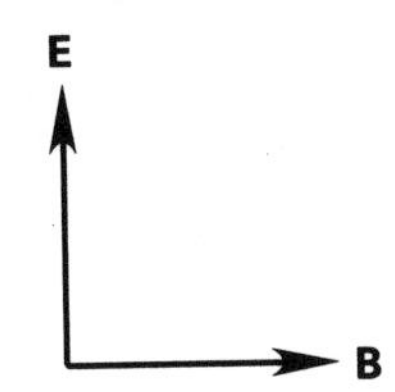

Determine the direction of the energy flow in the figure.

5. What fundamental action must occur to a charge in order to produce an electromagnetic wave capable of transporting energy in free space?

SOLUTIONS

1. The induced voltage resulting from a conductor moving in a uniform magnetic field is

$$v = -NBl\mathrm{v}$$

Rearranging and solving gives

$$N = \left| \frac{v}{Bl\mathrm{v}} \right|$$

$$= \frac{120 \text{ V}}{(1.2 \times 10^{-2} \text{ T})(0.4 \text{ m})\left(1450 \ \frac{\text{m}}{\text{s}}\right)} = 17.2$$

Therefore, 18 turns are required as a minimum.

The answer is (C).

2. Faraday's law of magnetic induction is used to determine the amount of induced voltage, also called induced electromotance.

The answer is (A).

3. The mmf is the integral of the magnetic field strength, **H**, along the path, $d\mathbf{l}$, where the magnetic flux is present. It is often given as NI, the number of turns times the current. The force produced per unit flux set up in the conductor is **H**, hence the name.

The answer is (D).

4. The direction of energy flow, or power density, is given by Poynting's vector, **S**.

$$\mathbf{S} = \mathbf{E} \times \mathbf{H}$$

In terms of the **B**-field, assuming a linear, isotropic medium,

$$\mathbf{S} = \frac{1}{\mu} \mathbf{E} \times \mathbf{B}$$

If the material were not isotropic, the permeability would remain with the **B**-field as part of the cross product. The direction of **S** would then depend on the product $\mu\mathbf{B}$ as well.

Using the right-hand rule, determine the direction of the cross product **S** is into the plane of the paper, ⊗.

Fields

5. The charge must be accelerated or decelerated by some means.

Note then that a stationary charge is related to an electric field, a charge in motion with uniform velocity is related to a magnetic field, and an accelerating charge is related to a radiated field.

25 Maxwell's Equations

PRACTICE PROBLEMS

There are no practice problems corresponding with Ch. 25 of the *Electrical Engineering Reference Manual.*

26 DC Circuit Fundamentals

PRACTICE PROBLEMS

1. What is the power dissipated in R_3 in the circuit shown?

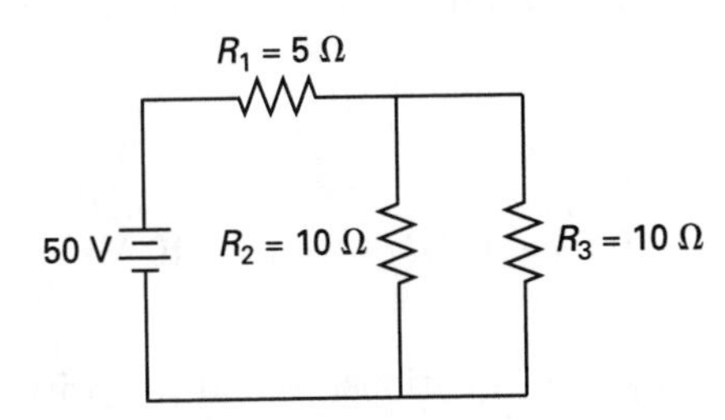

2. Calculate the resistance of the load necessary for maximum power transfer to occur in the circuit shown.

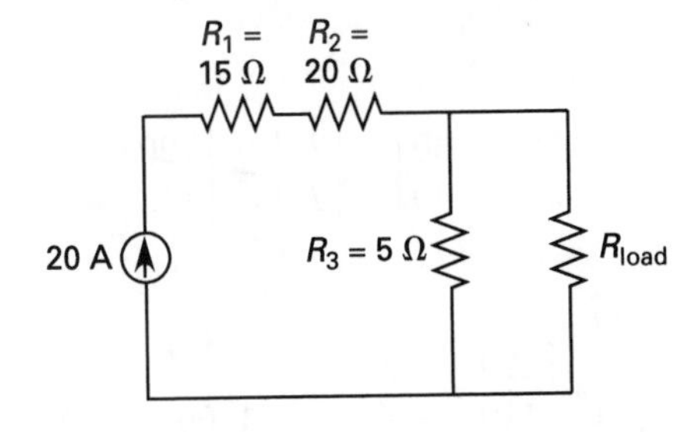

3. For the circuit shown, which method will result in the lesser number of equations: the loop-current method or the node-voltage method?

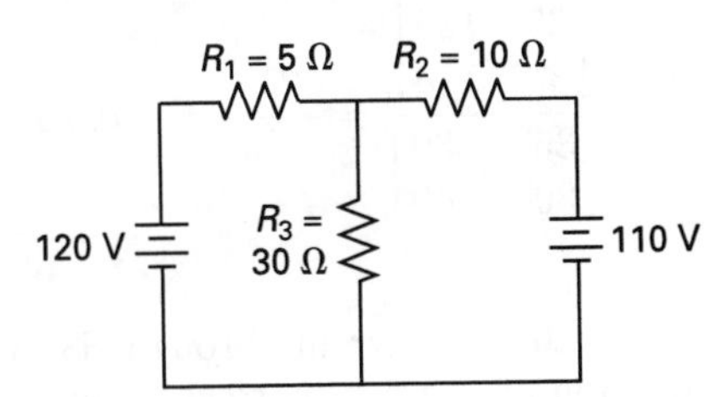

4. Using the loop-current method, solve for the current in R_3 in Prob. 3.

5. Using the node-voltage method, solve for the current in R_3 in Prob. 3.

SOLUTIONS

1. Power is given by

$$P = VI = I^2R = \frac{V^2}{R}$$

Using the last form of the equation, determine the voltage across R_3. Combine the parallel resistors R_2 and R_3.

$$R_{2/3} = \frac{R_2R_3}{R_2 + R_3} = \frac{(10\ \Omega)(10\ \Omega)}{10\ \Omega + 10\ \Omega} = 5\ \Omega$$

The circuit is transformed as follows.

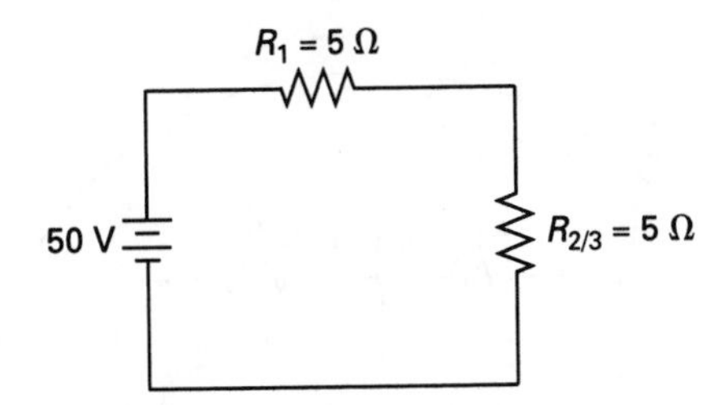

Using the voltage-divider concept, the voltage across $R_{2/3}$ (and R_3) is

$$V_{2/3} = V_3 = (50\ \text{V})\left(\frac{R_{2/3}}{R_1 + R_{2/3}}\right) = (50\ \text{V})\left(\frac{5\ \Omega}{10\ \Omega}\right)$$
$$= 25\ \text{V}$$

The power dissipated in R_3 is

$$P_3 = \frac{V_3^2}{R_3} = \frac{(25\ \text{V})^2}{10\ \Omega}$$
$$= \boxed{62\ \text{W}}$$

2. Maximum power transfer occurs when the load resistance equals the Thevenin or Norton equivalent resistance.

Open circuit the current source and determine R_{Th}.

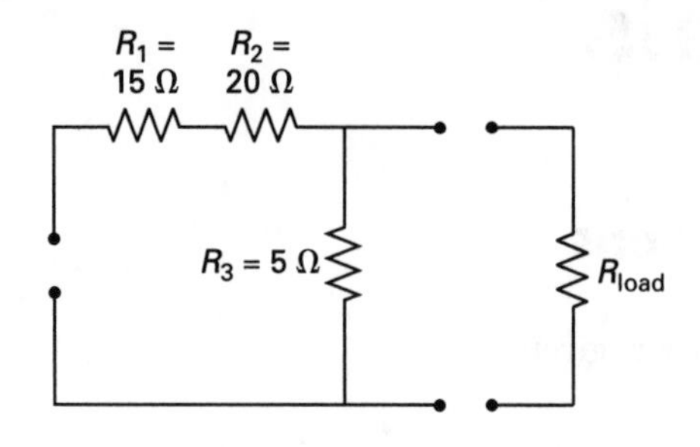

The 15 Ω and 20 Ω resistors are in an open-circuit path. Thus,

$$R_{\text{Th}} = R_3 = \boxed{5\ \Omega}$$

3. The loop-current method using KVL uses $n-1$ equations, where n is the number of loops. There are three loops present. The number of equations required is two.

The node-voltage method using KCL requires $n - 1$ equations, where n is the number of principal nodes. There are two principal nodes. Only one equation is required.

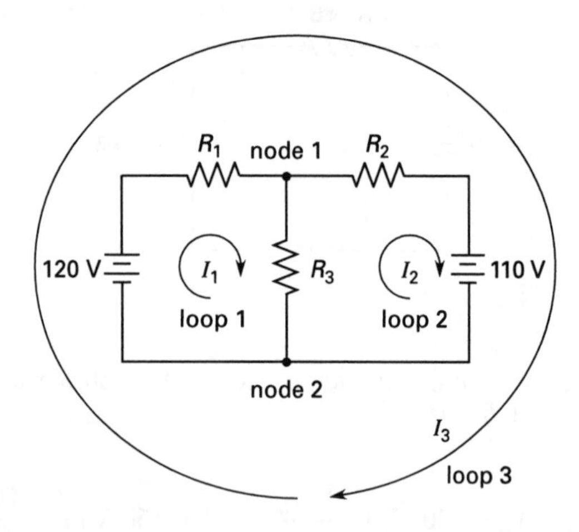

The node-voltage equation requires a lesser number of equations.

4. Consider the following circuit.

Using the selected loops with polarities for the resistive elements based on the loop-current direction, KVL for loop 1 is

$$120\ \text{V} - I_1R_1 - (I_1 - I_2)\,R_3 = 0$$

Rearranging gives

$$I_1\,(R_1 + R_3) - I_2R_3 = 120\ \text{V} \qquad \text{[I]}$$

KVL for loop 2 is

$$-\,(I_2 - I_1)\,R_3 - I_2R_2 - 110\ \text{V} = 0$$

Rearranging gives

$$I_1R_3 - I_2\,(R_3 + R_2) = 110\ \text{V} \qquad \text{[II]}$$

There are now two equations, [I] and [II], with two unknowns, I_1 and I_2. Use Cramer's rule and matrix algebra to solve for I_1 and I_2.

$$\begin{vmatrix} R_1 + R_3 & -R_3 \\ R_3 & -(R_3 + R_2) \end{vmatrix} \begin{vmatrix} I_1 \\ I_2 \end{vmatrix} = \begin{vmatrix} 120 \\ 110 \end{vmatrix}$$

$$\begin{vmatrix} 35 & -30 \\ 30 & -40 \end{vmatrix} \begin{vmatrix} I_1 \\ I_2 \end{vmatrix} = \begin{vmatrix} 120 \\ 110 \end{vmatrix}$$

$$I_1 = \frac{\begin{vmatrix} 120 & -30 \\ 110 & -40 \end{vmatrix}}{\begin{vmatrix} 35 & -30 \\ 30 & -40 \end{vmatrix}} = \frac{-1500}{-500} = 3\ \text{A}$$

$$I_2 = \frac{\begin{vmatrix} 35 & 120 \\ 30 & 110 \end{vmatrix}}{\begin{vmatrix} 35 & -30 \\ 30 & -40 \end{vmatrix}} = \frac{250}{-500} = -0.5\ \text{A}$$

Summing to obtain the current through R_3 and noting that the direction of I_2 was incorrect gives

$$I_3 = |I_1| + |I_2| = 3\ \text{A} + 0.5\ \text{A} = \boxed{3.5\ \text{A}}$$

Note that the direction of the current I_3 is identical to the direction of I_1, since it is positive.

5. Consider the following circuit.

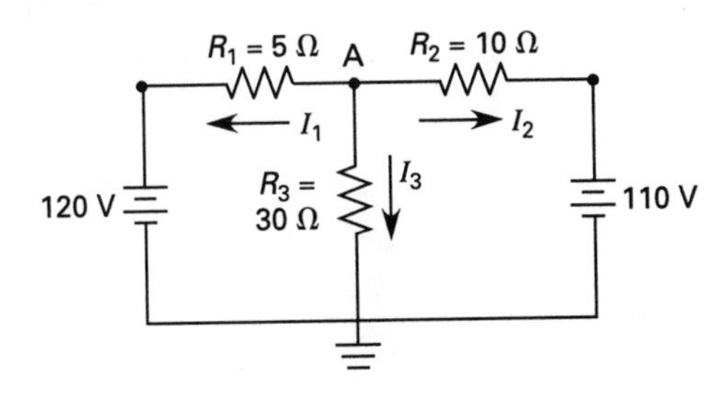

The principal node A is labeled. The node at the lower portion of the circuit is chosen as the reference or ground node. All currents are assumed to be positive.

Writing KCL for node A gives

$$\frac{V_A - 120\text{ V}}{R_1} + \frac{V_A - 0}{R_3} + \frac{V_A - 110\text{ V}}{R_2} = 0$$

Solve for V_A.

$$\frac{V_A}{R_1} - \frac{120\text{ V}}{R_1} + \frac{V_A}{R_3} + \frac{V_A}{R_2} - \frac{110\text{ V}}{R_2} = 0$$

$$\frac{V_A}{R_1} + \frac{V_A}{R_3} + \frac{V_A}{R_2} = \frac{120\text{ V}}{R_1} + \frac{110\text{ V}}{R_2}$$

$$V_A = \frac{\frac{120\text{ V}}{R_1} + \frac{110\text{ V}}{R_2}}{\frac{1}{R_1} + \frac{1}{R_3} + \frac{1}{R_2}}$$

$$= \frac{\frac{120\text{ V}}{5\ \Omega} + \frac{110\text{ V}}{10\ \Omega}}{\frac{1}{5\ \Omega} + \frac{1}{30\ \Omega} + \frac{1}{10\ \Omega}}$$

$$= 105\text{ V}$$

Use V_A to determine the desired current.

$$I_3 = \frac{V_A}{R_3} = \frac{105\text{ V}}{30\ \Omega} = \boxed{3.5\text{ A}}$$

Circuit Theory

27 AC Circuit Fundamentals

PRACTICE PROBLEMS

1. Express the phasor voltage $120\angle 30°$ as a function of time if the frequency is 60 Hz.

2. Express each of the following in rectangular form.

(a) $15\angle -180°$

(b) $10\angle 37°$

(c) $50\angle 120°$

(d) $21\angle -90°$

3. Express each of the following in phasor form.

(a) $(6 + j7)$

(b) $(50 - j60)$

(c) $(-75 + j45)$

(d) $(90 - j180)$

4. Calculate the following and express in phasor form.

(a) $\dfrac{7 + j5}{6}$

(b) $\dfrac{13 + j17}{15 - j10}$

(c) $\dfrac{0.020\angle 90°}{0.034\angle 56°}$

5. A sinusoidal waveform has a peak value at $t = 1$ ms and a period of 10 ms. Express this signal as a cosine and sine function.

6. A voltage of $10\cos(100t + 25°)$ is applied to a resistance of 25 Ω and an inductance of 0.50 H in parallel. What are the (a) phase angle and (b) rms value of the sum of the currents?

7. Draw the power triangle for the circuit in Prob. 6.

8. What is the capacitance, in units of farads, to completely correct the power factor of the circuit in Prob. 6, that is, to make $Q = 0$?

9. An 80 μF capacitor is in series with a 9 mH inductor. What is the total impedance (at 60 Hz), and is the circuit considered a leading or lagging circuit?

10. If the capacitor and inductor in Prob. 9 are connected in parallel, what is the total impedance, and is the circuit considered leading or lagging?

SOLUTIONS

1. As a function of time,

$$v(t) = V_p \sin(\omega t + \theta)$$

The peak voltage is

$$\begin{aligned} V_p &= V_{\text{eff}}\sqrt{2} \\ &= 120 \text{ V}\sqrt{2} = 169.71 \text{ V} \end{aligned}$$

The angular frequency is

$$\begin{aligned} \omega &= 2\pi f = 2\pi \left(60 \text{ s}^{-1}\right) \\ &= 376.99 \text{ rad/s} \end{aligned}$$

The given angle of 30° equals $\pi/6$ rad, thus

$$v(t) = \boxed{170 \sin\left(377t + \frac{\pi}{6}\right)}$$

2. (a) $15\cos(-180°) + j15\sin(-180°)$

$$= \boxed{-15 + j0}$$

(b) $10\cos(37°) + j10\sin(37°)$

$$= \boxed{7.99 + j6.02}$$

(c) $50\cos(120°) + j50\sin(120°)$

$$= \boxed{-25 + j43.3}$$

(d) $21\cos(-90°) + j21\sin(-90°)$

$$= \boxed{0 - j21}$$

3. (a) $\sqrt{(6)^2 + (7)^2}\angle \tan^{-1}\left(\frac{7}{6}\right)$

$$= \boxed{9.22\angle 49.4°}$$

(b) $\sqrt{(50)^2 + (-60)^2}\angle \tan^{-1}\left(\frac{-60}{50}\right)$

$$= \boxed{78.10\angle -50.19°}$$

(c) $\sqrt{(-75)^2 + (45)^2}\angle \tan^{-1}\left(\frac{45}{-75}\right)$

$$= \boxed{87.46\angle 149.04°}$$

(d) $\sqrt{(90)^2 + (-180)^2}\angle \tan^{-1}\left(\frac{-180}{90}\right)$

$$= \boxed{201.25\angle -63.43°}$$

4. (a) $\frac{1}{6}\sqrt{(7)^2 + (5)^2}\angle \tan^{-1}\left(\frac{5}{7}\right)$

$$= \boxed{1.4\angle 35.54°}$$

(b)

$$\frac{\sqrt{(13)^2 + (17)^2}\tan^{-1}\left(\dfrac{17}{13}\right)}{\sqrt{(15)^2 + (-10)^2}\tan^{-1}\left(\dfrac{-10}{15}\right)}$$

$$= \frac{21.4\angle 52.59°}{18.0\angle -33.69°}$$

$$= \boxed{1.19\angle 86.28°}$$

(c)

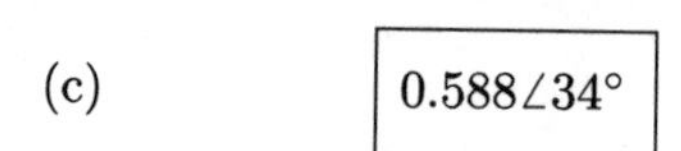

$$\boxed{0.588\angle 34°}$$

5.

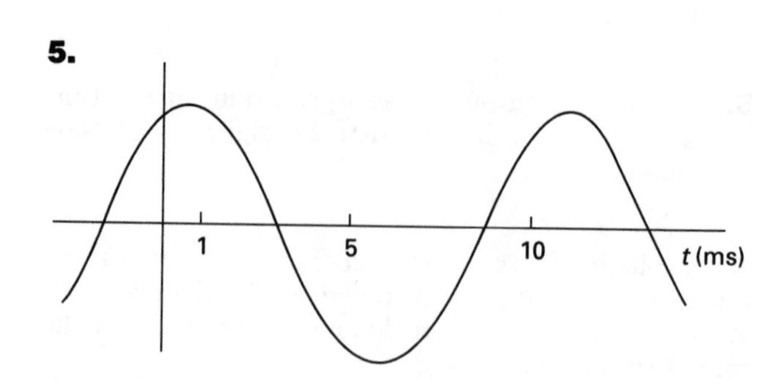

$$\omega = \frac{2\pi}{T} = \frac{2\pi}{0.01} = 200\pi \text{ rad/s}$$

$$10 \text{ ms} = 360°$$

$$1 \text{ ms} = 36°$$

For the sine, the positive zero crossing is at −1.5 ms, or −54°.

$$f = \boxed{F_p \sin(200\pi t + 54°) \text{ or } F_p \sin(200\pi t + 0.9)}$$

$$= \boxed{F_p \cos(200\pi t - 36°) \text{ or } F_p \cos(200\pi t - 0.6)}$$

6. Using cosine reference,

(a)

$$I_R = \frac{10\angle 25° \text{ V}}{25\ \Omega} = 0.4\angle 25° \text{ A}$$

$$= 0.3625 + j0.1690 \text{ A}$$

$$I_L = \frac{10\angle 25° \text{ V}}{(0.5)(100)\angle 90°\ \Omega}$$

$$= 0.2\angle -65° \text{ A}$$

$$= 0.0845 - j0.1813 \text{ A}$$

$$I_t = 0.4470 - j0.0123 \text{ A}$$

$$= 0.4472\angle - 1.6° \text{ A}$$

$$\text{phase angle} = \boxed{-1.6°}$$

(b) The calculations were made using peak values. Divide the total peak current by $\sqrt{2}$ to obtain the rms values.

$$I_{\text{rms}} = \frac{0.4472}{\sqrt{2}} \text{ A} = \boxed{0.316 \text{ A}}$$

7. Use rms values.

$$S = VI^* = \frac{(10\angle 25° \text{ V})(0.4472\angle 1.6° \text{ A})}{\sqrt{2}\sqrt{2}} \text{ VA}$$

$$= 2.236\angle 26.6° = 2 + j1 \text{ VA}$$

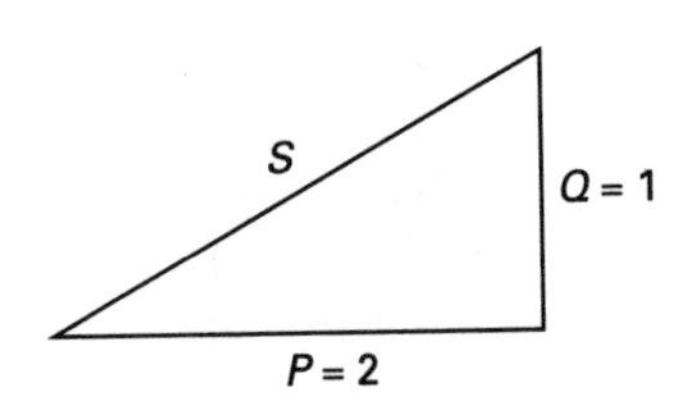

8. The capacitance current must cancel the inductance current, so its reactance must be $-j\omega L = -j50\ \Omega$.

$$X_C = \frac{1}{j\omega C} = \frac{-j}{100C} \rightarrow -j50\ \Omega$$

$$C = \frac{1}{(50\ \Omega)(100 \text{ rad/s})}$$

$$= \boxed{200 \times 10^{-6} \text{ F} \quad (200\ \mu\text{F})}$$

9. For the capacitor,

$$Z_C = \frac{1}{j\omega C} = \frac{1}{j\left(377\ \frac{\text{rad}}{\text{s}}\right)(80 \times 10^{-6} \text{ F})}$$

$$= -j\,(33.16\ \Omega) = 33.16\ \Omega\angle - 90°$$

For the inductor,

$$Z_L = j\omega L = j\left(377\ \frac{\text{rad}}{\text{s}}\right)(9 \times 10^{-3} \text{ H})$$

$$= j3.39\ \Omega = 3.39\ \Omega\angle 90°$$

For a series circuit, the impedance is

$$Z_{\text{total}} = Z_C + Z_L$$

$$= -j33.16\ \Omega + j3.39\ \Omega$$

$$= -j29.77\ \Omega = \boxed{29.77\ \Omega\angle -90°}$$

The impedance angle is negative. The circuit is capacitative and therefore leading.

10. For a parallel circuit, the impedance is

$$\frac{1}{Z_{\text{total}}} = \frac{1}{Z_C} + \frac{1}{Z_L}$$

$$= \frac{1}{33.16\ \Omega\angle -90°} + \frac{1}{3.39\ \Omega\angle 90°}$$

$$= 0.0302\ \Omega\angle 90° + 0.2950\ \Omega\angle -90°$$

$$= j0.0302\ \Omega - j0.2950\ \Omega$$

$$= -j0.2648$$

Therefore,

$$Z_{\text{total}} = \frac{1}{-j0.2648} = \frac{j}{0.2648}$$

$$= \boxed{3.78\angle 90°}$$

The impedance angle is positive. The circuit is inductive and therefore lagging.

28 Transformers

PRACTICE PROBLEMS

1. Which of the following describes the leakage flux?

(A) flux outside the core
(B) flux linking the primary
(C) flux linking the secondary
(D) flux other than the mutual flux

2. For the transformer shown, which of the following statement(s) is (are) true?

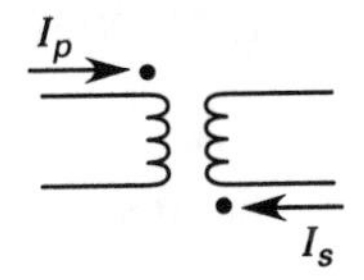

(A) $M > 0$
(B) $M < 0$
(C) $X_m > 0$ and $X_p > 0$
(D) Both (A) and (C) are true.

3. An ideal transformer has 90 turns on the primary and 2200 turns on the secondary. The load draws 8 A at 0.8 pf. What is the current in the primary?

(A) 90 ∠37° A
(B) 196 ∠37° A
(C) 320 ∠1° A
(D) 333 ∠0.2π A

4. An ideal transformer with a 120 V, 60 Hz source has a turns ratio of 10. What is the peak flux in the secondary?

5. A step-down transformer consists of 200 primary turns and 40 secondary turns. The primary voltage is 550 V. If an impedance of 4.2 Ω is the secondary load, what are the (a) secondary voltage, (b) primary current, and (c) secondary current?

6. If the amplifier and load impedances are matched, what is the turns ratio?

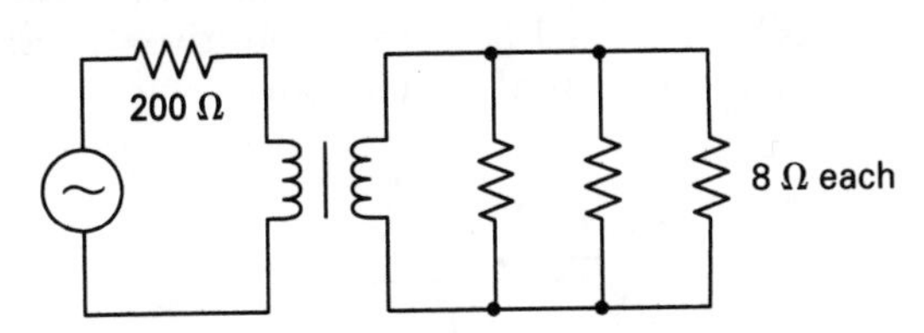

7. A 100 V (rms) source with an internal resistance of 8 Ω is connected to a 200/100-turn step-down transformer. An impedance of $6 + j8$ Ω is connected to the secondary. Find the (a) primary current and (b) secondary power.

SOLUTIONS

1. Leakage flux is defined as flux that goes beyond its intended path and does not serve its intended purpose—that is, the flux that links only one coil or the other, but not both, in a two-coil transformer. The flux linking both coils is the mutual flux.

The answer is (D).

2. Both assumed currents enter at the dotted terminals. Thus, the mutual inductor is positive. Further, X_m and X_p (or X_s) are the same sign.

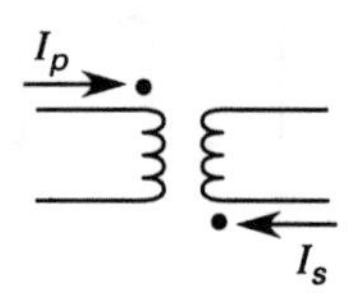

The answer is (D).

3. The relationship is

$$a = \frac{N_p}{N_s} = \frac{I_s}{I_p}$$

Therefore,

$$a = \frac{N_p}{N_s} = \frac{90}{2200} = 0.04091$$

The magnitude of the current is

$$a = \frac{I_s}{I_p}$$

$$I_p = \frac{I_s}{a} = \frac{8 \text{ A}}{0.04091} = 195.6 \text{ A}$$

In an ideal transformer, no phase shift occurs, thus

$$\text{pf} = \cos\psi$$

$$0.8 = \cos\psi$$

$$\psi = \cos^{-1}(0.8) = 36.87°$$

Thus,

$$\boxed{\mathbf{I}_p = 196\ \angle 37° \text{ A}}$$

The answer is (B).

4. The applicable formula is

$$v_s(t) = -4.44\, N_s f \Phi_m \cos\omega t$$

Since only the peak flux was requested

$$V_s = 4.44\, N_s f \Phi_m$$

$$\Phi_m = \frac{V_s}{4.44\, N_s f}$$

Now, from the turns ratio

$$a = \frac{N_p}{N_s} = \frac{V_p}{V_s}$$

$$V_s = \left(\frac{N_s}{N_p}\right) V_p = \tfrac{1}{10}\,(120 \text{ V}) = 12 \text{ V}$$

Substituting gives

$$\Phi_m = \frac{V_s}{4.44\, N_s f} = \frac{12 \text{ V}}{(4.44)(1)\,(60 \text{ s}^{-1})}$$

$$= \boxed{4.54 \times 10^{-2} \text{ Wb } (45 \text{ mWb})}$$

Note that this is the same flux present in the primary.

5. (a)

$$V_s = \left(\frac{N_s}{N_p}\right) V_p = \left(\frac{40}{200}\right)(550 \text{ V})$$

$$= \boxed{110 \text{ V}}$$

(b)

$$I_p = \frac{V_s^2}{V_p Z_s} = \frac{(110 \text{ V})^2}{(550 \text{ V})(4.2\ \Omega)}$$

$$= \boxed{5.24 \text{ A}}$$

(c)

$$I_s = \frac{V_s}{Z_s} = \frac{110 \text{ V}}{4.2\ \Omega}$$

$$= \boxed{26.2 \text{ A}}$$

6.

$$\frac{N_p}{N_s} = \sqrt{\frac{Z_p}{Z_s}} = \sqrt{\frac{200\ \Omega}{\frac{8\ \Omega}{3}}}$$

$$= \boxed{8.66}$$

7.

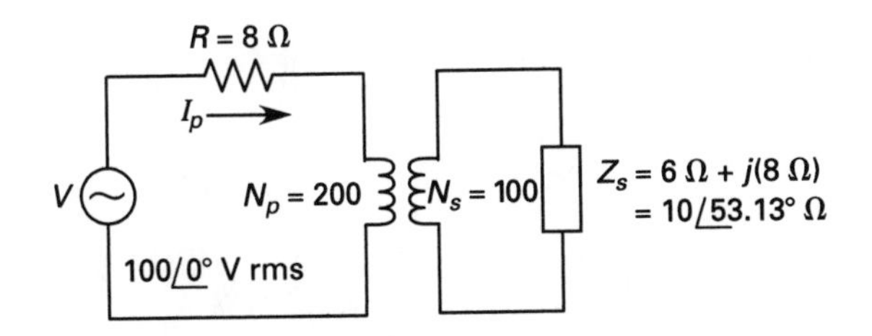

The circuit can be simplified to

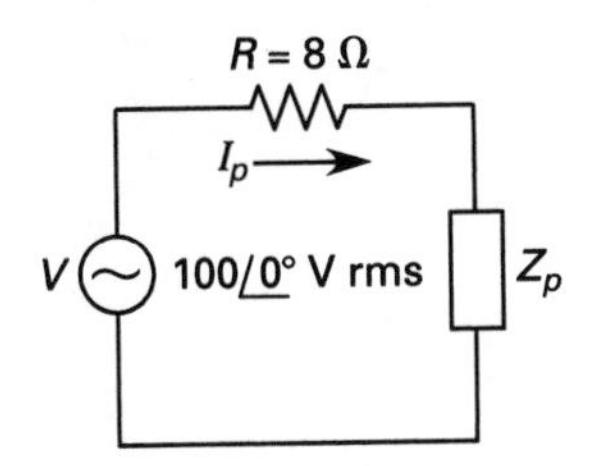

$$\mathbf{Z}_p = \left(\frac{N_p}{N_s}\right)^2 \mathbf{Z}_s = \left(\frac{200}{100}\right)^2 (10\angle 53.13^\circ\ \Omega)$$

$$= 40\angle 53.13^\circ\ \Omega = (24 + j32)\ \Omega$$

The circuit can be further simplified to

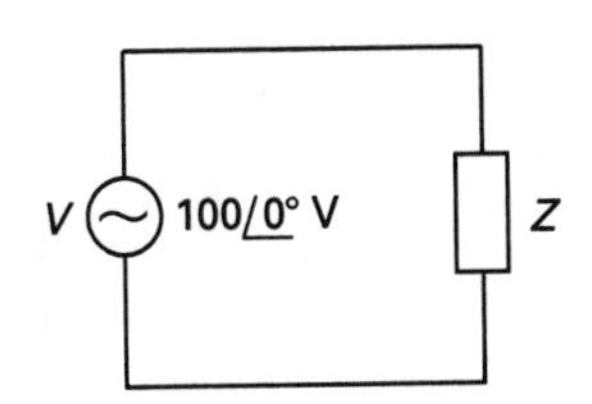

$$Z = \sqrt{(\Sigma R)^2 + (\Sigma X_L - \Sigma X_C)^2}$$

$$= \sqrt{(R + R_{Z_m})^2 + (X_{Z_m})^2}$$

$$= \sqrt{(8\ \Omega + 24\ \Omega)^2 + (32\ \Omega)^2} = 45.25\ \Omega$$

$$\phi\mathbf{z} = \tan^{-1}\left(\frac{\Sigma X_L - \Sigma X_C}{\Sigma R}\right)$$

$$= \tan^{-1}\left(\frac{X_{Z_m}}{R + R_{Z_m}}\right)$$

$$= \tan^{-1}\left(\frac{32\ \Omega}{8\ \Omega + 24\ \Omega}\right) = 45^\circ$$

$$\mathbf{Z} = 45.25\angle 45^\circ\ \Omega$$

(a) $$\mathbf{I}_p = \frac{\mathbf{V}}{\mathbf{Z}} = \frac{100\angle 0^\circ\ \text{V}}{45.25\angle 45^\circ\ \Omega}$$

$$= \boxed{2.21\angle{-45^\circ}\ \text{A rms}}$$

$$P_s = I_s^2 R_s = I_s^2 Z_s \cos\phi$$

(b) $$= \left(\frac{N_p I_p}{N_s}\right)^2 Z_s \cos\phi$$

$$= \left(\frac{(200)(2.21\ \text{A})}{100}\right)^2 (10\ \Omega)\cos 53.13^\circ$$

$$= \boxed{117.2\ \text{W}}$$

29 Linear Circuit Analysis

PRACTICE PROBLEMS

1. For which of the following circuit elements is superposition valid?

(A) independent source
(B) dependent source
(C) charged capacitor
(D) inductor with $\mathbf{I}_{\text{initial}} \neq 0$

2. A 5 Ω length of copper wire is used in a circuit designed for a maximum temperature of 50°C. What is the resistance of the wire at maximum design temperature specification?

3. Two square parallel plates (0.04 m × 0.04 m) are separated by a 0.1 cm thick insulator with a dielectric constant of 3.4. The plates are connected across 200 V. What is the capacitance?

4. For the parallel plates described in Prob. 3, what charge exists on the plates?

5. An 8 mH inductor carries 15 A of current. What is the average magnetic flux present?

6. For the circuit shown, which resistor(s) is (are) redundant?

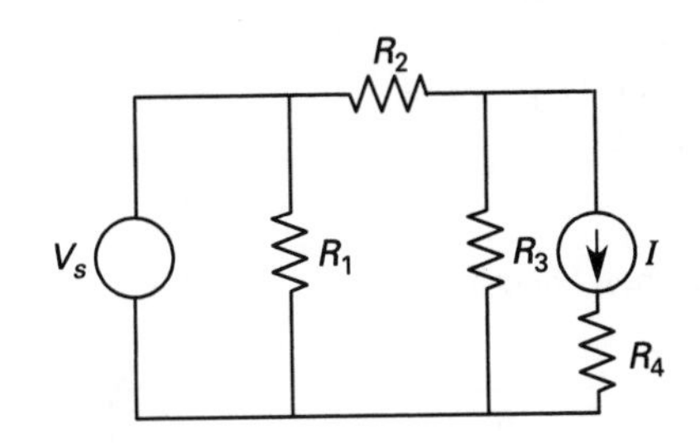

(A) R_1
(B) R_1 and R_3
(C) R_2 and R_4
(D) R_1 and R_4

7. For the circuit shown, what current is flowing in the 8 Ω resistor?

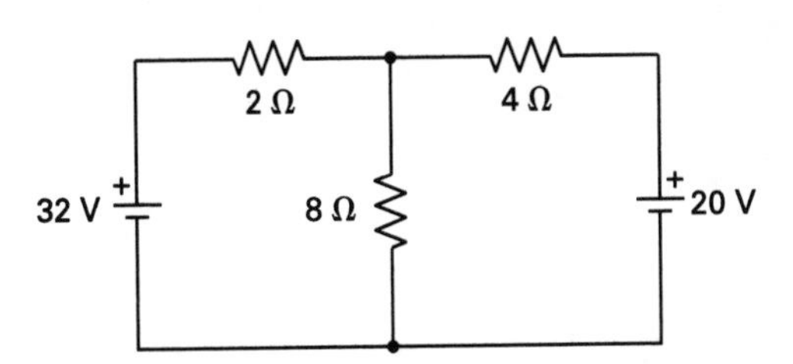

8. For the circuit shown, what current is flowing in the 6 Ω resistor?

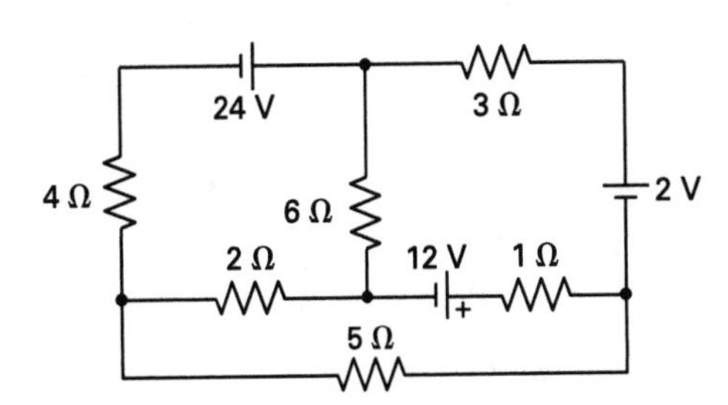

9. At $t = 0$, the switch connects to position A. At $t = 5 \times 10^{-4}$ s, the switch is moved to position B. Assume there is no current anywhere in the circuit prior to $t = 0$. (a) What is the current flowing in the circuit just prior to the switch moving to position B? (b) What is the equation of the current after the switch is moved to position B?

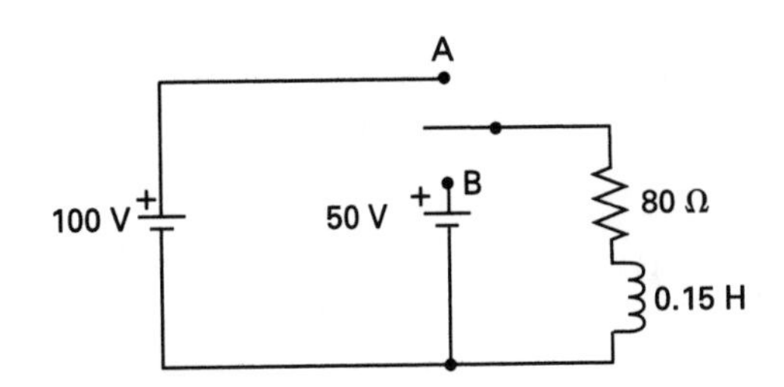

10. For the circuit shown, find the z parameters.

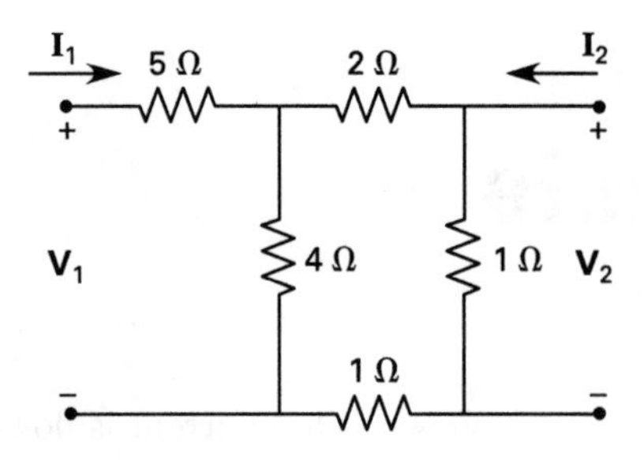

SOLUTIONS

1. Capacitors and inductors are linear elements for which superposition is valid only if initially uncharged or without an initial magnetic field, respectively. Therefore, neither (C) nor (D) is the answer. The output of dependent sources depends on electrical parameters elsewhere in the circuit and is therefore not linear. Therefore, (B) is not the answer.

Independent sources are linear and superposition can be applied.

The answer is (A).

2. Resistance is given by

$$R = \frac{\rho l}{A}$$

The length and cross-sectional area are unknown. Therefore, use the following ratio.

$$\frac{R_{50}}{R_{20}} = \frac{\rho_{50}\left(\frac{l}{A}\right)}{\rho_{20}\left(\frac{l}{A}\right)} = \frac{\rho_{50}}{\rho_{20}}$$

Thus,

$$R_{50} = R_{20}\left(\frac{\rho_{50}}{\rho_{20}}\right)$$

Solving for ρ_{50} gives

$$\begin{aligned}\rho_{50} &= \rho_{20}\big(1 + \alpha_{20}\,(T - 20)\big)\\ &= \left(1.8 \times 10^{-8}\ \Omega\text{·m}\right)\\ &\quad \times \left(1 + (3.9 \times 10^{-3}\,{}^\circ\text{C}^{-1})(50^\circ\text{C} - 20^\circ\text{C})\right)\\ &= 20.11 \times 10^{-9}\ \Omega\text{·m}\end{aligned}$$

Substituting the given and calculated information gives

$$\begin{aligned}R_{50} &= R_{20}\left(\frac{\rho_{50}}{\rho_{20}}\right) = (5\ \Omega)\left(\frac{20.11 \times 10^{-9}\ \Omega\text{·m}}{1.8 \times 10^{-8}\ \Omega\text{·m}}\right)\\ &= \boxed{5.6\ \Omega}\end{aligned}$$

3. The capacitance is

$$\begin{aligned}C &= \frac{\varepsilon A}{r} = \frac{\varepsilon_r \varepsilon_0 A}{r}\\ &= \frac{(3.4)\left(8.854 \times 10^{-12}\,\frac{\text{C}^2}{\text{N·m}^2}\right)(0.04\ \text{m})^2}{0.001\ \text{m}}\\ &= \boxed{4.817 \times 10^{-11}\ \text{F}}\end{aligned}$$

4. The charge on the plates is

$$Q = CV_{\text{plates}} = (4.817 \times 10^{-11}\ \text{F})(200\ \text{V})$$
$$= \boxed{9.63 \times 10^{-9}\ \text{C}}$$

5. The average energy in the inductor's magnetic field is

$$U = \tfrac{1}{2}LI^2 = \left(\tfrac{1}{2}\right)\left(8 \times 10^{-3}\ \text{H}\right)(15\ \text{A})^2$$
$$= 0.90\ \text{J}$$

The energy in terms of the magnetic flux is

$$U = \tfrac{1}{2}\left(\frac{\Psi^2}{L}\right)$$
$$\Psi = \sqrt{2UL} = \sqrt{(2)(0.90\ \text{J})(8 \times 10^{-3}\ \text{H})}$$
$$= \boxed{0.120\ \text{Wb}}$$

6. Resistors in parallel with voltage sources and in series with current sources do not affect the remainder of the circuit. Thus, R_1, and R_4 are redundant.

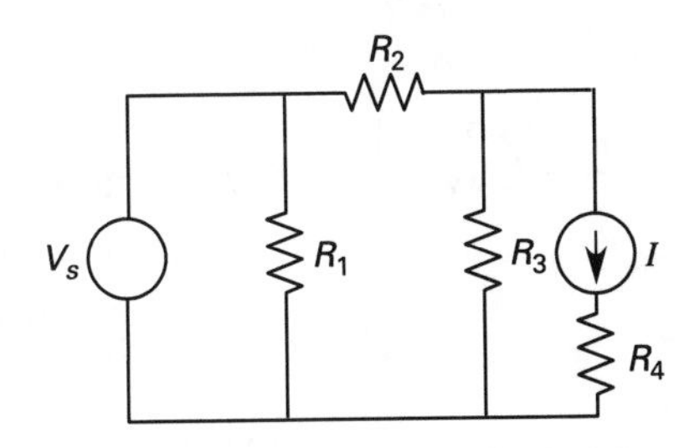

The answer is (D).

7. The circuit can be represented as

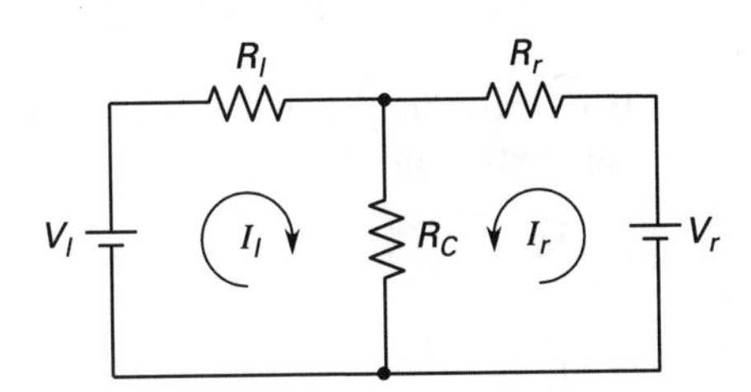

From the loop-current method,

$$V_l = R_lI_l + R_c(I_l + I_r)$$
$$V_r = R_rI_r + R_c(I_l + I_r)$$
$$I_l + I_r = \frac{R_rV_l + R_lV_r}{R_rR_l + R_rR_c + R_lR_c}$$
$$= \frac{(4\ \Omega)(32\ \text{V}) + (2\ \Omega)(20\ \text{V})}{(4\ \Omega)(2\ \Omega) + (4\ \Omega)(8\ \Omega) + (2\ \Omega)(8\ \Omega)}$$
$$= \boxed{3\ \text{A}\quad[\text{down}]}$$

If the first two equations are solved simultaneously,

$$I_l = 4\ \text{A}$$
$$I_r = -1\ \text{A}$$

8. The circuit can be represented as

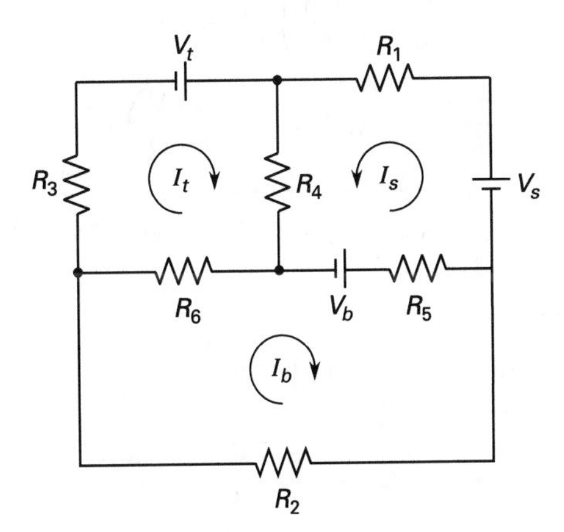

Using the loop-current method,

$$V_t = R_4(I_t + I_s) + R_6(I_t - I_b) + R_3I_t$$
$$V_s + V_b = R_1I_s + R_4(I_t + I_s) + R_5(I_s + I_b)$$
$$V_b + R_6(I_t - I_b) = R_5(I_b + I_s) + R_2I_b$$

After rearranging, substituting in the constants, and evaluating the coefficients, the previous three equations are

$$24\ \text{V} = (12\ \Omega)I_t + (6\ \Omega)I_s - (2\ \Omega)I_b$$
$$14\ \text{V} = (6\ \Omega)I_t + (10\ \Omega)I_s + (1\ \Omega)I_b$$
$$12\ \text{V} = -(2\ \Omega)I_t + (1\ \Omega)I_s + (8\ \Omega)I_b$$

Solving simultaneously,

$$I_t = 2.53\ \text{A}$$
$$I_s = -0.336\ \text{A}$$
$$I_t + I_s = \boxed{2.194\ \text{A}\quad[\text{down}]}$$

9. (a) Before the switch is moved to position B,

$$I(t) = \left(\frac{V_{100}}{R}\right)\left(1 - e^{\frac{-Rt}{L}}\right)$$

$$I\left(5 \times 10^{-4}\right) = \left(\frac{100\ \text{V}}{80\ \Omega}\right)\left(1 - e^{\frac{-(80\ \Omega)(5\times10^{-4}\ \text{s})}{0.15\ \text{H}}}\right)$$

$$= \boxed{0.2926\ \text{A}\quad[\text{clockwise}]}$$

(b) Move the switch to position B. Solve for currents I_1 and I_2, whose sum is I_N, the current flowing in the circuit.

$$I_1(t) = I_0 e^{\frac{-Rt}{L}}\quad[\text{clockwise}]$$

$$I_2(t) = \left(\frac{V_{50}}{R}\right)\left(1 - e^{\frac{-Rt}{L}}\right)\quad[\text{clockwise}]$$

$$I_N(t) = I_1 t + I_2 t$$

$$= I_0 e^{\frac{-Rt}{L}} + \left(\frac{V_{50}}{R}\right)\left(1 - e^{\frac{-Rt}{L}}\right)$$

$$= \frac{50\ \text{V}}{80\ \Omega} + \left(0.2926\ \text{A} - \frac{50\ \text{V}}{80\ \Omega}\right)e^{\frac{-(80\ \Omega)t}{0.15\ \text{H}}}$$

$$= \boxed{0.625 - 0.332e^{-(533s^{-1})t}\ \text{A}\quad[\text{clockwise}]}$$

10. Consider the circuit shown.

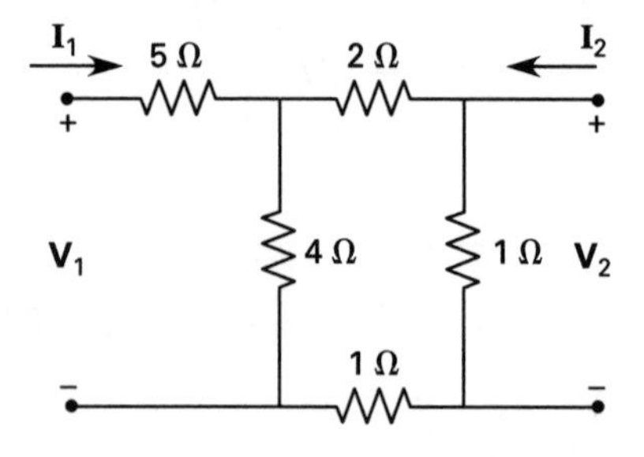

Since all the impedances are resistors, the phasor notation is dropped for convenience. From the deriving equations,

$$z_{11} = \left.\frac{V_1}{I_1}\right|_{I_2=0}$$

The circuit, redrawn with $I_2 = 0$, is

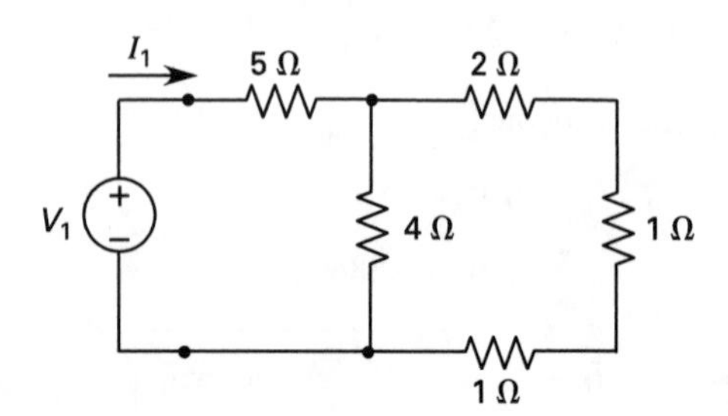

The equivalent impedance looking to the right, into the input terminals, is

$$z_e = R_e = 5\ \Omega + \frac{(4\ \Omega)(2\ \Omega + 1\ \Omega + 1\ \Omega)}{4\ \Omega + (2\ \Omega + 1\ \Omega + 1\ \Omega)}$$

$$= 7\ \Omega$$

Redrawing the circuit again gives

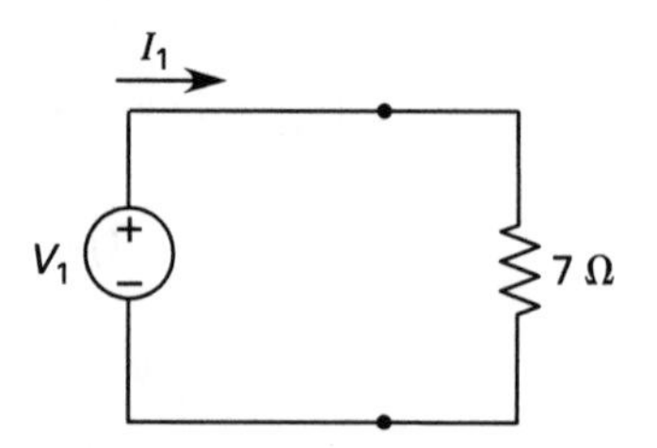

Using Ohm's law gives

$$V_1 = I_1(7\ \Omega)$$

$$\frac{V_1}{I_1} = (7\ \Omega)$$

Thus,

$$z_{11} = \boxed{7\ \Omega}$$

From the deriving equations,

$$z_{21} = \left.\frac{V_2}{I_1}\right|_{I_2=0}$$

The redrawn circuit is identical to that for z_{11}, since $I_2 = 0$.

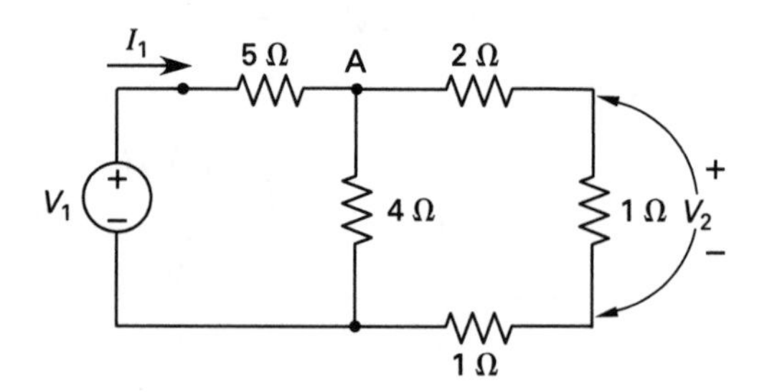

Using the concept of a current divider at node A gives the following for the current through the branch containing V_2.

$$I_{V_2} = I_1\left(\frac{4\ \Omega}{4\ \Omega + (2\ \Omega + 1\ \Omega + 1\ \Omega)}\right) = \tfrac{1}{2}I_1$$

Apply Ohm's law across the 1 Ω resistor associated with V_2, using the calculated current.

$$V_2 = \tfrac{1}{2}I_1(1\ \Omega) = \tfrac{1}{2}I_1$$

$$\frac{V_2}{I_1} = \tfrac{1}{2}\ \Omega$$

Thus,

$$z_{21} = \left.\frac{V_2}{I_1}\right|_{I_2=0} = \tfrac{1}{2}\ \Omega$$

From the deriving equations,

$$z_{12} = \left.\frac{V_1}{I_2}\right|_{I_1=0}$$

The circuit, redrawn with $I_1 = 0$, is

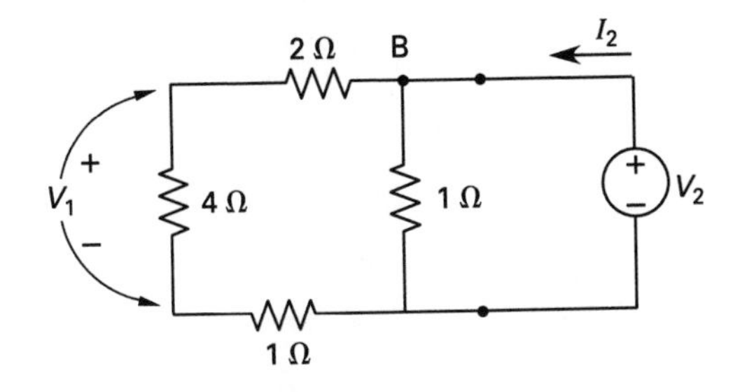

Using a current divider at node B gives the following for the branch current flowing through the 4 Ω resistor associated with V_1.

$$I_{V_1} = I_2\left(\frac{1\ \Omega}{1\ \Omega + (2\ \Omega + 4\ \Omega + 1\ \Omega)}\right) = \tfrac{1}{8}I_2$$

Applying Ohm's law for the 4 Ω resistor gives

$$V_1 = \tfrac{1}{8}I_2\,(4\ \Omega) = \tfrac{1}{2}I_2$$

$$\frac{V_1}{I_2} = \tfrac{1}{2}\ \Omega$$

Thus,

$$z_{12} = \left.\frac{V_1}{I_2}\right|_{I_1=0} = \tfrac{1}{2}\ \Omega$$

From the deriving equations,

$$z_{22} = \left.\frac{V_2}{I_2}\right|_{I_1=0}$$

The redrawn circuit is identical to that for z_{12}, since $I_1 = 0$.

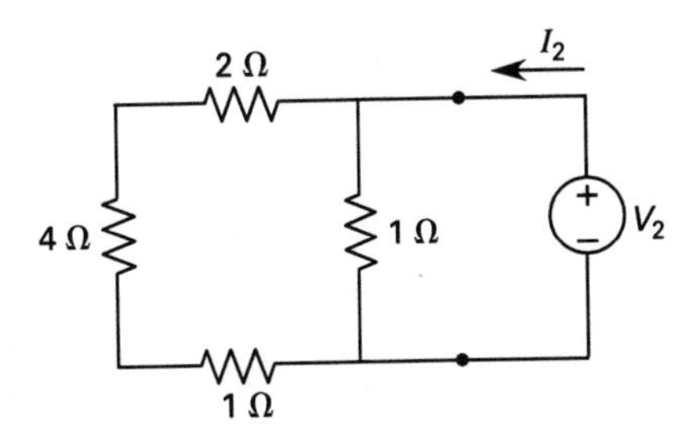

The equivalent impedance looking to the left, into the output terminals, is

$$z_e = R_e = \frac{(1\ \Omega)(2\ \Omega + 4\ \Omega + 1\ \Omega)}{1\ \Omega + (2\ \Omega + 4\ \Omega + 1\ \Omega)} = \tfrac{7}{8}\ \Omega$$

Redrawing the circuit again gives

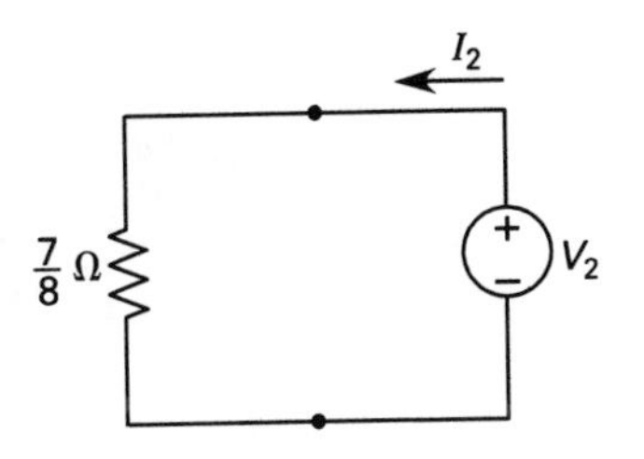

Using Ohm's law gives

$$V_2 = I_2\left(\tfrac{7}{8}\ \Omega\right)$$

$$\frac{V_2}{I_2} = \tfrac{7}{8}\ \Omega$$

Thus,

$$z_{22} = \tfrac{7}{8}\ \Omega$$

Circuit Theory

30 Transient Analysis

PRACTICE PROBLEMS

1. Consider the circuit shown. What is the time constant?

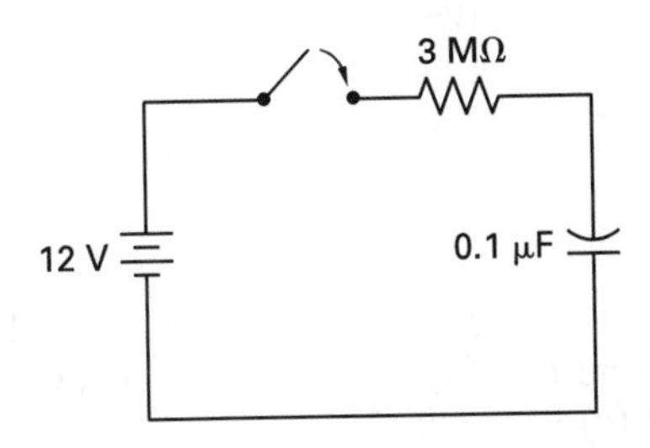

(A) 1.00 μs
(B) 36.0 μs
(C) 0.30 s
(D) 3.0 s

2. With an initially uncharged capacitor in the circuit in Prob. 1, what is the initial current?

(A) 0.0 A
(B) 4.0 μA
(C) 4.0 A
(D) 120 A

3. How long does it take the circuit shown in Prob. 1 to reach steady-state?

4. What is the rise time of the circuit in Prob. 1?

5. Consider the circuit shown. Write the equations for loops 1 and 2.

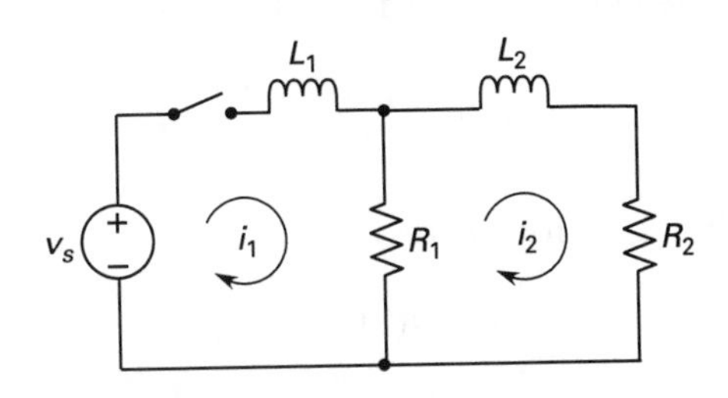

6. A series RC circuit contains a 4 Ω resistor and an uncharged 300 μF capacitor. The switch is closed at $t = 0$, connecting the circuit across a 500 V source. For the circuit shown, determine the current as a function of time.

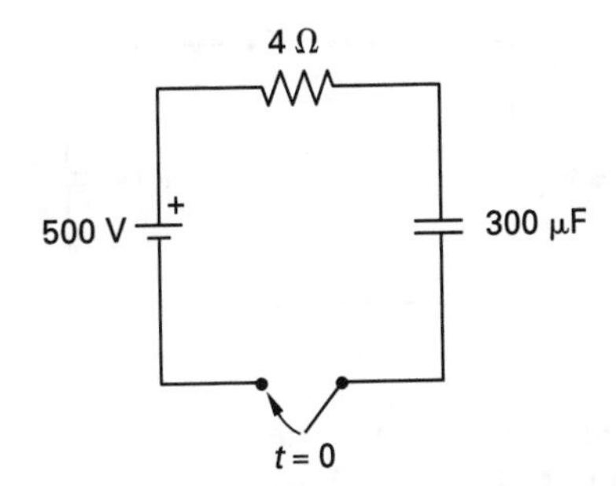

7. The switch in the circuit shown is closed at $t = 0$ and opened again at $t = 0.05$ s. How much energy is stored in the initially uncharged capacitor?

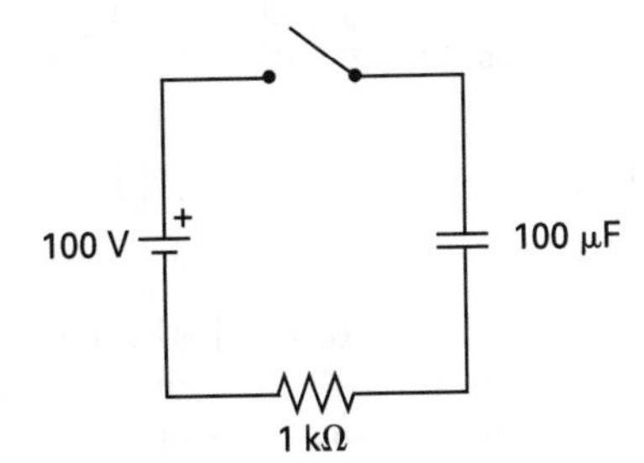

8. Find the current through the 80 Ω resistor (a) immediately after the switch is closed and (b) 2 s after the switch is closed. The switch has been open for a long time.

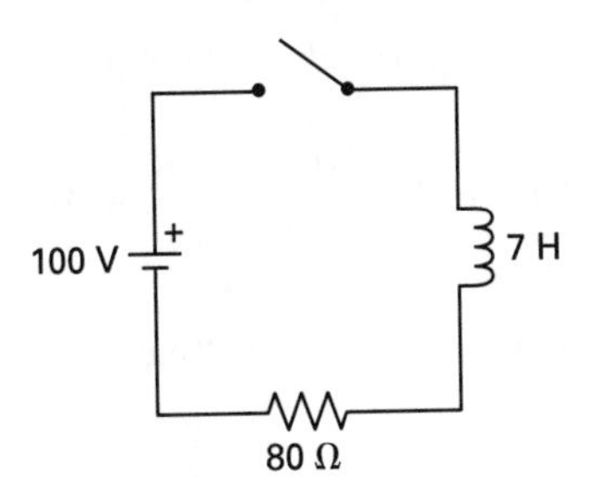

9. For the circuit shown, determine the steady-state inductance current, i_L, (a) with the switch closed and (b) with the switch open.

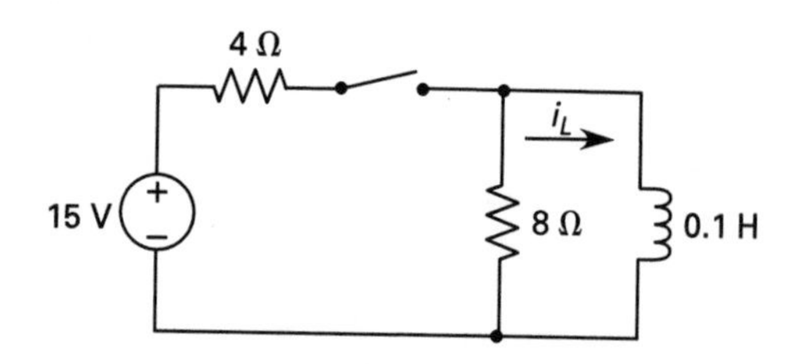

10. For the circuit shown, determine the steady-state capacitance voltage, V_C, (a) with the switch closed, and (b) with the switch open.

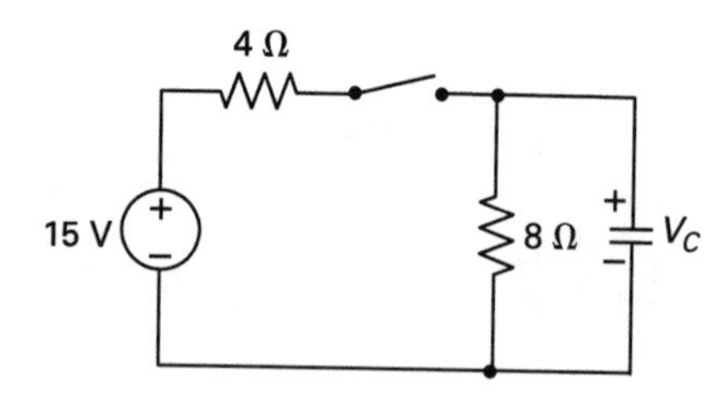

11. For the circuit shown, determine the steady-state inductance current, i_L, (a) with the switch closed and (b) with the switch open.

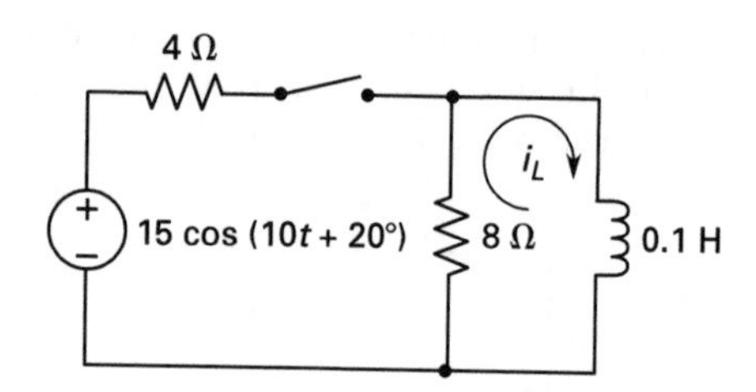

12. The circuit in Prob. 10 has had the switch closed for a long time. At $t = 0$, the switch opens. Determine the capacitance voltage, $v_C(t)$, for $t > 0$, for a capacitance of 1×10^{-6} F.

13. The circuit in Prob. 10 has had the switch open for a long time. At $t = 0$, the switch closes. Determine the capacitance voltage, $v_C(t)$, for $t > 0$.

14. A system is described by the equation

$$12 \sin 2t = \frac{d^2v}{dt^2} + 5\frac{dv}{dt} + 4v$$

The initial conditions on the voltage are

$$v(0) = -5$$

$$\frac{dv(0)}{dt} = 2$$

Determine $v(t)$ for $t > 0$.

15. For the circuit shown, the current source is

$$i_s(t) = 0.01 \cos 500t$$

Find the transfer function.

$$G(s) = \frac{V_C(s)}{I_s(s)}$$

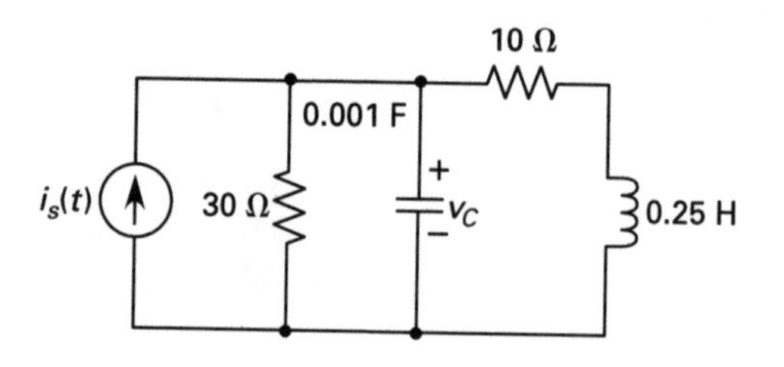

16. For a series RLC circuit of 10 Ω, 0.5 H, and 100 μF, determine the (a) resonant frequency, (b) quality factor, and (c) bandwidth.

17. For a parallel RLC circuit of 10 Ω, 0.5 H, and 100 μF, determine the (a) resonant frequency, (b) quality factor, and (c) bandwidth.

18. The switch of the circuit shown has been open for a long time and is closed at $t = 0$. Determine the currents flowing in the two capacitances at the instant the switch closes. Give the magnitudes and directions.

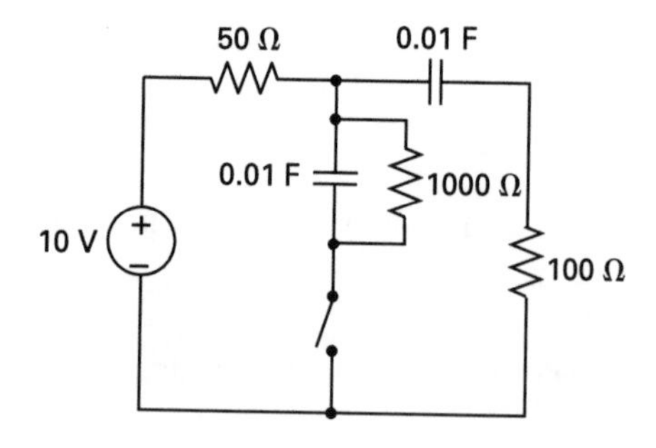

19. A system is described by the following differential equation.

$$20 \sin 4t = \frac{d^2i}{dt^2} + 4\frac{di}{dt} + 4i$$

The initial conditions are

$$i(0) = 0$$

$$\frac{di(0)}{dt} = 4$$

Determine $i(t)$ for $t > 0$.

20. The circuit shown has the 1 μF capacitance charged to 100 V and the 2 μF capacitance uncharged before the switch is closed. Determine the (a) energy stored in each capacitance at the instant the switch closes, (b) energy stored in each capacitance a long time after the switch closes, and (c) energy dissipated in the resistance.

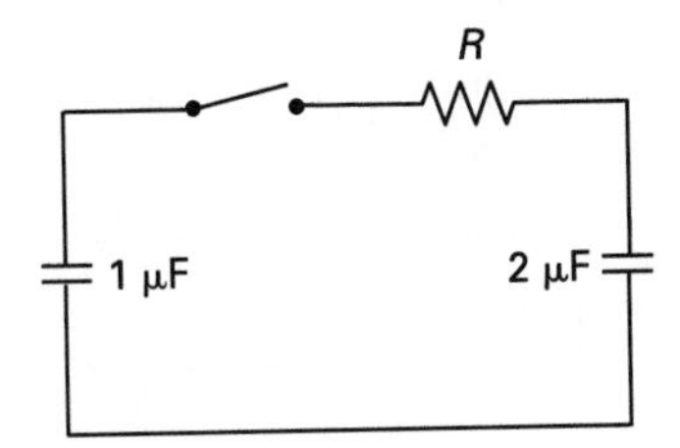

SOLUTIONS

1.

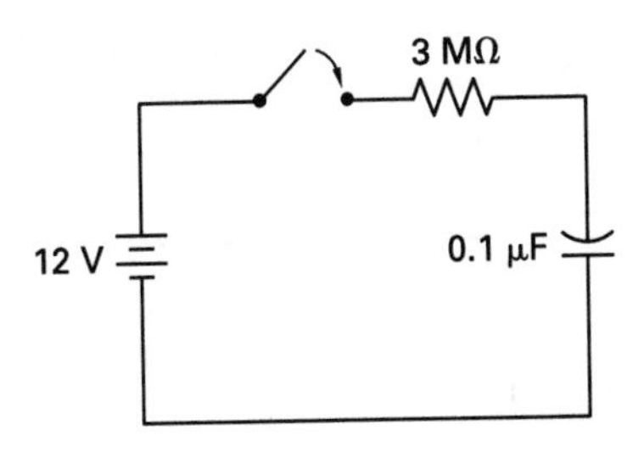

The time constant is

$$\tau = RC = \left(3 \times 10^6\ \Omega\right)\left(0.1 \times 10^{-6}\ \text{F}\right)$$
$$= 0.3\ \text{s}$$

The answer is (C).

2. The capacitor initially acts as a short. Thus, from Ohm's law,

$$V = IR$$
$$I = \frac{V}{R} = \frac{12\ \text{V}}{3 \times 10^6\ \Omega} = 4.0 \times 10^{-6}\ \text{A}$$

The answer is (B).

3. Steady-state is reached in approximately 5 time constants. Thus,

$$t = 5\tau$$

The time constant is $RC = 0.3$ s. Substituting,

$$t = (5)(0.3\ \text{s}) = \boxed{1.5\ \text{s}}$$

4. The rise time is most often determined by measurements in the lab. Nevertheless, the rise time is

$$t_r = 2.2\tau$$

The time constant was found in Prob. 1. Substituting, gives

$$t_r = (2.2)(0.3\ \text{s}) = \boxed{0.660\ \text{s}}$$

Circuit Theory

5.

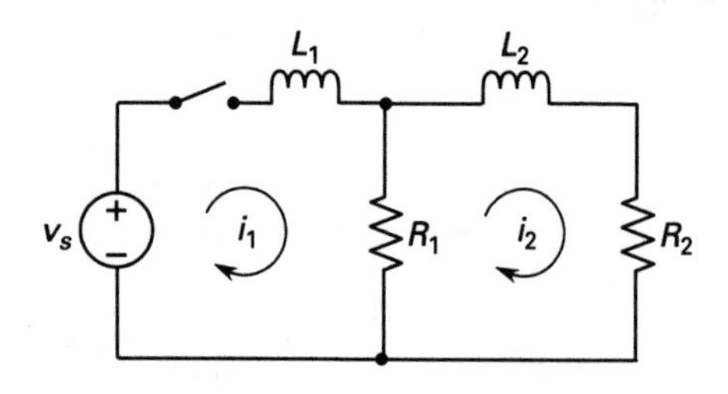

KVL for loop 1 gives

$$v_s - L_1\left(\frac{di_1}{dt}\right) - (i_1 - i_2)R_1 = 0$$

$$v_s = \boxed{L_1\left(\frac{di_1}{dt}\right) + (i_1 - i_2)R_1}$$

KVL for loop 2 gives

$$-(i_2 - i_1)R_1 - L_2\left(\frac{di_2}{dt}\right) - i_2R_2 = 0$$

$$-i_2R_1 + i_1R_1 - L_2\left(\frac{di_2}{dt}\right) - i_2R_2 = 0$$

$$\boxed{L_2\left(\frac{di_2}{d_t}\right) + R_1(i_2 - i_1) + i_2R_2 = 0}$$

6. From Table 30.1, the current response for a series RC circuit is

$$i(t) = \left(\frac{V_{\text{bat}} - V_0}{R}\right)e^{\frac{-t}{\tau}} = \left(\frac{500\text{ V} - 0}{4\ \Omega}\right)e^{\frac{-t}{1.2\times10^{-3}}}$$

$$= \boxed{125e^{-833t}\text{ A}}$$

7.

$$Q(t) = CV_C(t) = CV\left(1 - e^{\frac{-t}{RC}}\right)$$

$$E_C(t) = \tfrac{1}{2}CV_C^2(t) = \tfrac{1}{2}CV^2\left(1 - e^{\frac{-t}{RC}}\right)^2$$

$$E_C(0.05) = \left(\tfrac{1}{2}\right)(100\times10^{-6}\text{ F})(100\text{ V})^2 \times\left(1 - e^{\frac{-0.05\text{ s}}{(10^3\ \Omega)(100\times10^{-6}\text{ F})}}\right)^2$$

$$= \boxed{0.0774\text{ J}}$$

8. (a) Since the current in the inductor cannot change instantaneously,

$$I(0) = \boxed{0}$$

(b)

$$I(t) = \left(\frac{V}{R}\right)\left(1 - e^{\frac{-Rt}{L}}\right)$$

$$I(2) = \left(\frac{100\text{ V}}{80\ \Omega}\right)\left(1 - e^{\frac{-(80\ \Omega)(2\text{ s})}{7\text{ H}}}\right)$$

$$= \boxed{1.25\text{ A}\quad\text{[down]}}$$

9. (a) Use Kirchhoff's voltage law around the outer loop. With the switch closed, KVL gives

$$15 = 4i_{\text{ss}} + v_L$$

$$v_L = L\left(\frac{di}{dt}\right) = 0\quad\text{[in steady-state]}$$

$$i_{\text{ss}} = \frac{15}{4}$$

> The current through the 8 Ω resistor is $v_L/8 = 0$, therefore all of i_{ss} flows through the inductance in steady-state.

(b) With the switch open, the inductor dissipates its energy in the 8 Ω resistor.

$$\boxed{i_{\text{ss}} = 0}$$

10. (a) With the switch closed,

$$i_C = C\left(\frac{dv_C}{dt}\right)$$

In steady-state,

$$\frac{dv_C}{dt} = 0$$

Therefore, $i_C \to 0$. By voltage division,

$$v_C = \left(\frac{8\ \Omega}{12\ \Omega}\right)(15\text{ V}) = \boxed{10\text{ V}}$$

With the switch open, the capacitor dissipates its energy in the 8 Ω resistor.

(b)

$$\boxed{v_C = 0}$$

11. (a) A Thevenin equivalent circuit seen from the inductance is obtained.

$$v_{oc} = v_{Th} = \left(\frac{8\ \Omega}{12\ \Omega}\right) v_s = 10\cos(10t + 20°)\ \text{V}$$

$$i_{sc} = \frac{15\cos(10t+20°)\ \text{V}}{4\ \Omega} = 3.75\cos(10t+20°)\ \text{A}$$

$$R_{Th} = \frac{v_{oc}}{i_{sc}} = \frac{10\ \text{V}}{3.75\ \text{A}} = 8/3\ \Omega$$

Taking $\cos(10t + 20°)$ as the phasor reference,

$$V_{Th} = 10\angle 0°\ \text{V}$$

$$Z_L = j\omega L = \left(j10\ \frac{\text{rad}}{\text{s}}\right)(0.1\ \text{H}) = j1\ \Omega$$

$$I_L = \frac{V_{Th}}{R_{Th} + j1\ \Omega}$$

$$= \left(\frac{10\ \text{V}}{\frac{8}{3}\ \Omega + j1\ \Omega}\right)\left(\frac{\frac{8}{3}\ \Omega - j1\ \Omega}{\frac{8}{3}\ \Omega - j1\ \Omega}\right)$$

$$= \left(\frac{90}{73}\ \text{A}\right)\left(\frac{8}{3} - j1\right) = \boxed{3.51\angle{-20.6°}\ \text{A}}$$

As a time function,

$$i_L(\infty) = 3.51\cos(10t + 20° - 20.6°)$$

$$= 3.51\cos(10t - 0.6°)\ \text{A}$$

(b) $\boxed{i_L(\infty) \text{ is zero with the switch open.}}$

12. With the switch closed for a long time, the capacitance voltage is 10 V (see Prob. 10). The initial voltage when the switch opens is therefore 10 V.

With the switch open, KVL is

$$v_C = 8i$$

$$i_C = 10^{-6}\left(\frac{dv_C}{dt}\right)$$

$$i_C = -i$$

$$v_C + (8 \times 10^{-6})\left(\frac{dv_C}{dt}\right) = 0$$

$$\frac{dv_C}{dt} + (1.25 \times 10^5)v_C = 0$$

Any appropriate mathematical method may now be used to solve the differential equation. Let $\alpha = 1.25 \times 10^5$, the solution is of the form

$$v_C = V(t_1)e^{-\alpha(t-t_1)}$$

Taking $t_1 = 0$, $v_C(0) = 10$ V.

$$\frac{dv_C}{dt} = -\alpha v_C$$

$$\alpha = 1.25 \times 10^5$$

$$v_C(t) = \boxed{10e^{-(1.25\times 10^5)t}\ \text{V}}$$

13. The initial capacitance voltage is zero. Kirchhoff's current law at the upper node, with the lower node as reference, is

$$\frac{15\ \text{V} - v_C}{4\ \Omega} = \frac{v_C}{8\ \Omega} + (10^{-6}\ \text{F})\left(\frac{dv_C}{dt}\right)$$

This is manipulated to

$$\left(\frac{8}{3} \times 10^{-6}\right)\left(\frac{dv_C}{dt}\right) + v_C = 10\ \text{V}$$

$$V_{ss} = 10\ \text{V}$$

$$\tau = \frac{8}{3} \times 10^{-6}\ \text{s} = 2.67 \times 10^{-6}\ \text{s}$$

$$V_0 = 0$$

$$v_C(t) = V_{ss} + (V_0 - V_{ss})e^{\frac{-t}{\tau}}$$

$$= \boxed{(10)\left(1 - e^{-(3.75\times 10^5)t}\right)u(t)\ \text{V}}$$

14. The homogeneous differential equation is

$$\frac{d^2v}{dt^2} + 5\frac{dv}{dt} + 4v = 0$$

This has solutions e^{st},

$$\frac{d^2v}{dt^2} \rightarrow s^2V$$

$$\frac{dv}{dt} \rightarrow sV$$

$$s^2 + 5s + 4 = 0$$

$$(s+4)(s+1) = 0$$

Transient solutions are of the form

$$Ae^{-4t} + Be^{-t}$$

The steady-state sinusoidal solution must be found before A and B can be evaluated.

Circuit Theory

The phasor solution is

$$12 \sin 2t \Rightarrow 12\angle 0°$$

$$\omega = 2$$

$$\frac{d}{dt} \rightarrow j\omega = j2$$

$$\frac{d^2}{dt^2} \rightarrow -\omega^2 = -4$$

The phasor equation is

$$12 \text{ V} = -4V_{ss} + (5)(j2)V_{ss} + 4V_{ss}$$

$$= j10V_{ss}$$

$$V_{ss} = \frac{1.2}{j} = 1.2\angle -90° \text{ V}$$

$$v_{ss} = 1.2 \sin(2t - 90°) \text{ V}$$

$$= -1.2 \cos 2t \text{ V}$$

$$v(t) = Ae^{-4t} + Be^{-t} - 1.2 \cos 2t$$

$$v(0) = A + B - 1.2 \text{ V} = -5 \text{ V}$$

$$A + B = -3.8 \text{ V}$$

$$\frac{dv}{dt} = -4Ae^{-4t} - Be^{-t} + 2.4 \sin 2t$$

$$\frac{dv}{dt}(0) = -4A - B = 2$$

$$A = 0.6 \text{ V}$$

$$B = -4.4 \text{ V}$$

$$v(t) = \boxed{(0.6e^{-4t} - 4.4e^{-t} - 1.2 \cos 2t)u(t) \text{ V}}$$

15. The impedance of the inductance branch is

$$Z_{LR} = 10 + 0.25s \ \Omega$$

KCL applied to the upper node is

$$I_s = \frac{V_C}{30 \ \Omega} + (0.001 \text{ F})sV_C + \frac{V_C}{10 \ \Omega + (0.25 \text{ H})s}$$

Solving,

$$G = \frac{V_C}{I_s} = \left(\frac{3 \times 10^4 + 750s}{0.75s^2 + 55s + 4000}\right) \Omega$$

$$= \boxed{(1000)\left(\frac{s + 40}{s^2 + 73.33s + 5333}\right) \Omega}$$

16. (a)

$$Z = 10 + 0.5s + \frac{10^4}{s} \ \Omega$$

$$= \left(\frac{1}{2}\right)\left(\frac{s^2 + 20s + 2 \times 10^4}{s}\right) \Omega$$

$$= \left(\frac{1}{2s}\right)\left(s^2 + \frac{\omega_o}{Q}s + \omega_o^2\right) \Omega$$

$$\omega_o = \boxed{\sqrt{2} \times 10^2 \text{ rad/s}}$$

(b)

$$Q = \frac{\omega_o}{20} = \boxed{5\sqrt{2}}$$

(c)

$$\text{BW} = \frac{\omega_o}{Q} = \boxed{20 \text{ rad/s}}$$

17.

$$Y = 0.1 + \frac{1}{0.5s} + 10^{-4}s$$

$$= \left(\frac{10^{-4}}{s}\right)(s^2 + 10^3 s + 2 \times 10^4)$$

$$= \left(\frac{10^{-4}}{s}\right)\left(s^2 + \frac{\omega_0}{Q}s + \omega_o^2\right)$$

$$\omega_0 = \sqrt{2 \times 10^4} = \boxed{100\sqrt{2} \text{ rad/s}}$$

$$\text{BW} = \frac{\omega_0}{Q} = \boxed{10^3 \text{ rad/s}}$$

$$Q = \frac{\omega_0}{\text{BW}} = \frac{\sqrt{2}}{10} = \boxed{0.1\sqrt{2}}$$

18. With the switch open for a long time, the upper-right capacitor is charged to 10 V and the lower (center) capacitor is discharged to 0 V. At the instant of switching, the capacitors can be modeled as ideal voltage sources with their initial voltages. The resulting equivalent circuit is shown.

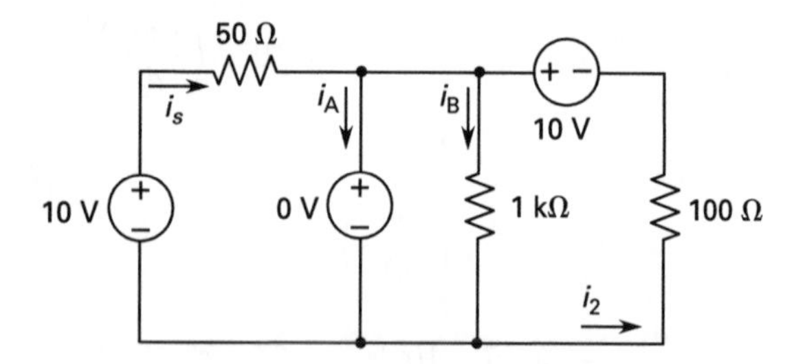

Kirchhoff's voltage law on the left loop is

$$10 \text{ V} - (50 \ \Omega)i_s + 0.0 = 0$$

$$i_s = \frac{1}{5} \text{ A}$$

Kirchhoff's voltage law on the right loop is

$$0.0 + 10\ \text{V} - (100\ \Omega)i_2 = 0$$

$$i_2 = \frac{1}{10}\ \text{A}$$

The current in the right (upper) capacitor is i_2, 0.1 A flowing left.

As the voltage across the 1 kΩ resistor is zero, $i_B = 0$. By Kirchhoff's current law, the other capacitor current is

$$i_A = i_s + i_2 = \frac{1}{5}\ \text{A} + \frac{1}{10}\ \text{A} = \boxed{0.3\ \text{A}}$$

19. The homogeneous equation is

$$s^2 + 4s + 4 = 0$$

The roots are $(s+2)^2$. The homogeneous solution is

$$Ae^{-2t} + Bte^{-2t}$$

The steady-state solution is

$$s \rightarrow j\omega$$
$$\omega = 4$$
$$\frac{d}{dt} \rightarrow j\omega$$
$$\frac{d^2}{dt^2} \rightarrow -\omega^2 = -16$$

The sine reference is used.

$$20\ \text{V} = (-16\ \Omega + 16j\ \Omega + 4\ \Omega)I_{\text{ss}}$$

$$I_{\text{ss}} = \frac{20}{-12 + j16} = -0.6 + j(-0.8)\ \text{A}$$

$$i_{\text{ss}} = -0.6 \sin 4t - 0.8 \sin(4t + 90°)\ \text{A}$$
$$= -0.6 \sin 4t - 0.8 \cos 4t\ \text{A}$$

$$i = i_{\text{ss}} + Ae^{-2t} + Bte^{-2t}$$

$$i(0) = -0.8 + A = 0$$

$$A = 0.8\ \text{A}$$

$$\frac{di}{dt} = -2.4 \cos 4t + 3.2 \sin 4t - 2Ae^{-2t} + Be^{-2t} - 2Bte^{-2t}$$

$$\frac{di}{dt}(0) = -2.4 - 2A + B = 4$$

$$B = 2.4 + 2A + 4 = 8\ \text{A}$$

$$i(t) = 0.8e^{-2t} + 8te^{-2t} - 0.6 \sin 4t - 0.8 \cos 4t\ \text{A}$$

$$= \boxed{0.8e^{-2t} + 8te^{-2t} - \sin(4t + 53.1°)\ \text{A}}$$

20. The initial charge on the 1 μF capacitor is

$$Q_0 = CV_0$$
$$= (10^{-6}\ \text{F})(100\ \text{V}) = 10^{-4}\ \text{C}$$

The initial energy, W, is

(a)

$$W_0 = \tfrac{1}{2}C_1V_0^2$$
$$= \left(\frac{10^{-6}\ \text{F}}{2}\right)(100\ \text{V})^2 = 0.005\ \text{J}$$

$$W_{1,\text{initial}} = 5\ \text{mJ}$$

In the final condition, the voltage across the two capacitors is the same.

$$W_{2,\text{initial}} = 0\ \text{mJ}$$

$$V_f = \frac{Q_1}{C_1} = \frac{Q_2}{C_2}$$

$$Q_2 = 2Q_1$$

As $Q_1 + Q_2 = Q_0$,

$$Q_1 = \frac{1}{3} \times 10^{-4}\ \text{C}$$

$$Q_2 = \frac{2}{3} \times 10^{-4}\ \text{C}$$

$$W_{\text{final}} = \left(\frac{1}{2}\right)\left(\frac{Q_1^2}{C_1}\right) + \left(\frac{1}{2}\right)\left(\frac{Q_2^2}{C_2}\right)\ \text{J}$$
$$= \left(\frac{1}{2}\right)(10^6\ \text{F}^{-1})\left(\frac{1}{9} \times 10^{-8}\ \text{C}^2\right) + \left(\frac{1}{2}\right)\left(\frac{10^6\ \text{F}^{-1}}{2}\right)\left(\frac{4}{9} \times 10^{-8}\ \text{C}^2\right)$$

$$W_{1,\text{final}} = 0.556\ \text{mJ}$$
$$W_{2,\text{final}} = 1.111\ \text{mJ}$$
$$W_{\text{loss}} = 3.333\ \text{mJ}$$

(b) $W_{\text{final}} = \frac{3}{18} \times 10^{-2} = \boxed{0.00167\ \text{J}}$

(c) $W_{\text{loss}} = 0.005 - 0.00167 = \boxed{0.00333\ \text{J}}$

Circuit Theory

31 Time Response

PRACTICE PROBLEMS

1. Which of the following is in the form of the solution to a first-order circuit?

(A) $\kappa + Ae^{\frac{-t}{\tau}}$
(B) $\kappa + Ae^{s_1 t} + Be^{s_2 t}$
(C) $\kappa + Ae^{st} + Bte^{st}$
(D) $\kappa + e^{\alpha t}(A\cos\beta t + B\sin\beta t)$

2. In the circuit shown, how long will it take for the capacitor voltage to reach 98% of its final value, assuming the capacitor is initially uncharged?

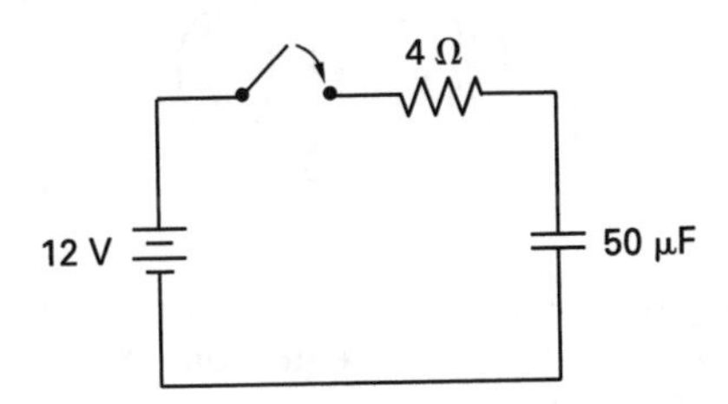

3. Determine the time domain equation for the circuit shown.

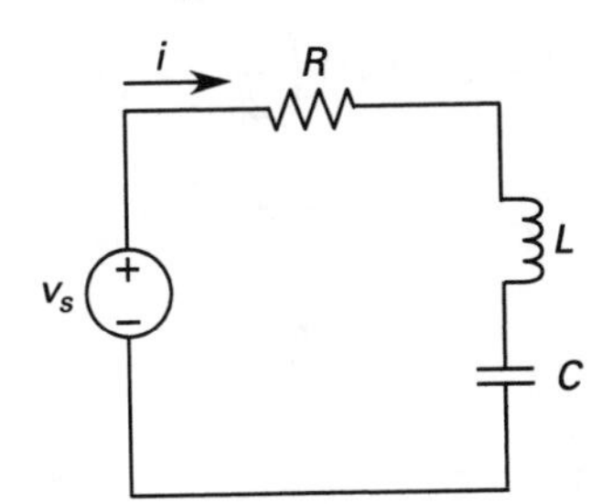

4. In the circuit shown in Prob. 3, the inductance is 80 mH and the capacitance is 20 μF. What is the minimum value of the resistance required for the circuit to be overdamped?

(A) 25 Ω
(B) 50 Ω
(C) 100 Ω
(D) 150 Ω

5. Complete the following table for a sine reference.

function	$A_p\angle\phi$	s
5		
$5e^{-200t}$		
$10\sin 377t$		
$10\sin(500t + 30°)$		
$20\cos(1000t + 60°)$		
$20e^{-200t}\sin 1000t$		

SOLUTIONS

1. The form of a first-order circuit solution is

$$x_t = \kappa + Ae^{-\frac{t}{\tau}}$$

The answer is (A).

2. The time constant, τ, is

$$\tau = RC = (4\ \Omega)\left(50 \times 10^{-6}\ \text{F}\right) = 200 \times 10^{-6}\ \text{s}$$
$$= 200\ \mu\text{s}$$

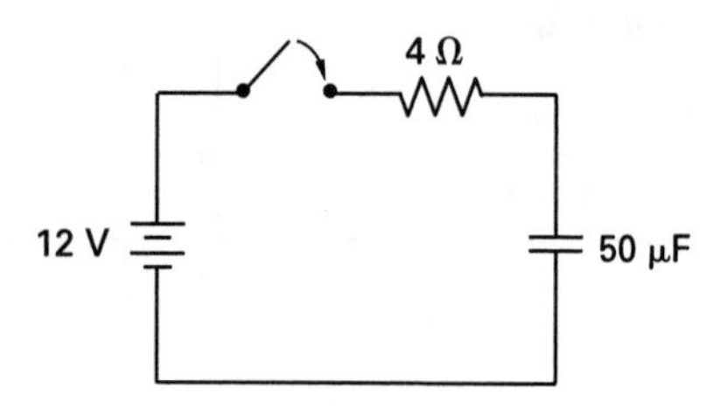

The build-up factor after a step change takes the form

$$1 - e^{-\frac{t}{\tau}}$$

Therefore, the time to 98% of the final value, regardless of the actual value, is

$$0.98 = 1 - e^{-\frac{t}{\tau}}$$

For the given circuit, $\tau = 200\ \mu\text{s}$. Substituting and solving for the time gives

$$0.98 = 1 - e^{\frac{-t}{200\times 10^{-6}\ \text{s}}}$$
$$-0.02 = -e^{-\frac{t}{200\times 10^{-6}\ \text{s}}}$$
$$\ln(0.02) = -\frac{t}{200 \times 10^{-6}\ \text{s}}$$
$$t = -\left(\ln(0.02)\right)\left(200 \times 10^{-6}\ \text{s}\right)$$
$$= 782.4 \times 10^{-6}\ \text{s}$$
$$= \boxed{782\ \mu\text{s}\quad [\approx 4\tau]}$$

3.

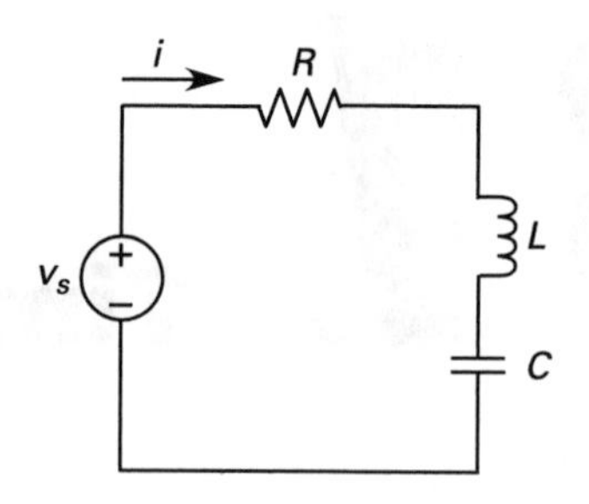

Using KVL around the circuit gives

$$v_s - iR - v_L - v_C = 0$$
$$v_s = iR + v_L + v_C$$
$$= iR + L\left(\frac{di}{dt}\right) + \frac{1}{C}\int idt$$

Differentiating gives

$$\frac{dv_s}{dt} = \left(\frac{di}{dt}\right)R + L\left(\frac{d^2i}{dt^2}\right) + \frac{i}{C}$$
$$= \boxed{L\left(\frac{d^2i}{dt^2}\right) + R\left(\frac{di}{dt}\right) + \frac{1}{C}i}$$

4. The overdamped condition is determined by the coefficients of the differential equation. Specifically, for the overdamped condition, $b^2 > 4ac$.

From the solution of Prob. 31.3

$$R^2 > 4L\left(\frac{1}{C}\right)$$

$$R < \sqrt{(4)\left(80 \times 10^{-3}\ \text{H}\right)\left(\frac{1}{20 \times 10^{-6}\ \text{F}}\right)} = 126.5\ \Omega$$

The answer is (D).

5.

function	$A_p\angle\phi$	s
5	$5\angle 0°$	$0 + j0$
$5e^{-200t}$	$5\angle 0°$	$-200 + j0$ Np/s
$10\sin 377t$	$10\angle 0°$	$0 \pm j377$ rad/s
$10\sin(500t + 30°)$	$10\angle 30°$	$0 \pm j500$ rad/s
$20\cos(1000t + 60°)$	$20\angle{-30°}$	$0 \pm j1000$ rad/s
$20e^{-200t}\sin 1000t$	$20\angle 0°$	$-200 \pm j1000\ \text{s}^{-1}$

32 Frequency Response

PRACTICE PROBLEMS

1. For the circuit shown, determine the transfer function with $V(s)$ as the input and $I(s)$ as the output.

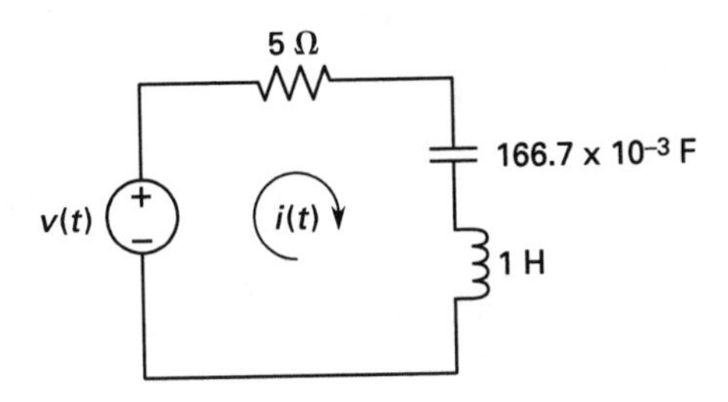

2. Plot the zeros and poles on the s domain for the circuit in Prob. 1.

3. Plot the magnitude Bode plot for the circuit in Prob. 1.

4. For the circuit shown, plot the idealized Bode plot of magnitude and phase.

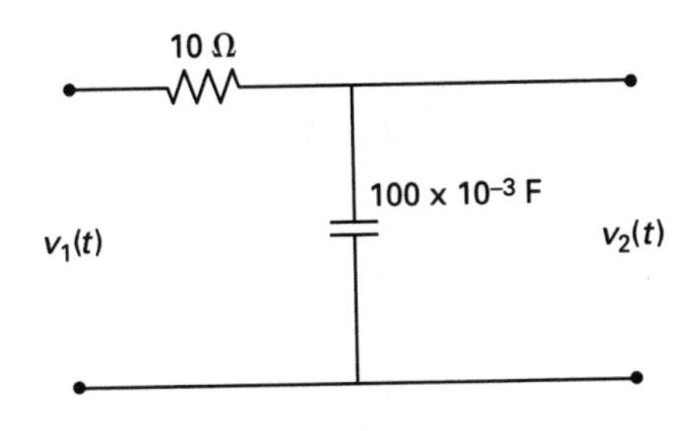

5. In the Bode plot for the circuit in Prob. 4, what type of filter is represented?

(A) low-pass filter
(B) high-pass filter
(C) bandpass filter
(D) notch filter

6. For the circuit shown, determine the (a) voltage transfer function, (b) frequency response, and (c) unit step voltage response.

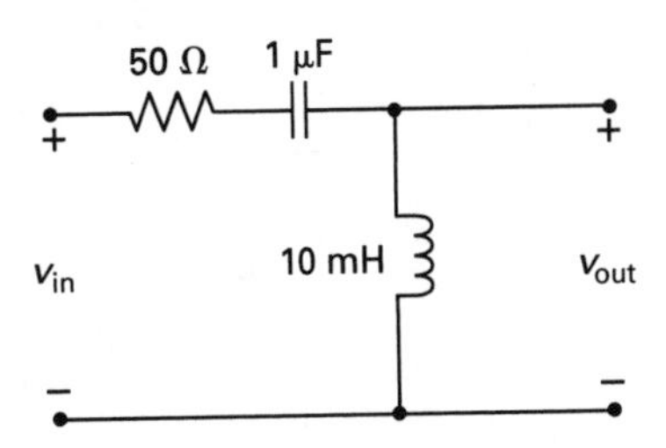

7. For the circuit shown, determine the (a) voltage transfer function, (b) frequency response, and (c) unit step voltage response.

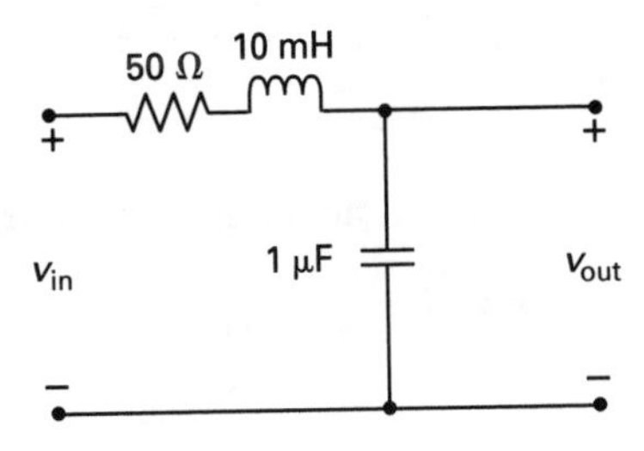

Circuit Theory

SOLUTIONS

1. The transfer function is

$$T_{\text{net}}(s) = \frac{I(s)}{V(s)} = \frac{1}{Z(s)}$$

First, transfer the circuit into the s domain.

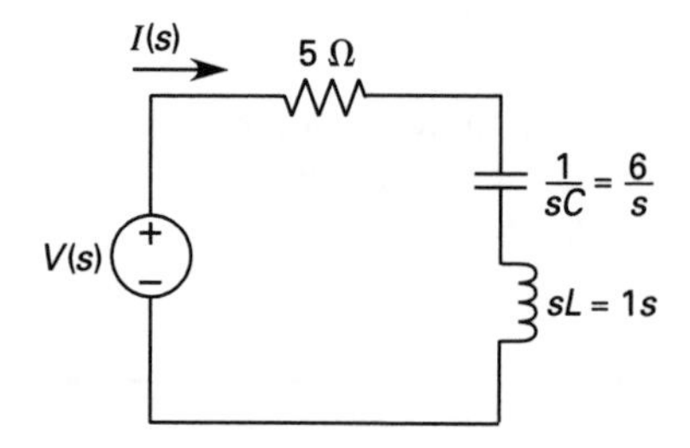

The impedance for this series circuit is

$$Z(s) = R + Z_C + Z_L = 5 + \frac{6}{s} + 1s$$

$$= \frac{5s + 6 + s^2}{s} = \frac{s^2 + 5s + 6}{s}$$

The transfer function is

$$\boxed{T_{\text{net}} = \frac{1}{Z(s)} = \frac{s}{s^2 + 5s + 6}}$$

2. Factor the transfer function into standard form.

$$T_{\text{net}}(s) = \frac{s}{s^2 + 5s + 6}$$

$$= \frac{s}{(s+2)(s+3)}$$

One zero at the origin is given by the numerator term. Two poles, one at -2 and the other at -3, exist. Plotting gives

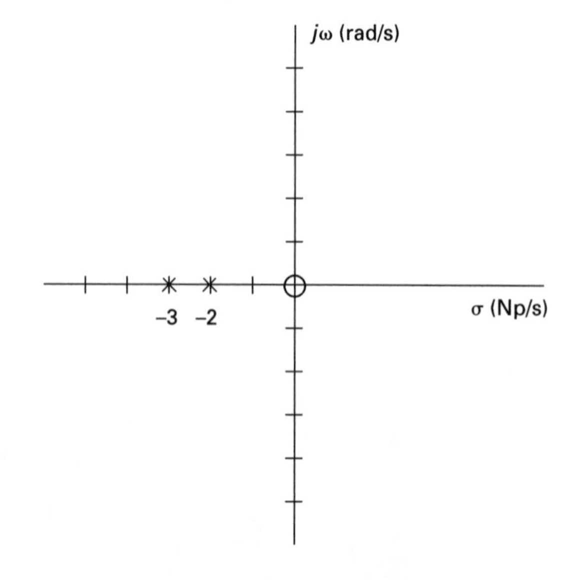

3. The transfer function was determined in Prob. 2 to be

$$T_{\text{net}}(s) = \frac{s}{(s+2)(s+3)}$$

The zero break frequency is $\omega = 1$. The break frequency for the poles is $\omega = 2$ and $\omega = 3$. These points are plotted on the diagram below.

For the zero, a +20 dB/decade line through the origin is plotted. For the poles, the low-frequency asymptotes are plotted on the frequency axis extending to lower frequencies from $\omega = 2$ and $\omega = 3$. The high-frequency asymptotes break down at -20 dB per decade from these points. Summing the asymptotic lines gives the net magnitude Bode plot as shown.

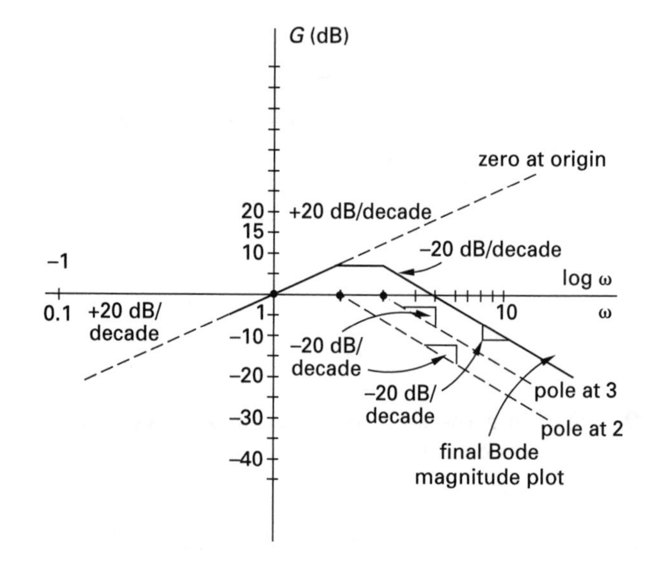

The value of K is determined from

$$T_{\text{net}}(j\omega) = K\left(\frac{j\omega}{\left(1 + \frac{j\omega}{2}\right)\left(1 + \frac{j\omega}{3}\right)}\right)$$

$$K = A\left(\frac{\prod_1^n z_n}{\prod_1^d p_d}\right)$$

The transfer $T_{\text{net}}(s)$ gives $A = 1$. For a zero at the origin in the s domain, the break frequency is $\omega = 1$. Therefore $z_n = 1$. Thus,

$$K = (1)\left(\frac{(1)}{(2)(3)}\right) = \frac{1}{6}$$

This makes $20 \log K = -15.563$ dB. Either the Bode plot should be shifted downward by -15.563 or the zero dB point should be changed to -15.563 to obtain the final solution.

4. Since the calculations will take place in the s domain, transform the circuit.

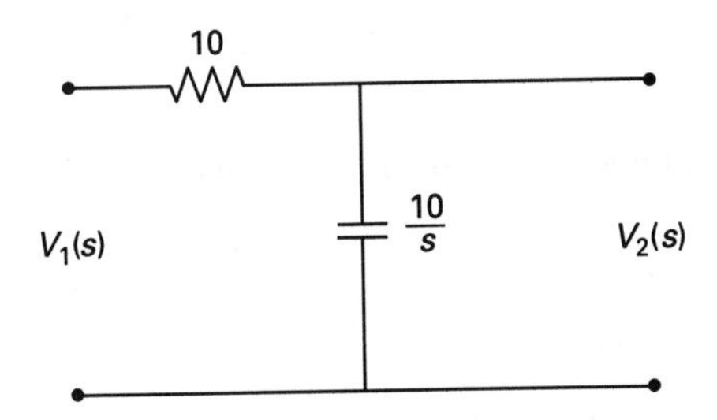

The transfer function using the voltage divider concept is

$$T_{\text{net}}(s) = \frac{V_2(s)}{V_1(s)} = \frac{\frac{10}{s}}{\frac{10}{s} + 10}$$

$$= \frac{\frac{10}{s}}{\frac{10 + s10}{s}} = \frac{10}{10 + s10} = \frac{1}{s+1}$$

The break frequency, or pole, occurs at $\omega = 1$. At this point the phase is $-45°$. Using the standard method, the resulting plot is

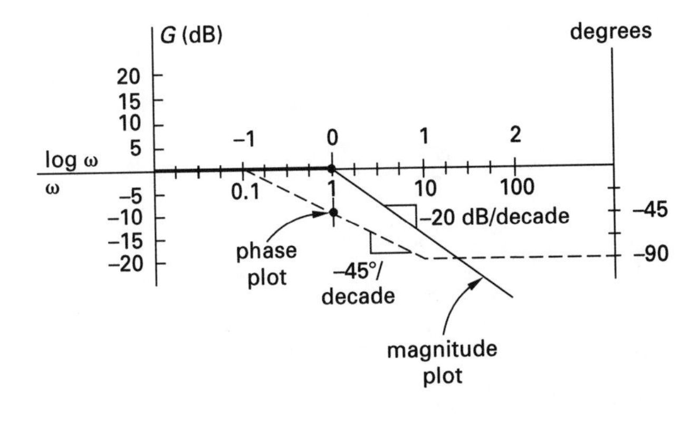

5. The Bode plot of Prob. 4 indicates that low frequencies are passed with a gain of 0 dB. At high frequencies, the output is attenuated and the gain steadily drops. The circuit is called a low-pass filter.

The answer is (A).

6. (a) By voltage division,

$$\left(\frac{V_{\text{out}}}{V_{\text{in}}}\right)(s) = \frac{0.01s}{0.01s + 50 + \frac{10^6}{s}}$$

$$= \boxed{\frac{s^2}{s^2 + 5000s + 10^8}}$$

(b) The denominator requires an underdamped response, so it is put into standard form to take advantage of the Laplace transform pairs table.

$$\left(\frac{V_{\text{out}}}{V_{\text{in}}}\right)(s) = \frac{s^2}{(s+2500)^2 + (9682)^2}$$

For the frequency response $s \to j\omega$,

$$\left(\frac{V_{\text{out}}}{V_{\text{in}}}\right)(j\omega) = \frac{-\omega^2}{10^8 - \omega^2 + j5000\omega}$$

At low frequency,

$$\frac{V_{\text{out}}}{V_{\text{in}}} \to \frac{-\omega^2}{10^8}$$

This has a slope of 40 dB per decade and a phase of 180° on a Bode diagram.

At high frequency,

$$\frac{V_{\text{out}}}{V_{\text{in}}} \to 1$$

At $\omega = 10^4$ rad/s,

$$\frac{V_{\text{out}}}{V_{\text{in}}} = 2\angle 90°$$

The value is 6 dB.

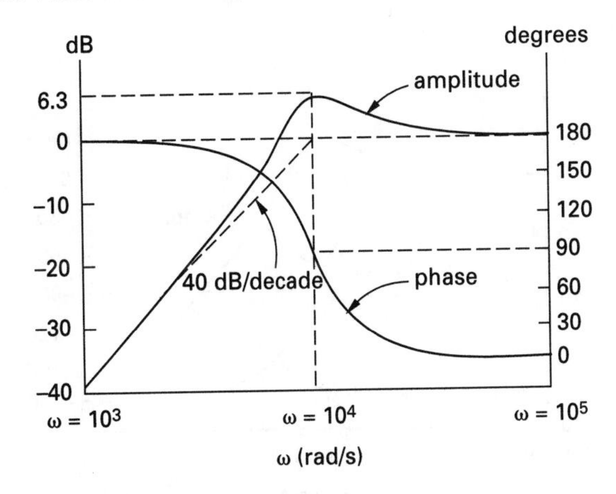

(c) The step response with $V_{\text{in}} = 1/s$ is

$$V_{\text{out}} = \frac{s}{(s+2500)^2 + (9682)^2} \text{ V·s}$$

$$= \frac{s+2500}{(s+2500)^2 + (9682)^2} - \left(\frac{2500}{9682}\right)\left(\frac{9682}{(s+2500)^2 + (9682)^2}\right)$$

$$v_{\text{out}}(t) = e^{-2500t}(\cos 9682t - 0.258 \sin 9682t)$$

$$= \boxed{1.03e^{-2500t}\cos(9682t + 14.5°) \text{ V}}$$

7. (a) By voltage division,

$$\frac{V_{\text{out}}}{V_{\text{in}}} = \frac{\frac{10^6}{s}}{10^{-2}s + 50 + \frac{10^6}{s}}$$

$$= \boxed{\frac{10^8}{s^2 + 5000s + 10^8}}$$

$$\left(\frac{V_{\text{out}}}{V_{\text{in}}}\right)(j\omega) = \frac{10^8}{10^8 - \omega^2 + j5000\omega}$$

(b) At low frequency,

$$\frac{V_{\text{out}}}{V_{\text{in}}} \to 1\angle 0°$$

At high frequency,

$$\frac{V_{\text{out}}}{V_{\text{in}}} \to \frac{10^8}{-\omega^2}$$

On the Bode diagram, this has a slope of -40 dB per decade and an angle of $-180°$. At $\omega = 10^4$ rad/s,

$$\frac{V_{\text{out}}}{V_{\text{in}}} = 2\angle -90°$$

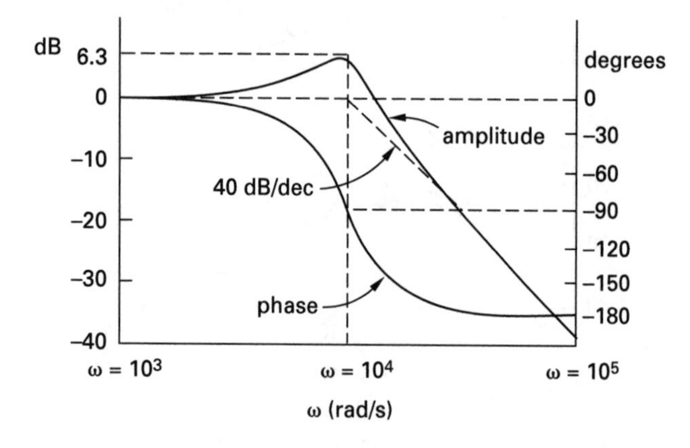

(c) The step response with $V_{\text{in}} = 1/s$ is

$$V_{\text{out}} = \frac{10^8}{s(s^2 + 5000s + 10^8)} \text{ V·s}$$

This is expanded in partial fractions to

$$V_{\text{out}} = \frac{1}{s} - \frac{s + 5000}{(s + 2500)^2 + (9682)^2}$$

This is recast to fit the Laplace transform pairs table.

$$V_{\text{out}} = \frac{1}{s} - \frac{s + 2500}{(s + 2500)^2 + (9682)^2} - \frac{\left(\frac{2500}{9682}\right)(9682)}{(s + 2500)^2 + (9682)^2}$$

$$v_{\text{out}}(t) = 1 - e^{-2500t}(\cos 9682t + 0.258 \sin 9682t) \text{ V}$$

$$= \boxed{1 - 1.033e^{-2500t} \cos(9682t - 14.5°) \text{ V}}$$

33 Generation Systems

PRACTICE PROBLEMS

1. What term describes the process of minimizing irreversibilities by raising the temperature of the water returning to the steam generator?

(A) superheat
(B) reheat
(C) regeneration
(D) both (A) and (C)

2. What fission rate (fissions/s) is required for a nuclear plant to have a 1000 MW thermal output?

3. What type of first stage, or control stage, is normally used on large turbines used for electric power generation?

(A) impulse
(B) Rateau
(C) Curtis
(D) Parsons

4. In the load sharing diagram shown, the speed droop for both generators is 0.6 Hz/1000 kW. If generator A is to carry all the load, what no-load frequency setpoint is required on generator A while maintaining the system frequency at 60 Hz?

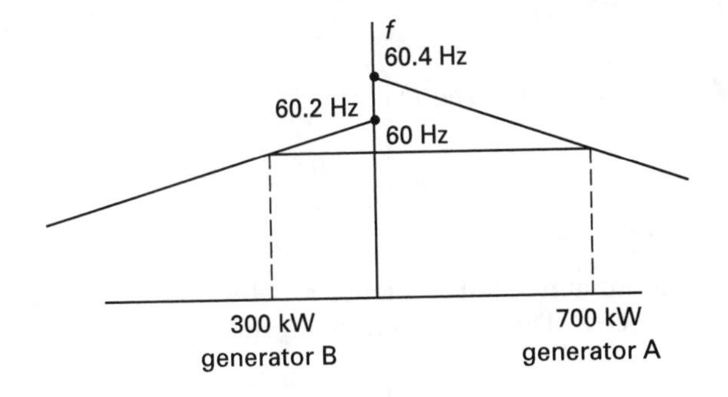

5. Draw the load sharing diagram for Prob. 4 with generator A carrying all the load. Label all significant points.

6. The field of a DC generator is normally located on the

(A) stator
(B) rotor
(C) armature
(D) round rotor

7. What is the maximum speed for 60 Hz AC generator operation?

8. What power quality term describes a transient of short duration and high magnitude?

(A) overvoltage
(B) harmonic
(C) sag
(D) surge

SOLUTIONS

1. Superheat raises the temperature of the steam supplied to the turbine. Reheat raises the temperature of the steam at the input of the low-temperature stages of a turbine. Regeneration raises the temperature of the water returning to the steam generator.

The answer is (C).

2. The fission rate is in units of fissions/s. The average energy released per fission is 200 MeV. Unit analysis gives

$$P = \left(\frac{\text{fissions}}{\text{s}}\right)\left(\frac{\text{energy}}{\text{fission}}\right) = \text{fr}\left(200\ \frac{\text{MeV}}{\text{fission}}\right)$$

$$\text{fr} = \frac{P}{200\ \frac{\text{MeV}}{\text{fission}}} = \frac{1000\times 10^6\ \text{W}}{\left(200\times 10^6\ \frac{\text{eV}}{\text{fission}}\right)\left(\frac{1.602\times 10^{-19}\,\text{J}}{1\ \text{eV}}\right)} = \boxed{3.12\times 10^{19}\ \text{fissions/s}}$$

3. Large turbines use impulse control stages since no pressure drop occurs across such stage, thereby allowing partial arc admission of the steam without significant forces being exerted on the turbine rotor.

The answer is (A).

4.

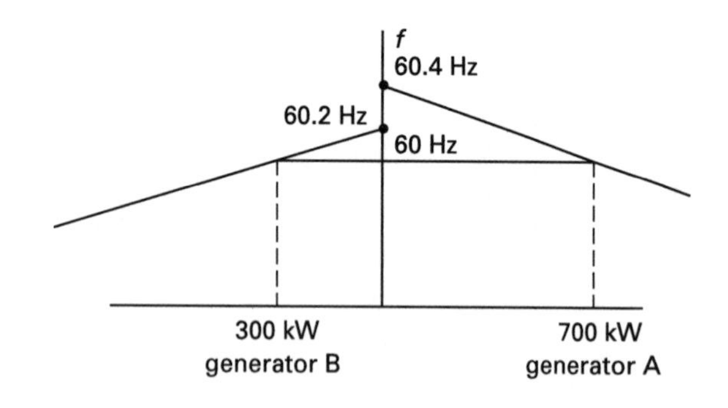

The power carried by an AC generator is given by

$$P = \frac{f_{\text{nl}} - f_{\text{sys}}}{f_{\text{droop}}}$$

Solving for the no-load frequency gives

$$f_{\text{nl}} = P f_{\text{droop}} + f_{\text{sys}}$$

Solving for the no-load frequency increase necessary to carry an additional 300 kW gives

$$\Delta f_{\text{nl}} = \Delta P f_{\text{droop}} = (300\ \text{kW})\left(\frac{0.6\ \text{Hz}}{1000\ \text{kW}}\right) = 0.18\ \text{Hz}$$

The required no-load frequency is

$$f_{\text{nl}} = f_{\text{nl,current}} + \Delta f_{\text{nl}} = 60.4\ \text{Hz} + 0.18\ \text{Hz} \approx \boxed{60.6\ \text{Hz}}$$

Generator B's frequency will have to be lowered by 0.2 Hz to maintain the system frequency at 60 Hz.

5. The real load sharing diagram with generator A carrying all the load is

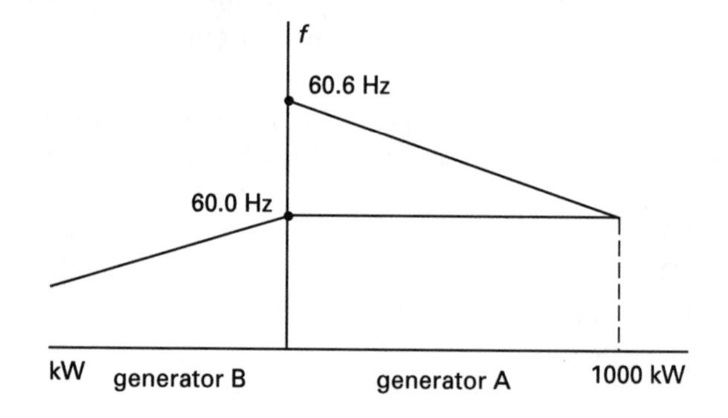

6. The field of a DC machine is normally on the stationary portion, or stator. This allows commutation to be accomplished on the rotor.

The answer is (A).

7. Synchronous speed is derived by

$$N_s = \frac{120f}{P}$$

The minimum number of poles determines the maximum speed. Two is the minimum number. Thus,

$$N_s = \frac{(120)(60\ \text{Hz})}{2} = \boxed{3600\ \text{rpm}}$$

8. A transient condition of short duration and high magnitude is called a *surge*.

The answer is (D).

34 Three-Phase Electricity and Power

PRACTICE PROBLEMS

1. A three-phase 208 V (rms) system supplies heating elements connected in wye. What is the resistance of each element if the total balanced load is 3 kW?

2. A three-phase system has a balanced delta load of 5000 kW at 84% power factor. If the line voltage is 4160 V (rms), what is the line current?

3. A 120 V (per phase, rms) three-phase system has a balanced load consisting of three 10 Ω resistances. What total power is dissipated if the connection is (a) wye and (b) delta?

4. A balanced delta load consists of three $20\angle 25^\circ$ Ω impedances. The 60 Hz line voltage is 208 V (rms). Find the (a) phase current, (b) line current, (c) phase voltage, (d) power consumed by each phase, and (e) total power.

5. A balanced wye load consist of three $3 + j4$ Ω impedances. The system voltage is 110 V (rms). Find the (a) phase voltage, (b) line current, (c) phase current, and (d) total power.

6. Consider the one-line diagram of a three-phase distribution system. What is the generator base voltage?

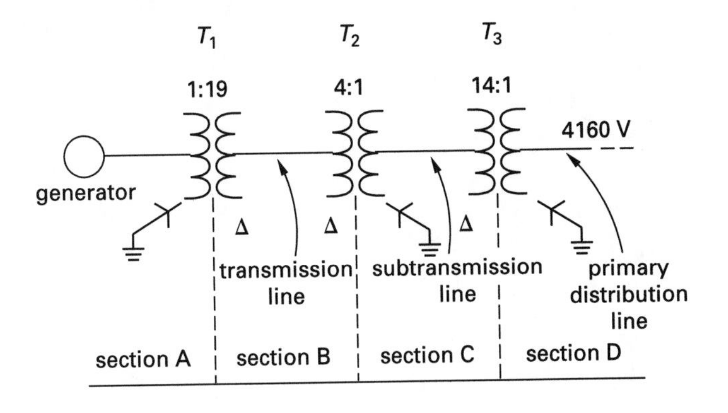

SOLUTIONS

1.
$$R = \frac{V_l^2}{P} = \frac{(208\text{ V})^2}{3\times 10^3\text{ W}} = \boxed{14.42\ \Omega}$$

2.
$$I_l = \left(\frac{1}{\sqrt{3}}\right)\left(\frac{P_\Delta}{V_l\cos\phi}\right)$$
$$= \left(\frac{1}{\sqrt{3}}\right)\left(\frac{5000\times 10^3\text{ W}}{(4160\text{ V})(0.84)}\right) = \boxed{826\text{ A}}$$

3. (a)
$$P_p = \frac{V_l^2}{R} = \frac{(120\text{ V})^2}{10\ \Omega} = 1440\text{ W}$$
$$P_t = (3)(1440\text{ W}) = \boxed{4320\text{ W}}$$

(b)
$$P_t = (3)\left(\frac{V_l^2}{R}\right) = (3)\left(\frac{(120\text{ V})^2}{10\ \Omega}\right)$$
$$= \boxed{4320\text{ W}}$$

4. (a)
$$\mathbf{I}_p = \frac{\mathbf{V}_l}{\mathbf{Z}} = \frac{208\angle 0^\circ\text{ V}}{20\angle 25^\circ\ \Omega}$$
$$= \boxed{10.4\ \angle{-25^\circ}\text{ A}}$$

(b)
$$\mathbf{I}_l = \sqrt{3}\ \angle{-30^\circ}\ \mathbf{I}_p$$
$$= \left(\sqrt{3}\ \angle{-30^\circ}\right)(10.4\ \angle{-25^\circ}\text{ A})$$
$$= \boxed{18\ \angle{-55^\circ}\text{ A}}$$

(c)
$$\mathbf{V}_p = \mathbf{V}_l$$
$$= \boxed{208\ \angle 0^\circ\text{ V}}$$

Power–Generation

(d)

$$P_p = V_p I_p \cos\phi$$
$$= (208 \text{ V})(10.4 \text{ A})(\cos 25°)$$
$$= \boxed{1960.5 \text{ W}}$$

(e)

$$P_\Delta = 3P_p = (3)(1960.5 \text{ W})$$
$$= \boxed{5882 \text{ W}}$$

5. (a)

$$V_p = \frac{110 \text{ V}}{\sqrt{3}} = \boxed{63.51 \text{ V}}$$

(b)

$$Z = \sqrt{(3\ \Omega)^2 + (4\ \Omega)^2} = 5\ \Omega$$
$$\phi = \arctan\left(\frac{3\ \Omega}{4\ \Omega}\right) = 53.15°$$
$$\mathbf{I}_l = \frac{\mathbf{V}_p}{\mathbf{Z}} = \frac{63.51\ \angle 0° \text{ V}}{5\ \angle 53.13°\ \Omega}$$
$$= \boxed{12.70\ \angle{-53.13°} \text{ A}}$$

(c)

$$\mathbf{I}_p = \mathbf{I}_l = \boxed{12.70\ \angle{-53.13°} \text{ A}}$$

(d)

$$P_t = \sqrt{3} V I \cos\phi$$
$$= \left(\sqrt{3}\right)(110 \text{ V})(12.7 \text{ A})(\cos 53.13°)$$
$$= \boxed{1451 \text{ W}}$$

6. For a three-phase system, the line voltage is the base. The base in the primary distribution line is given as 4160 V, or $V = 1.0 \text{ pu } \angle 0°$.

Since the transformers scale the base values in each section, the base voltage value of 4160 V will be determined at the generator. The $\sqrt{3}$ is used to account for the wye connection line-to-phase and phase-to-line conversions.

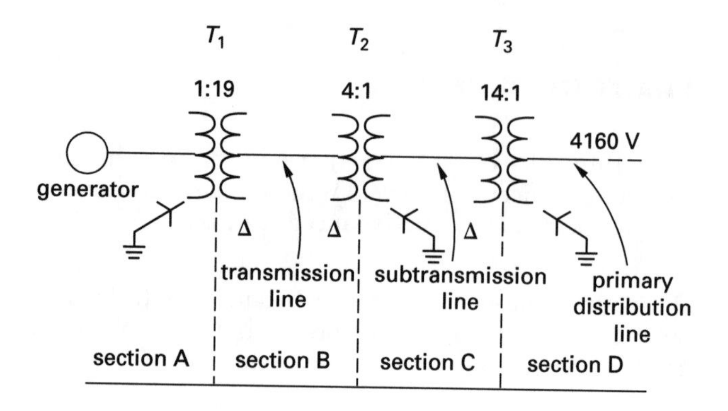

$$V_{\text{gen,base}} = V_{\text{gen,line}}$$
$$= \left(\frac{V_{\text{line,D}}}{\sqrt{3}}\right)\left(\frac{N_{3,\text{pri}}}{N_{3,\text{sec}}}\right)\left(\frac{1}{\sqrt{3}}\right)$$
$$\times \left(\frac{N_{2,\text{pri}}}{N_{2,\text{sec}}}\right)\left(\frac{N_{1,\text{pri}}}{N_{1,\text{sec}}}\right)\left(\sqrt{3}\right)$$
$$= \left(\frac{4160 \text{ V}}{\sqrt{3}}\right)\left(\frac{14}{1}\right)\left(\frac{1}{\sqrt{3}}\right)\left(\frac{4}{1}\right)\left(\frac{1}{19}\right)\left(\sqrt{3}\right)$$
$$= \boxed{7.07 \text{ kV}}$$

This generator line voltage of 7.07 kV is the 1.0 pu $\angle 0°$ value in section A of the distribution system.

35 Power Distribution

PRACTICE PROBLEMS

1. Which of the following voltages is not commonly used in the subtransmission portion of the power distribution system?

(A) 25 kV
(B) 69 kV
(C) 115 kV
(D) 138 kV

2. A single-conductor cable with an outside diameter of 0.5 cm is to be used in an underground installation. What is the optimal conductor radius?

(A) 0.05 cm
(B) 0.10 cm
(C) 0.20 cm
(D) 0.30 cm

3. A single-conductor capacitance-graded cable with an outside diameter of 10 cm can withstand a maximum electric field of 700 kV/m before insulation breakdown occurs. What is the maximum operating voltage of the conductor?

4. What is the total capacitance of 100 m of single-conductor cable with the optimal radii ratio and polyethylene insulation ($\epsilon_r = 2.25$)?

5. A three-phase 11.5 kV generator drives a 500 kW, 0.866 lagging power factor load. Determine the (a) line current and (b) necessary generator rating.

6. A transmission line with 8% reactance on a 200 MVA base connects two substations. At the first substation the voltage is $1.03\angle 5°$ pu, and at the second the voltage is $0.98\angle{-2.5°}$ pu. Find the (a) power and (b) volt-amperes reactive flowing out of the first substation.

SOLUTIONS

1. The trend in transmission voltages is toward higher voltages. The most common are 115 kV, 69 kV, 138 kV, and 238 kV. Thus, 25 kV is not common.

The answer is (A).

2. The optimal ratio is given by

$$\frac{r_2}{r_1} = 2.718$$

The outside diameter is given as 0.5 cm. Thus, the radius, r_2, is 0.25 cm. Substituting gives

$$r_1 = \frac{r_2}{2.718} = \frac{0.25 \text{ cm}}{2.718} = \boxed{0.0920 \text{ cm}}$$

The answer is (B).

3. The operating voltage for a capacitance-graded cable is

$$\begin{aligned} V &= E_{\text{max}}\left(r_1 \ln\left(\frac{r_2}{r_1}\right) + r_2 \ln\left(\frac{r_3}{r_2}\right)\right) \\ &= E_{\text{max}}\left(r_1 \ln\xi + r_2 \ln\xi\right) \end{aligned}$$

The values of r_1 and r_2 are the only unknowns. They are determined from

$$\xi = \frac{r_3}{r_2}$$

$$r_2 = \frac{r_3}{\xi} = \frac{\left(\frac{10 \text{ cm}}{2}\right)}{2.718} = 1.84 \text{ cm}$$

$$\xi = \frac{r_2}{r_1}$$

$$r_1 = \frac{r_2}{\xi} = \frac{1.84 \text{ cm}}{2.718} = 0.677 \text{ cm}$$

Substituting known and calculated values gives

$$\begin{aligned} V &= E_{\text{max}}\left(r_1 \ln\xi + r_2 \ln\xi\right) \\ &= \left(700 \ \frac{\text{kV}}{\text{m}}\right)\left(\begin{array}{l}(0.677 \text{ cm})(\ln 2.718) \\ \quad + (1.84 \text{ cm})(\ln 2.718)\end{array}\right)\left(\frac{1 \text{ m}}{100 \text{ cm}}\right) \\ &= \boxed{17.6 \text{ kV}} \end{aligned}$$

Power–Transmission

4. The capacitance per unit length is given by

$$\begin{aligned} C_l &= \frac{Q}{V} = \frac{2\pi\epsilon}{\ln \xi} \\ &= \frac{2\pi\epsilon_r\epsilon_0}{\ln e} \\ &= \frac{2\pi\epsilon_r\epsilon_0}{1} \\ &= 2\pi\,(2.25)\left(8.854\times 10^{-12}\ \frac{\text{F}}{\text{m}}\right) \\ &= 125\times 10^{-12}\ \text{F/m} \end{aligned}$$

The total capacitance is given by

$$\begin{aligned} C &= C_l L \\ &= \left(125\times 10^{-12}\ \frac{\text{F}}{\text{m}}\right)(100\ \text{m}) \\ &= \boxed{125\times 10^{-10}\ \text{F}} \end{aligned}$$

5. (a)

$$\begin{aligned} P_{\text{phase}} &= I_p V_p\ (\text{pf}) \\ I_p &= \frac{P_{\text{phase}}}{V_p\ (\text{pf})} \\ &= \frac{\frac{P_{\text{total}}}{3\ \text{phases}}}{\frac{V_p}{\sqrt{3}}} \\ &= \frac{\frac{500\ \text{kW}}{3\ \text{phases}}}{\left(\frac{11.5\ \text{kV}}{\sqrt{3}}\right)(0.866)} \\ &= \boxed{29\ \text{A}} \end{aligned}$$

(b)

$$\begin{aligned} P &= S\ (\text{pf}) \\ S &= \frac{P}{\text{pf}} \\ &= \frac{500\ \text{kW}}{0.866} \\ &= \boxed{577\ \text{kVA}} \end{aligned}$$

6.

j0.08

I

1.03∠5°

0.98∠−2.5°

(a)

$$\begin{aligned} I_{\text{pu}} &= \frac{V_{\text{pu}}}{Z_{\text{pu}}} = \frac{1.03\angle 5^\circ - 0.98\angle -2.5^\circ}{j0.08} \\ &= 1.758\angle -19.5^\circ\ \text{pu} \\ S_{\text{pu}} &= V_{\text{pu}} I^*_{\text{pu}} \\ &= (1.03\angle 5^\circ)(1.758\angle 19.5^\circ) \\ &= 1.811\angle 24.5^\circ\ \text{pu} \\ &= 1.648 + j0.751\ \text{pu} \\ P &= P_{\text{pu}} P_{\text{base}} = (1.648)(200\ \text{MW}) \\ &= \boxed{330\ \text{MW}} \end{aligned}$$

(b)

$$\begin{aligned} Q &= Q_{\text{pu}} Q_{\text{base}} = (0.751)(200\ \text{MVAR}) \\ &= \boxed{150\ \text{MVAR}} \end{aligned}$$

36 Power Transformers

PRACTICE PROBLEMS

1. Which electrical element in the exact (real) transformer model accounts for the eddy current and hysteresis losses?

(A) B_c
(B) R_c
(C) R_p
(D) R_s

2. Which electrical element in the exact (real) transformer model accounts for magnetic flux leakage?

(A) B_c
(B) X_p
(C) X_s
(D) (B) or (C)

3. What is the eddy current power loss for a 500 kVA rated 200 kg iron-core transformer with a coupling coefficient of 4×10^{-4} in a 1.4 T peak magnetic field?

(A) 3.0 W
(B) 9.0 W
(C) 400 W
(D) 575 W

4. A transformer is rated for 34.5 kV. The measured secondary voltage at no load is 35.02 kV. What is the percent voltage regulation?

(A) 0.5%
(B) 1.4%
(C) 1.5%
(D) 2.0%

5. An open-circuit test is conducted on a 120 V transformer winding. The results are $V_{1oc} = 120$ V, $V_{2oc} = 240$ V, $I_{1oc} = 0.25$ A, and $P_{oc} = 20$ W. What apparent power is necessary to supply core losses in this transformer?

(A) 20 VA
(B) 30 VA
(C) 60 VA
(D) 120 VA

6. A 13.8 kV single-phase transformer is subjected to an open-circuit test with the following results: $P_{in} = 900$ W, $I_{in} = 0.2$ A, and $V_{out} = 460$ V (secondary). Determine the values of the equivalent circuit parameters that can be found from this test.

7. A 32 kVA single-phase transformer is rated at 1.5 kV on the primary. A 60 Hz short-circuit test on the primary indicates $P_{in} = 1$ kW, $I_{in} = 20$ A, $I_s = 100$ A, and $V_p = 80$ V. Determine the (a) winding resistances and (b) leakage inductances.

8. A single-phase transformer is rated at 115 kV at 500 kVA. A short-circuit test on the high-voltage side at rated current indicates $P_{sc} = 435$ W and $V_{sc} = 2.5$ kV. Determine the per-unit impedance of the transformer.

SOLUTIONS

1. Core losses include those due to eddy currents and hysteresis. In the exact transformer model, these are accounted for by $G_c = 1/R_c$.

The answer is (B).

2. Magnetic flux leakage is a term used to describe any flux that is not mutual. The primary and secondary self-inductance represented by X_p and X_s are not mutual.

The answer is (D).

3. The eddy current loss is found from

$$P_e = k_e B_m^2 f^2 m$$

Assume 60 Hz is the frequency. Substitute the given values.

$$\begin{aligned} P_e &= (4 \times 10^{-4})(1.4\ \text{T})^2(60\ \text{Hz})^2(200\ \text{kg}) \\ &= \boxed{565\ \text{W}} \end{aligned}$$

The answer is (D).

4. The voltage regulation is given by

$$\begin{aligned} \text{VR} &= \frac{V_{\text{nl}} - V_{\text{fl}}}{V_{\text{fl}}} \\ &= \frac{\frac{V_p}{a} - V_{\text{s,rated}}}{V_{\text{s,rated}}} \end{aligned}$$

Since the voltage was measured at the output of the secondary, it is equivalent to V_p/a. Thus,

$$\text{VR} = \frac{35.02\ \text{V} - 34.5\ \text{V}}{34.5\ \text{V}} = 15.07 \times 10^{-3}$$

To express as a percentage, multiply by 100.

$$\text{VR} = \boxed{1.5\%}$$

The answer is (C).

5. The open-circuit test determines the parameters associated with core losses. The apparent power is

$$\begin{aligned} S_{\text{oc}} &= I_{1\text{oc}} V_{1\text{oc}} \\ &= (0.25\ \text{A})(120\ \text{V}) \\ &= \boxed{30\ \text{VA}} \end{aligned}$$

The answer is (B).

6. The open-circuit test obtains the core parameters G and B.

$$G = \frac{P_{\text{oc}}}{V_{\text{oc}}^2} = \frac{900\ \text{W}}{(13{,}800\ \text{V})^2} = \boxed{4.73 \times 10^{-6}\ \text{S}}$$

$$Q^2 = S^2 - P^2 = ((13{,}800\ \text{VA})(0.2))^2 - (900\ \text{W})s^2$$

$$Q = 2609\ \text{VAR}$$

$$B = \frac{Q}{V^2} = \frac{2609\ \text{VAR}}{(13{,}800\ \text{V})^2} = \boxed{13.7 \times 10^{-6}\ \text{S}}$$

The turns ratio is

$$\begin{aligned} \frac{V_p}{N_p} &= \frac{V_s}{N_s} \\ \frac{N_p}{N_s} &= \frac{V_p}{V_s} \\ &= \frac{13{,}800\ \text{V}}{460\ \text{V}} = \boxed{30} \end{aligned}$$

7. (a)

$$\begin{aligned} N_p I_p &= N_s I_s \\ a &= \frac{N_p}{N_s} = \frac{I_s}{I_p} \\ &= \frac{100\ \text{A}}{20\ \text{A}} = 5 \end{aligned}$$

$$\begin{aligned} P_{\text{sc}} &= I_p^2 R_p + I_s^2 R_s = 1000\ \text{W} \\ I_p^2(R_p + a^2 R_s) &= 1000\ \text{W} \\ R_p + a^2 R_s &= \frac{1000\ \text{W}}{(20\ \text{A})^2} = 2.5\ \Omega \end{aligned}$$

Assume $R_p = a^2 R_s = \boxed{1.25\ \Omega}$

$$R_s = \frac{1.25\ \Omega}{(5)^2} = \boxed{0.05\ \Omega}$$

(b)

$$S = V_pI_p = (80 \text{ V})(20 \text{ A}) = 1600 \text{ VA}$$
$$= P + jQ$$
$$Q^2 = S^2 - P^2$$
$$Q = 1249 \text{ VAR}$$
$$= I_1^2(X_p + a^2X_s)$$
$$X_p + a^2X_s = \frac{Q}{I_1^2} = \frac{1249 \text{ VAR}}{(20 \text{ A})^2} = 3.12 \ \Omega$$

Assume $X_p = a^2X_s$, then

$$2X_p = 3.12$$
$$X_p = \frac{3.12}{2} = 1.56 \ \Omega = \omega L_p = 377L_p$$
$$L_p = \frac{1.56 \ \Omega}{377 \ \frac{\text{rad}}{\text{s}}} = \boxed{4.14 \text{ mH}}$$
$$X_s = \frac{X_p}{a^2} = \frac{1.56}{a^2} = 0.0624 \ \Omega = \omega L_s = 377L_s$$
$$L_s = \frac{0.0624 \ \Omega}{377 \ \frac{\text{rad}}{\text{s}}} = \boxed{0.166 \text{ mH}}$$

8. The per-unit impedance is found from

$$Z_{\text{pu}} = \frac{Z_{\text{actual}}}{Z_{\text{base}}} = \frac{R + jX}{Z_{\text{base}}} \qquad \text{[I]}$$

Find the unknown quantities.

$$Z_{\text{base}} = \frac{V_{\text{base}}}{I_{\text{base}}} \qquad \text{[II]}$$

The short-circuit test is performed at rated current, which is I_{base}. Thus,

$$I_{\text{rated}} = I_{\text{base}} = \frac{S_{\text{rated}}}{V_{\text{rated}}}$$
$$= \frac{500 \text{ kVA}}{115 \text{ kV}} = 4.35 \text{ A}$$

Substitute the rated current into Eq. II.

$$Z_{\text{base}} = \frac{115 \text{ kV}}{4.35 \text{ A}} = 26{,}440 \ \Omega$$

Find R from

$$P_{sc} = I_{sc}^2R = I_{\text{rated}}^2R$$
$$R = \frac{P_{sc}}{I_{\text{rated}}^2} = \frac{435 \text{ W}}{(4.35 \text{ A})^2} = 23 \ \Omega$$

Find X from

$$Q_{sc} = I_{\text{rated}}^2X$$
$$X = \frac{Q}{I_{\text{rated}}^2} \qquad \text{[III]}$$

The reactive power is found using

$$S_{sc}^2 = P_{sc}^2 + Q_{sc}^2$$
$$Q_{sc} = \sqrt{S_{sc}^2 - P_{sc}^2}$$
$$= \sqrt{(I_{\text{rated}}V_{sc})^2 - (435 \text{ W})^2}$$
$$= \sqrt{\left((4.35 \text{ A})(2.5 \times 10^3 \text{ V})\right)^2 - (435 \text{ W})^2}$$
$$= 10{,}870 \text{ VAR}$$

Substitute into Eq. III giving

$$X = \frac{10{,}870 \text{ VAR}}{(4.35 \text{ A})^2} = 574 \ \Omega$$

Substitute the calculated values for R, X, and Z_{base} into Eq. I.

$$Z_{\text{pu}} = \frac{23 \ \Omega + j574 \ \Omega}{26{,}440 \ \Omega} = 0.0009 + j0.022 \text{ pu}$$
$$= \boxed{0.001 + j0.022 \text{ pu}}$$

37 Power Transmission Lines

PRACTICE PROBLEMS

1. As frequency increases, the skin effect causes which of the following parameters to increase?

(A) AC resistance
(B) DC resistance
(C) internal inductance
(D) external inductance

2. What distributed parameter is normally insignificant and ignored in calculations?

(A) line resistance
(B) line inductance
(C) shunt resistance
(D) shunt capacitance

3. The velocity factor is defined as the ratio of the velocity of propagation, v_w, to the speed of light, c. What is the velocity factor in a single waxwing conductor at 60 Hz?

(A) 0.29
(B) 0.61
(C) 0.90
(D) 0.97

4. What is parameter A of a 220 km falcon line with $Z_l = 0.85\angle 70°\ \Omega/\text{mi}$ and $Y_l = 7 \times 10^{-6}\angle 90°\ \text{S/mi}$?

(A) $4.0 \times 10^{-4}\angle 160°$
(B) $0.948\angle 1.2°$
(C) $0.996\angle 0.008°$
(D) $1.0\angle 20°$

5. A three-phase 215 kV transmission line supplies a 100 MW wye-connected load at 0.8 pf lagging. The ABCD parameters are as follows: $\text{A} = 0.9\angle 2.0°$, $\text{B} = 150\angle 80°$, $\text{C} = 0.001\angle 90°$, and $\text{D} = 0.9\angle 2°$. What is the sending-end voltage?

SOLUTIONS

1. The skin effect causes AC resistance to increase and internal inductance to decrease.

The answer is (A).

2. The shunt resistance is normally very large and thus has a minimal effect on current and voltage.

The answer is (C).

3. The velocity of wave propagation is given by

$$v_w = \frac{1}{\sqrt{L_l C_l}}$$

From App. 37.A, a waxwing conductor has an inductive reactance of

$$X_{L,l} = 0.476\ \Omega/\text{mi}$$

Solving for the inductance per unit length gives

$$X_{L,l} = \omega L = 2\pi f L$$

$$L = \frac{X_{L,l}}{2\pi f} = \frac{0.476\ \frac{\Omega}{\text{mi}}}{2\pi\,(60\ \text{s}^{-1})} = 1.26 \times 10^{-3}\ \text{H/mi}$$

The capacitive reactance, from App. 37.A, is

$$X_C = 0.1090\ \text{M}\Omega/\text{mi}$$

Solving for C_l gives

$$|X_C| = \frac{1}{\omega C_l} = \frac{1}{2\pi f C_l}$$

$$C_l = \frac{1}{2\pi f\,|X_C|} = \frac{1}{2\pi\,(60\ \text{s}^{-1})\left(0.1090 \times 10^6\ \frac{\Omega}{\text{mi}}\right)} = 2.43 \times 10^{-8}\ \text{F/mi}$$

Substituting the calculated values gives

$$\begin{aligned} \mathrm{v}_w &= \frac{1}{\sqrt{L_l C_l}} \\ &= \frac{1}{\sqrt{\left(1.26\times10^{-3}\ \frac{\mathrm{H}}{\mathrm{mi}}\right)\left(2.43\times10^{-8}\ \frac{\mathrm{F}}{\mathrm{mi}}\right)}} \\ &= \left(1.81\times10^{5}\ \frac{\mathrm{mi}}{\mathrm{s}}\right)\left(\frac{1.609\ \mathrm{km}}{1\ \mathrm{mi}}\right)\left(\frac{1000\ \mathrm{m}}{1\ \mathrm{km}}\right) \\ &= 2.91\times10^{8}\ \mathrm{m/s} \end{aligned}$$

The velocity factor is

$$\begin{aligned} \mathrm{v}_{\mathrm{factor}} &= \frac{\mathrm{v}_w}{C} \\ &= \frac{2.91\times10^{8}\ \frac{\mathrm{m}}{\mathrm{s}}}{3.00\times10^{8}\ \frac{\mathrm{m}}{\mathrm{s}}} \\ &= \boxed{0.97} \end{aligned}$$

The answer is (D).

4. This is a medium-length transmission line. Regardless of the model chosen,

$$\begin{aligned} \mathrm{A} &= 1 + \tfrac{1}{2}YZ \\ &= 1 + \tfrac{1}{2}\left(7\times10^{-6}\angle 90^\circ\ \frac{\Omega^{-1}}{\mathrm{mi}}\right)\left(0.85\angle 70^\circ\ \frac{\Omega}{\mathrm{mi}}\right) \\ &= 1 + \left(2.98\times10^{-6}\angle 160^\circ\ \mathrm{mi}^{-2}\right)(220\ \mathrm{km})^2 \\ &\quad \times \left(\frac{1\ \mathrm{mi}}{1.609\ \mathrm{km}}\right)^2 \\ &= 1\angle 0^\circ + 5.56\times10^{-2}\angle 160^\circ \\ &= (1+0j) + (-0.0523 + j0.0190) \\ &= 0.9477 + j0.0190 \\ &= \boxed{0.948\angle 1.2^\circ} \end{aligned}$$

The answer is (B).

The type of cable determines the electrical parameters, but these were given. Also, the first term of the parameter, that is, one, can be represented as $1\angle 0^\circ$.

5. The sending-end voltage is given by (see Fig. 37.3)

$$V_S = \mathrm{A}V_R + \mathrm{B}I_R$$

V_R and I_R must be determined. Since the two-part model represents a single phase, the receiving-end phase voltage for a wye-connected load is

$$V_R = \frac{215\ \mathrm{kV}}{\sqrt{3}} = 124\ \mathrm{kV}$$

Since the power factor is given for the load, let V_R be the reference voltage. Thus,

$$V_R = 124\angle 0^\circ\ \mathrm{kV}$$

The receiving-end current is found from

$$\begin{aligned} P &= 3I_R V_R(\mathrm{pf}) \\ I_R &= \frac{P}{3V_R(\mathrm{pf})} = \frac{100\times10^{6}\ \mathrm{W}}{(3)(124\times10^{3}\ \mathrm{V})(0.8)} \\ &= 336\ \mathrm{A} \end{aligned}$$

Since the power factor is 0.8 lagging, the current angle is -36.8°.

$$I_R = 336\angle -36.8^\circ\ \mathrm{A}$$

Substituting gives

$$\begin{aligned} V_S &= \mathrm{A}V_R + \mathrm{B}I_R \\ &= (0.9\angle 2.0^\circ)(124\times10^{3}\angle 0^\circ\ \mathrm{V}) \\ &\quad + (150\angle 80^\circ)(336\angle -36.8^\circ\ \mathrm{A}) \\ &= 1.116\times10^{5}\angle 2.0^\circ\ \mathrm{V} + 5.04\times10^{4}\angle 43.2^\circ\ \mathrm{V} \\ &= \boxed{153\angle 14.5^\circ\ \mathrm{kV}} \end{aligned}$$

Note that the power could have been calculated using the quantities, in which case the current would be given by

$$I_R = \frac{P}{\sqrt{3}V_{ll}(\mathrm{pf})}$$

38 Batteries and Fuel Cells

PRACTICE PROBLEMS

1. A cell that undergoes a chemical reaction in such a manner that it cannot be recharged is called a

(A) dry cell
(B) primary cell
(C) reserve cell
(D) secondary cell

2. Within a battery, conventional current flows from

(A) anion to anode
(B) anion to cation
(C) anode to cathode
(D) positive electrode to negative electrode

3. A series of fuel cells is connected to produce a no-load voltage of 320 V. At rated voltage of 300 V the cells provide 2.0 MW to a DC-AC converter. Assuming the cell can be modeled as a standard battery, what is the total internal resistance of the fuel cells?

(A) 0.0001 Ω
(B) 0.003 Ω
(C) 0.050 Ω
(D) 3.0 Ω

4. A wristwatch mercury cell has a capacity of 200 mA·h. How many months can the watch function without replacement if it draws 15 μA?

(A) 3 months
(B) 6 months
(C) 12 months
(D) 18 months

5. Increasing the current withdrawn from a battery has what effect on the battery capacity and why?

(A) increases capacity due to less internal heat loss
(B) increases capacity due to lower internal resistance
(C) decreases capacity due to higher load voltage drop
(D) decreases capacity due to increased polarization

SOLUTIONS

1. A cell in which the electrodes are consumed and in general cannot be recharged is called a primary cell.

The answer is (B).

2. Within a battery, the conventional current flows from anode to cathode. (The anode is labeled as the positive terminal. Externally, the anode is the terminal into which positive current flows.)

The answer is (C).

3. The circuit is

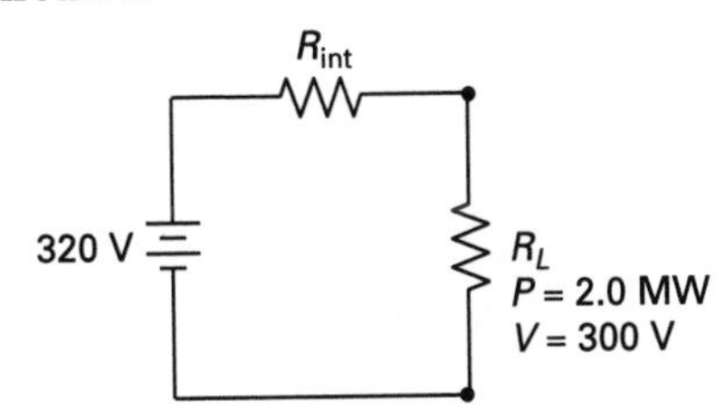

The voltage across the internal resistance is given (20 V). To find the resistance, the current must be determined. From the rated conditions, the load resistance is

$$P = \frac{V^2}{R_L}$$

$$R_L = \frac{V^2}{P} = \frac{(300 \text{ V})^2}{2.0 \times 10^6 \text{ W}} = 0.045 \ \Omega$$

The rated current is

$$P = I^2 R_L$$

$$I = \sqrt{\frac{P}{R_L}} = \sqrt{\frac{2.0 \times 10^6 \text{ W}}{0.045 \ \Omega}} = 6.67 \times 10^3 \text{ A}$$

Using Ohm's law on the internal resistor gives

$$V = IR_{\text{int}}$$

$$R_{\text{int}} = \frac{V}{I} = \frac{320 \text{ V} - 300 \text{ V}}{6.67 \times 10^3 \text{ A}} = \boxed{0.003 \ \Omega}$$

The answer is (B).

Power–Transmission

4. Amp-hour capacity is given by

$$Ah = It$$

Thus,

$$t = \frac{Ah}{I} = \frac{200 \times 10^{-3}\ \text{A·h}}{15 \times 10^{-6}\ \text{A}}$$
$$= 1.33 \times 10^{4}\ \text{h}$$

Assuming 30-day months, the capacity is

$$\text{total months} = \left(1.33 \times 10^{4}\ \text{h}\right)\left(\frac{1\ \text{d}}{24\ \text{h}}\right)\left(\frac{1\ \text{month}}{30\ \text{d}}\right)$$
$$= \boxed{18.5\ \text{months}}$$

The answer is (D).

5. Battery capacity drops as the discharge current increases due to increased losses on the internal resistance and polarization of electrodes, that is, hydrogen bubble insulation of the electrodes.

The answer is (D).

39 Rotating DC Machinery

PRACTICE PROBLEMS

1. A four-pole generator is turned at 3600 rpm. Each of its poles has a flux of 3.0×10^{-4} Wb. The armature has a total of 200 conductors and is simplex lap wound. What average voltage is generated at no load?

2. A DC generator has an armature resistance of 0.31 Ω. The shunt field resistance is 134 Ω. No-load voltage is 121 V at 1775 rpm. Full-load current is 30 A at 110 V. Assume a constant flux. What are the (a) no-load shunt field current, (b) full-load speed, and (c) copper power losses at full load?

3. (a) What is the back emf of a 10 hp shunt motor operating on 110 V terminals if 90 A flow through a 0.05 Ω armature resistance? (b) What is the line current if the field resistance is 60 Ω?

4. The nameplate of a 240 V DC motor states that the full-load line current is 67 A. During a no-load test at rated voltage, 3.35 A armature current and 3.16 A field current are drawn. The no-load armature resistance is 0.207 Ω. The no-load brush drop is 2 V. What are the (a) horsepower and (b) efficiency at full load?

5. A DC generator has a 250 V output at a 50 A load current and 260 V at no load. Estimate the armature resistance.

6. A separately excited DC generator is rated at 10 kW, 240 V, 2800 rpm for an input power of 15 hp. No-load voltage is 260 V. Determine the (a) armature resistance and (b) mechanical power loss at 2800 rpm.

7. A DC shunt motor is running at rated values when the line voltage suddenly drops by 10%. Determine the resulting changes in input power and motor speed if (a) the load torque is constant and (b) the load torque is proportional to speed.

8. A DC series motor is running at rated speed, voltage, and current when the load torque increases by 25%. Determine the effects of this change, assuming that the line voltage remains the same.

9. A DC shunt motor takes 200 V and 25 A at 1000 rpm and 5 hp. It has an armature resistance of 0.5 Ω and a shunt field resistance of 100 Ω. Determine the (a) rotational losses and (b) no-load speed at 200 V.

10. A DC shunt motor is rated at 7.5 hp, 60 A, 1000 rpm, and 120 V. Assume the load torque is constant, regardless of speed. What field resistance values are required to change the speed to (a) 1150 rpm and (b) 750 rpm?

SOLUTIONS

1.

$$V = \left(\frac{z}{a}\right) p\Phi \left(\frac{n}{60}\right)$$

For simplex lap winding, $a = p$.

$$V = \left(\frac{200}{4}\right)(4)(3.0 \times 10^{-4}\ \text{Wb})\left(\frac{3600\ \text{rpm}}{60\ \frac{\text{sec}}{\text{min}}}\right)$$

$$= \boxed{3.6\ \text{V}}$$

2. (a)

$$I_{f,\text{nl}} = \frac{V_{\text{nl}}}{R_f} = \frac{121\ \text{V}}{134\ \Omega}$$

$$= \boxed{0.903\ \text{A}}$$

(b) Voltage is proportional to rotational speed.

$$n_{\text{fl}} = \left(\frac{V_{\text{fl}} + R_a\left(I_{\text{fl}} + \frac{V_{\text{fl}}}{R_f}\right)}{V_{\text{nl}} + R_a\left(\frac{V_{\text{nl}}}{R_f}\right)}\right) n_{\text{nl}}$$

$$= \left(\frac{110V + (0.31\ \Omega)\left(30\ \text{A} + \frac{110\ \text{V}}{134\ \Omega}\right)}{121\ \text{V} + (0.31\ \Omega)\left(\frac{121\ \text{V}}{134\ \Omega}\right)}\right)(1775\ \text{rpm})$$

$$= \boxed{1750\ \text{rpm}}$$

(c) $$P = \frac{V_{\text{fl}}^2}{R_f} + \left(I_{\text{fl}} + \frac{V_{\text{fl}}}{R_f}\right)^2 R_a$$

$$= \frac{(110\ \text{V})^2}{134\ \Omega} + \left(30\ \text{A} + \frac{110\ \text{V}}{134\ \Omega}\right)^2 (0.31\ \Omega)$$

$$= \boxed{385\ \text{W}}$$

3. (a)

$$E = V_{\text{fl}} - I_{a,\text{load}} R_a$$

$$= 110\ \text{V} - (90\ \text{A})(0.05\ \Omega)$$

$$= \boxed{105.5\ \text{V}}$$

(b)

$$I_{\text{fl}} = \frac{V_{\text{fl}}}{R_f} + I_{a,\text{load}} = \frac{110\ \text{V}}{60\ \Omega} + 90\ \text{A}$$

$$= \boxed{91.8\ \text{A}}$$

4.

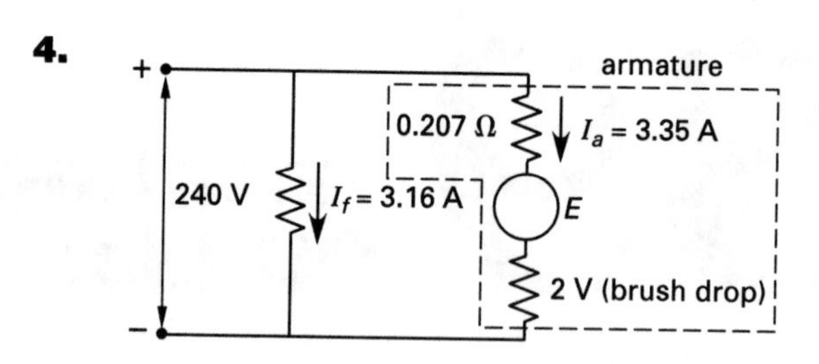

The total no-load power input to the armature is

$$P_a = I_a V = (3.35\ \text{A})(240\ \text{V})$$

$$= 804\ \text{W}$$

The I^2R power loss in the armature resistance is

$$P_{\text{resistance}} = I_a^2 R = (3.35\ \text{A})^2(0.207\ \Omega)$$

$$= 2.323\ \text{W}$$

The armature brush power loss is

$$P_{\text{brush}} = I_a V_{\text{brush}}$$

$$= (3.35\ \text{A})(2\ \text{V}) = 6.7\ \text{W}$$

The no-load rotational loss is

$$P_{\text{rotational}} = P_a - P_{\text{resistance}} - P_{\text{brush}}$$

$$= 804\ \text{W} - 2.323\ \text{W} - 6.7\ \text{W}$$

$$\approx 795.0\ \text{W}$$

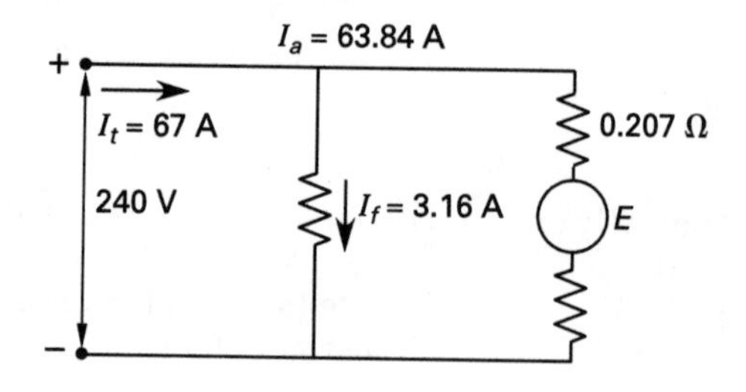

At full load, the total input power is

$$P_{\text{in}} = I_{\text{fl}} V_{\text{nameplate}}$$

$$= (67\ \text{A})(240\ \text{W}) = 16{,}080\ \text{W}$$

The field current is the same as for the no-load condition.

$$I_f = 3.16\ \text{A}$$

The field power loss is

$$P_f = I_f V = (3.16 \text{ A})(240 \text{ V})$$
$$= 758.4 \text{ W}$$
$$I_a = I_{\text{fl}} - I_f = 67 \text{ A} - 3.16 \text{ A}$$
$$= 63.84 \text{ A}$$

$$P_{\text{resistance}} = I_a^2 R = (63.84 \text{ A})^2 (0.207 \ \Omega)$$
$$= 843.6 \text{ W}$$

Using the no-load data, the equivalent brush resistance is

$$R_{\text{brush}} = \frac{V_{\text{brush}}}{I} = \frac{2 \text{ V}}{3.35 \text{ A}} = 0.597 \ \Omega$$

The armature brush power loss is

$$P_{\text{brush}} = I_a^2 R_{\text{brush}} = (63.84)^2 (0.597 \ \Omega)$$
$$= 2433.1 \text{ W}$$

The rotational loss is the same as for the no-load condition.

$$P_{\text{rotational}} = 795.0 \text{ W}$$

The net power to the load is

$$P_{\text{net}} = P_{\text{in}} - P_f - P_{\text{resistance}} - P_{\text{brush}} - P_{\text{rotational}}$$
$$= 16{,}080 \text{ W} - 758.4 \text{ W} - 843.6 \text{ W}$$
$$- 2433.1 \text{ W} - 795.0 \text{ W}$$
$$= 11{,}249.9 \text{ W}$$

(a) The motor output horsepower is

$$P = \frac{(11{,}249.9 \text{ W})\left(1.341 \ \frac{\text{hp}}{\text{kW}}\right)}{1000 \ \frac{\text{W}}{\text{kW}}} = 15.09 \text{ hp}$$

(b) The efficiency is

$$\eta = \frac{P_{\text{net}}}{P_{\text{in}}} = \frac{11{,}249.9 \text{ W}}{16{,}080 \text{ W}} = 0.70 \quad (70.0\%)$$

5.

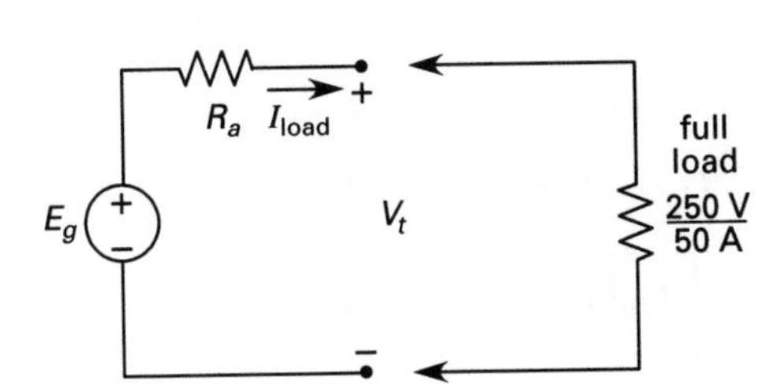

$$V_t = E_g - R_a I_{\text{load}}$$

At no load,

$$V_t = E_g = 260 \text{ V}$$

At full load,

$$V_t = E_g - (50 \text{ A}) R_a$$
$$250 \text{ V} = E_g - (50 \text{ A}) R_a$$
$$R_a = \frac{260 \text{ V} - 250 \text{ V}}{50 \text{ A}}$$
$$= \boxed{0.2 \ \Omega}$$

6. (a)

$$I_{\text{fl}} = \frac{10 \text{ kW}}{240 \text{ V}} = 41.67 \text{ A}$$
$$R_a = \frac{V_{\text{nl}} - V_{\text{fl}}}{I_{\text{fl}}} \quad \text{[see Prob. 5]}$$
$$= \frac{260 \text{ V} - 240 \text{ V}}{41.67 \text{ A}}$$
$$= \boxed{0.48 \ \Omega}$$

(b)

$$P_{\text{in}} = P_{\text{mech. loss}} + I_a^2 R_a + p_{\text{out}}$$
$$= (15 \text{ hp})\left(\frac{746 \text{ W}}{\text{hp}}\right) = 11.190 \text{ kW}$$
$$I_a^2 R_a = (41.67 \text{ A})^2 (0.48 \ \Omega)$$
$$= 0.833 \text{kW}$$
$$P_{\text{out}} = 10 \text{ kW}$$
$$P_{\text{mech. loss}} = 11.190 \text{ kW} - 10.833 \text{ kW}$$
$$= \boxed{357 \text{ W}}$$

7. For field flux, Φ is proportional to line voltage.

$$\Phi = aV$$
$$\Omega_r = \text{rotor speed} \quad \text{[rad/s]}$$
$$E_g = k\Phi\Omega_r$$
$$= kaV\Omega_r$$
$$T = k\Phi I_a = kaVI_a \quad \text{[torque in N·m]}$$
$$V = E_g + R_a I_a$$
$$P_{\text{in}} \approx VI_a$$
$$P_{\text{out}} \approx T\Omega_r$$

(a) For $T =$ constant, if V is decreased by 10%, to keep T constant, I_a is increased by approximately 10% and E_g is decreased by 10% with V.

$$P_{\text{in}} = \boxed{VI_a \approx \text{constant}}$$

$$P_{\text{out}} = P_{\text{in}} - I_a^2 R_a - \text{mech. loss}$$

Since I_a has increased by 10%, I_a^2 is increased by 20%, but $I_a^2 R_a$ is on the order of 5% of the load power, so it only causes a 1% drop in P_{out}.

E_g drops 10% with V, but

$\boxed{\Omega_r \text{ drops only 1\%.}}$

(b) For $T \propto \Omega_r$, in a first approximation,

$$E_g \approx V$$

$$E_g = kaV\Omega_r$$

$ka\Omega_r \approx 1$, and there is no change in speed.

In a second approximation,

$$I_a = \frac{V - E_g}{R_a}$$

$$T = kaVI_a$$

The air gap power is

$$P_{\text{air gap}} = E_g I_a = T\Omega_r$$

$$T = m\Omega_r$$

$$T\Omega_r = m\Omega_r^2$$

$$E_g = kaV\Omega_r$$

$$E_g I_a = \frac{kaV^2\Omega_r(1 - ka\Omega_r)}{R_a} = m\Omega_r^2$$

$$\frac{kaV^2}{R_a} = \frac{m\Omega_r}{1 - ka\Omega_r} = \frac{m}{\frac{1}{\Omega_r} - ka}$$

When V goes down 10%, $1\ \Omega_r - ka$ must go up by 21%, so Ω_r must decrease some amount. If under normal conditions $I_a R_a = 0.05V$,

$$I_a R_a = V - E_{g,\text{nl}} = V(1 - ka\Omega_{r,\text{nl}})$$

$$1 - ka\Omega_{r,\text{nl}} = 0.05$$

$$ka = \frac{0.95}{\Omega_{r,\text{nl}}}$$

When V goes down 10%,

$$\frac{m\Omega_r}{1 - ka\Omega_r} \rightarrow (0.81)\left(\frac{m\Omega_{r,\text{nl}}}{1 - ka\Omega_{r,\text{nl}}}\right)$$

$$\frac{\Omega_r}{1 - (0.95)\left(\frac{\Omega_r}{\Omega_{r,\text{nl}}}\right)} = \left(\frac{0.81}{0.05}\right)\Omega_{r,\text{nl}} = 16.2\Omega_{r,\text{nl}}$$

$$\frac{\Omega_r}{\Omega_{r,\text{nl}}} = (16.2)\left(1 - 0.95\left(\frac{\Omega_r}{\Omega_{r,\text{nl}}}\right)\right)$$

$$= \frac{16.2}{1 + (0.95)(16.2)} = 0.99$$

$\boxed{\Omega_r \text{ drops by about 1\%.}}$

$$I_a R_a = V - E_g = V - ka\Omega_r$$

$$T = m\Omega_r = kaVI_a$$

Eliminating Ω_r gives

$$I_a R_a = \frac{V}{1 + \frac{(kaV)^2}{mR_a}}$$

The original value of $I_a R_a$ is

$$I_{a,\text{nl}} R_a \approx 0.05V_{\text{nl}}$$

V_0 is the initial voltage. Then,

$$1 + \frac{(kaV_{\text{nl}})^2}{mR_a} \approx 20$$

When the voltage drops 10%,

$$1 + \frac{\left(ka(0.9)V_{\text{nl}}\right)^2}{mR_a} = 16.39$$

The resulting armature IR drop is

$$I_a R_a = \frac{0.9V_{\text{nl}}}{16.39} = 0.0549V_{\text{nl}}$$

The current changes by

$$\frac{I_a R_a - I_{a,\text{nl}} R_a}{I_{a,\text{nl}} R_a} = \frac{0.0549 - 0.05}{0.05}$$

$$= 0.098 \text{ pu}$$

The input power changes also.

$$P_{\text{in}} = (0.9V_{\text{nl}})(1.098I_{a,\text{nl}}) = 0.988P_{\text{nl}}$$

$\boxed{\text{The power drops by about 1\%.}}$

8.

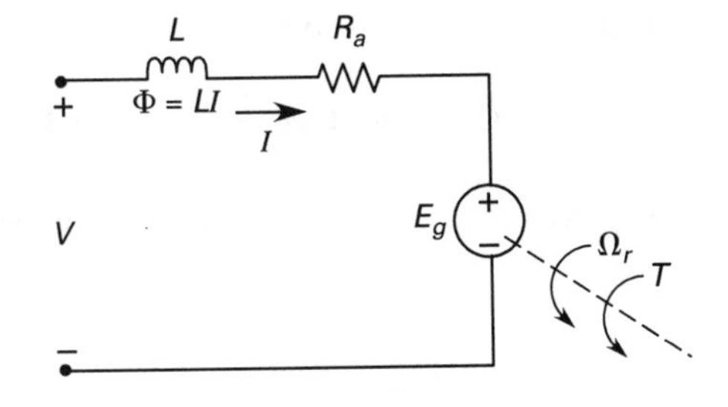

$$V = RI + E_g$$

$$E_g = k\Phi\Omega_r = kLI\Omega_r$$

$$T = k\Phi I = kLI^2$$

If T is up 25%, so is I^2, and $\sqrt{1.25} = 1.11$, so

I is up 11%.

Using subscript zero for the initial values,

$$E_{g0} = V - I_0 R_a$$

When I increases by 11%,

E_g decreases by about 1%.

$$E_{g1} = 0.99E_{g0}$$

$$E_{g0} = kLI_0\Omega_{r0}$$

$$E_{g1} = 0.99kLI_0\Omega_{r0}$$

$$= kLI_1\Omega_{r1}$$

$$\Omega_{r1} = \left(\frac{0.99I_0}{I_1}\right)\Omega_{r0}$$

$$= \left(\frac{0.99}{1.11}\right)\Omega_{r0} = 0.89\Omega_{r0}$$

The speed also drops by about 11%.

The output power is

$$P_2 = (1.25T_0)(0.89\Omega_{r0}) = 1.11\ P_0$$

The air-gap power increases by 11%.

9. (a)

$$I_t = 25\text{ A} = I_a + I_f$$

$$I_f = \frac{200\text{ V}}{100\ \Omega} = 2\text{ A}$$

$$I_a = It - I_f = 23\text{ A}$$

$$E_g = 200\text{ V} - (23\text{ A})R_a = 188.5\text{ V}$$

$$P_{\text{air gap}} = E_g I_a = 4335.5\text{ W}$$

$$P_{\text{load}} = (5\text{ hp})\left(\frac{746\text{ W}}{\text{hp}}\right) = 3730\text{ W}$$

$$P_{\text{rotational losses}} = P_{\text{air gap}} - P_{\text{load}} = \boxed{605.5\text{ W}}$$

(b) The speed (n) is proportional to the counter voltage as seen from

$$E_g = k\Phi n$$

The term $k\Phi$ is determined from the information in Part (a) to be

$$k\Phi = \frac{188.5\text{ V}}{1000\text{ rpm}}$$

Assuming the rotational losses are fixed at 605.5 W,

$$E_g I_a = 605.5\text{ W}$$

$$I_a = \frac{605.5\text{ W}}{E_g}$$

The armature current at the specified 200 V is given by

$$I_a = \frac{200\text{ V} - E_g}{0.5\ \Omega}$$

Substitute for the armature current giving

$$\frac{605.5\text{ W}}{E_g} = \frac{200\text{ V} - E_g}{0.5\ \Omega}$$

Rearranging gives the quadratic equation

$$E_g^2 - 200E_g + 302.75 = 0$$

Solving the quadratic and using the positive sign gives

$$E_g = \frac{-(-200\text{ V}) \pm \sqrt{(-200\text{ V})^2 - (4)(1)(302.75\text{ V}^2)}}{(2)(1)}$$

$$= 198.5\text{ V}$$

Comparing voltage to rpm ratios gives

$$\frac{n}{198.5\text{ V}} = \frac{1000\text{ rpm}}{188.5\text{ V}}$$

$$n = \boxed{1053\text{ rpm}}$$

Power–Machinery

10.

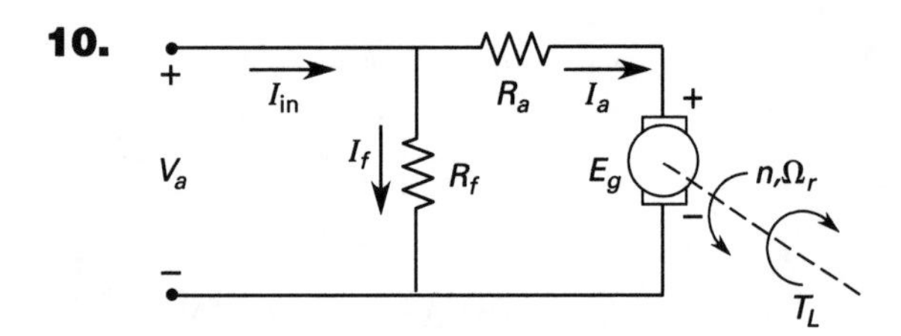

$$P_{in} = V_A I_{in} = (60\text{ A})(120\text{ V}) = 7200\text{ W}$$

$$P_{out} = (7.5\text{ hp})\left(746\ \frac{\text{W}}{\text{hp}}\right) = 5595\text{ W}$$

Since load torque is constant,

$$P_{out} = k_T n$$

$$k_T = \frac{5595\text{ W}}{1000\text{ rpm}} = 5.595n \quad [n \text{ in rpm}]$$

The losses are the difference between P_{in} and P_{out}.

$$P_{loss} = P_{in} - P_{out} = 7200\text{ W} - 5595\text{ W} = 1605\text{ W}$$

With no information given, it is necessary to assign these losses to rotational losses, $I_a^2 R_a$ and $I_f^2 R_f$. Make them equal.

$$I_a^2 R_a = I_f^2 R_f = P_{\text{rotational losses}} = 535\text{ W}$$

$$I_f^2 R_f = \frac{V_a^2}{R_f} = 535\text{ W}$$

$$R_f = \frac{(120\text{ V})^2}{535\text{ W}} = 26.9\ \Omega$$

$$I_f = \frac{120\text{ V}}{26.9\ \Omega} = 4.46\text{ A}$$

Therefore,

$$I_a = 60\text{ A} - 4.46\text{ A} = 55.54\text{ A}$$

$$R_a = \frac{535\text{ W}}{(55.54\text{ A})^2} = 0.173\ \Omega$$

Because of rounding, the powers are recalculated and the remainder assigned to rotational losses.

$$P_{\text{rotational losses}} = 1605\text{ W} - \frac{(120\text{ V})^2}{26.9\ \Omega} - (0.173\ \Omega)(55.54\text{ A})^2 = 536\text{ W}$$

The air-gap power must then be

$$P_{\text{air gap}} = P_{out} + P_{\text{rotational losses}} = 5.595n + 536\text{ W}$$

The air-gap power is also $E_g I_a$, and E_g is proportional to I_f and n.

$$E_g = k' I_f n = k''\left(\frac{n}{R_f}\right)$$

At $n = 1000$ rpm and $R_f = 26.9\ \Omega$,

$$E_g = 120\text{ V} - (55.54\text{ A})(0.173\ \Omega) = 110.4\text{ V}$$

$$k'' = \frac{(110.4\text{ V})(26.9\ \Omega)}{1000\text{ rpm}} = 2.97\text{ V}\cdot\Omega/\text{rpm}$$

$$E_g = 2.97\left(\frac{n}{R_f}\right)$$

$$I_a = \frac{120\text{ V} - E_g}{R_a} = \frac{120\text{ V} - (2.97)\left(\frac{n}{R_f}\right)}{0.173\ \Omega} = 694\text{ A} - (17.2)\left(\frac{n}{R_f}\right)$$

$$P_{\text{air gap}} = E_g I_a = (2061)\left(\frac{n}{R_f}\right) - (51.08)\left(\frac{n}{R_f}\right)^2$$

Equating with the mechanical side,

$$(51.08)\left(\frac{n}{R_f}\right)^2 + \left(5.595 - \frac{2061}{R_f}\right)n + 536\text{ W} = 0$$

$$0 = R_f^2 - \left(\frac{2061n}{5.595n + 536\text{ W}}\right)(R_f) + \frac{51.08n^2}{5.595n + 536\text{ W}}$$

(a) For $n = 1150$ rpm,

$$R_f^2 - (340.0\ \Omega)R_f + 9692\ \Omega^2 = 0$$

$$R_f = \boxed{31.4\ \Omega}$$

(b) For $n = 750$ rpm,

$$R_f^2 - (326.6\ \Omega)R_f + 6072\ \Omega^2 = 0$$

$$R_f = \boxed{19.8\ \Omega}$$

40 Rotating AC Machinery

PRACTICE PROBLEMS

1. An old single-phase generator has four poles. The armature is simplex lap wound and has a total of 240 armature conductors. The effective armature length and radius are 16.7 cm and 5 cm, respectively. The armature rotates at 1800 rpm. (a) If the uniform magnetic flux density per pole is 1.0 T, what is the maximum voltage induced? (b) What is the horsepower of the driving motor if the rated output is 1 kW and the conversion efficiency is 90%?

2. (a) What is the rotational speed of a 24-pole alternator that produces a 60 Hz sinusoid? (b) If the effective emf is 2200 V and each phase coil has 20 turns in series, what is the maximum flux?

3. Calculate the full-load phase current drawn by a 440 V (rms), 20 hp (per phase) induction motor having a full-load efficiency of 86% and a full-load power factor of 76%.

4. A 200 hp, three-phase, four-pole, 60 Hz, 440 V (rms) squirrel-cage induction motor operates at full load with an efficiency of 85%, a power factor of 91%, and 3% slip. Find the (a) speed in rpm, (b) torque developed, and (c) line current.

5. A factory's induction motor load draws 550 kW at 82% power factor. What size synchronous motor is required to carry 250 hp and raise the power factor to 95%? The line voltage is 220 V (rms).

6. The nameplate of an induction motor lists 960 rpm as the full-load speed. For what frequency was the motor designed?

7. A three-phase, 440 V (line-to-line) synchronous motor has a synchronous reactance (X_s) of 20 Ω and an armature resistance of 0.5 Ω per phase. The synchronous speed is 1800 rpm. The no-load field current is adjusted for minimum armature current at a field current (I_f) of 5 A DC, at which time the input power is 900 W. This machine is to provide 10 kVAR of power factor correction when running lightly loaded. Determine the rotor current required.

8. A 440 V, three-phase synchronous motor is driving a 45 hp pump. Assume the motor is 93% efficient. The rotor current is adjusted until the line current is at a minimum, occurring when the rotor current is 8.2 A DC. Determine the rotor current necessary to provide a leading power factor of 0.8.

9. A 50 hp, 50 kVA, 1746 rpm squirrel-cage induction motor has phase windings rated at 440 V. With the windings connected in a delta configuration, a blocked rotor test is carried out under rated current conditions. The result in per-phase volt-amperes is found to be $P + jQ = 1097 + j2769$ VA. This machine is fed from a three-phase, 50 kVA, 440 V transformer that has 10% reactance per phase (on its own base). Determine the dip in the line voltage when the motor is started. The transformer can be modeled as a wye connection of 254 V sources in series with the 10% reactance on each phase.

10. A 50 hp, 50 kVA, 1746 rpm squirrel-cage induction motor has phase windings rated at 440 V and is normally connected in delta. Its starting current is 367 A and 38.3 kW per phase when started directly across the line, and under those conditions it develops a starting torque of 190 ft-lbf. A blocked rotor test at rated current results in a voltage of 78.6 V and a power of 1.1 kW per phase. A no-load test at rated voltage results in a current of 25.9 A and 1.1 kW per phase. If this machine is started in the wye configuration, determine its starting current.

SOLUTIONS

1. (a)

$$V_m = \left(\frac{\pi}{2}\right) V_{\text{ave}} = \left(\frac{\pi}{2}\right) pNAB\left(\frac{n}{60}\right)$$
$$= \frac{\pi pnNAB}{60}$$
$$= \pi p\left(\frac{n}{60}\right)\left(\frac{z}{a}\right)AB$$

For a simplex lap-wound armature, $a = p$.

$$V_m = (\pi)(4)\left(\frac{1800\ \frac{\text{rev}}{\text{min}}}{60\ \frac{\text{s}}{\text{min}}}\right)\left(\frac{240}{4}\right) \times \left((0.10\ \text{m})(0.167\ \text{m})\right)\left(1.0\ \frac{\text{Wb}}{\text{m}^2}\right)$$
$$= \boxed{377.8\ \text{V}}$$

(b)

$$P = \frac{(1\ \text{kW})\left(\frac{1\ \text{hp}}{0.7457\ \text{kW}}\right)}{0.9} = \boxed{1.49\ \text{hp}}$$

2. (a)

$$n = (2)\left(60\ \frac{\text{s}}{\text{min}}\right)\left(\frac{f}{p}\right)$$
$$= (2)\left(60\ \frac{\text{s}}{\text{min}}\right)\left(\frac{60\ \text{Hz}}{24}\right)$$
$$= \boxed{300\ \text{rpm}}$$

(b)

$$BA = \frac{\sqrt{2}\,V_{\text{eff}}}{N\omega} = \frac{(\sqrt{2})\,(2200\ \text{V})}{(20)(2\pi)(60\ \text{Hz})}$$
$$= \boxed{0.413\ \text{Wb}}$$

3.

$$I_p = \frac{P_p}{\eta V \cos\phi}$$
$$= \frac{(20\ \text{hp})\left(\frac{0.7457\times 10^3\ \text{W}}{1\ \text{hp}}\right)}{(0.86)(440\ \text{V})(0.76)}$$
$$= \boxed{51.86\ \text{A}}$$

4. (a)

$$n_r = \left(\frac{(2)\left(60\ \frac{\text{s}}{\text{min}}\right)f}{p}\right)(1-\text{s})$$
$$= \left(\frac{(2)\left(60\ \frac{\text{s}}{\text{min}}\right)(60\ \text{Hz})}{4}\right)(1-0.03)$$
$$= \boxed{1746\ \text{rpm}}$$

(b)

$$T = \frac{P}{\omega} = \frac{(200\ \text{hp})\left(\frac{550\ \frac{\text{ft-lbf}}{\text{s}}}{\text{hp}}\right)}{2\pi\left(1746\ \frac{\text{rev}}{\text{min}}\right)\left(\frac{1\ \text{min}}{60\ \text{s}}\right)}$$
$$= \boxed{602\ \text{ft-lbf}}$$

(c)

$$I_l = \left(\frac{1}{\sqrt{3}}\right)\left(\frac{P}{\eta V_l \cos\phi}\right)$$
$$= \left(\frac{1}{\sqrt{3}}\right)\left(\frac{(200\ \text{hp})\left(\frac{0.7457\times 10^3\ \text{W}}{1\ \text{hp}}\right)}{(0.85)(440\ \text{V})(0.91)}\right)$$
$$= \boxed{253\ \text{A}}$$

5.

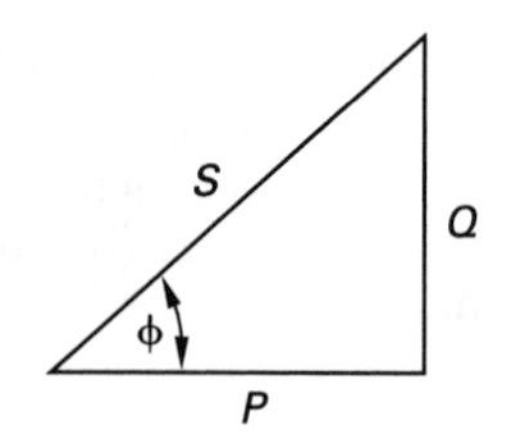

The original power angle is

$$\phi_1 = \arccos 0.82 = 34.92^\circ$$
$$P_1 = 550\ \text{kW}$$
$$Q_i = P_1 \tan\phi_1$$
$$= (550\ \text{kW})(\tan 34.92^\circ)$$
$$= 384.0\ \text{kVAR}$$

The new conditions are

$$P_2 = (250\ \text{hp})\left(0.7457\ \frac{\text{kW}}{\text{hp}}\right)$$
$$= 186.4\ \text{kW}$$
$$\phi_f = \arccos 0.95 = 18.19^\circ$$

Since both motors perform real work,

$$P_f = P_1 + P_2 = 550 \text{ kW} + 186.4 \text{ kW}$$
$$= 736.4 \text{ kW}$$

The new reactive power is

$$Q_f = P_f \tan\phi_f = (736.4 \text{ kW})(\tan 18.19°)$$
$$= 242.0 \text{ kVAR}$$

The change in reactive power is

$$\Delta Q = 384.0 \text{ kVAR} - 242.0 \text{ kVAR} = 142 \text{ kVAR}$$

Synchronous motors used for power factor correction are rated by apparent power.

$$S = \sqrt{(\Delta P)^2 + (\Delta Q)^2}$$
$$= \sqrt{(186.4 \text{ kW})^2 + (142 \text{ kVAR})^2}$$
$$= \boxed{234.3 \text{ kVA}}$$

6.

$$f = \frac{pn_s}{(2)\left(60 \frac{\text{s}}{\text{min}}\right)}$$
$$= \frac{pn}{(2)\left(60 \frac{\text{s}}{\text{min}}\right)(1-s)}$$

The slip and number of poles are unknown. Assume $s = 0$ and $p = 4$.

$$f = \frac{(4)\left(960 \frac{\text{rev}}{\text{min}}\right)}{(2)\left(60 \frac{\text{s}}{\text{min}}\right)} = 32 \text{ Hz}$$

This is not close to anything in commercial use. Try $p = 6$.

$$f = \frac{(6)\left(960 \frac{\text{rev}}{\text{min}}\right)}{(2)\left(60 \frac{\text{s}}{\text{min}}\right)} = 48 \text{ Hz}$$

With a 4% slip, $f = 50$ Hz.

$$\boxed{50 \text{ Hz (European)}}$$

7. The required volt-amps per phase are determined by

$$\frac{Q}{3 \text{ phases}} = V_a I_a = \frac{V_a(E_g - V_a)}{X_s}$$
$$E_g = \left(\frac{Q}{3 \text{ phases}}\right)\left(\frac{X_s}{V_a}\right) + V_a$$
$$= \left(\frac{10^4 \text{ VA}}{3 \text{ phases}}\right)\left(\frac{20\ \Omega}{254 \text{ V}}\right) + 254 \text{ V} = 516 \text{ V}$$
$$I_f = \left(\frac{5 \text{ A}}{254 \text{ V}}\right)(516 \text{ V})$$
$$= \boxed{10.16 \text{ A}}$$

8.

$$P_{\text{in}} = \frac{(45 \text{ hp})\left(746 \frac{\text{W}}{\text{hp}}\right)}{0.93 \text{ (efficiency)}} = 36.1 \text{ kW}$$

At minimum line current, the current is in phase with the voltage.

$$I = \frac{\frac{36.1 \times 10^3 \text{ W}}{3 \text{ phases}}}{\frac{440 \text{ V}}{\sqrt{3}}} = 47.37 \text{ A}$$

$$Z_{\text{base}} = \frac{254 \text{ V}}{47.37 \text{ A}} = 5.36\ \Omega$$

A synchronous reactance of 100% is common and is therefore assumed.

$$X_s = 5.36\ \Omega$$
$$I_a X_s = (5.36\ \Omega)(47.37 \text{ A}) = 254 \text{ V}$$
$$E_g = V_a - jX_s I_a = (254 \text{ V})(1 - j)$$
$$|E_g| = 359 \text{ V}$$

The rotor current is given as $I_r = 8.2$ A.

To provide a leading power factor of 0.8, it is necessary to retain the same power.

$$|I_a| \cos\theta = 47.37 \text{ A}$$
$$|I_a| = \frac{47.37 \text{ A}}{0.8} = 59.2 \text{ A}$$
$$I_a = (59.2 \text{ A})(\cos\theta + j\ \sin\theta)$$
$$= 47.37 \text{ A} + j35.52 \text{ A}$$
$$= 59.2\angle 36.87° \text{ A}$$
$$E_g = 254 \text{ V} - (j5.36\ \Omega)I_a$$
$$= 512\angle{-29.74°} \text{ V}$$

Using a ratio gives

$$I_r = \left(\frac{8.2\ \text{A}}{359\ \text{V}}\right)(512\ \text{V})$$
$$= \boxed{11.7\ \text{A}}$$

9. The bases for the motor and transformer are the same.

$$I_{\text{rated}} = \frac{\frac{50\ \text{kVA}}{3\ \text{phases}}}{254\ \text{V}} = 65.62\ \text{A}$$

$$Z_{\text{base}} = \frac{254\ \text{V}}{65.62\ \text{A}} = 3.87\ \Omega$$

From the blocked rotor test,

$$P + jQ = I^2(R_e + jX_e)$$
$$1097\ \text{W} + j2791\ \text{VAR} = (65.62\ \text{A})^2(R_e + jX_e)$$
$$R_e + jX_e = 0.2548 + j0.6482\ \Omega$$

On a per-unit basis,

$$R_e + jX_e = 0.0658 + j0.1675\ \text{pu}$$

The per-unit circuit for starting is

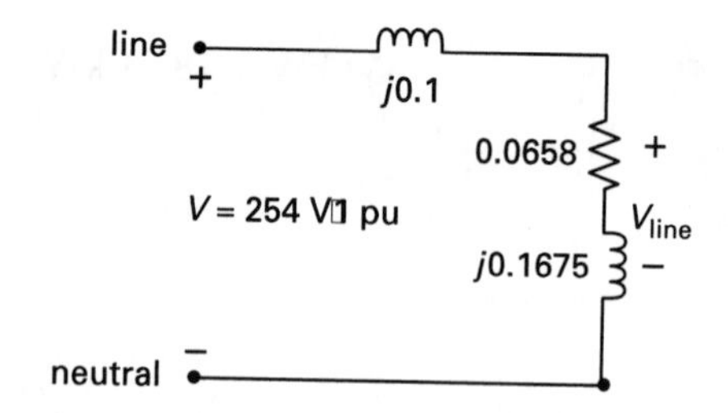

By voltage division,

$$|V_{\text{line}}| = \left|\frac{0.0658 + j0.1675}{0.0658 + j0.2675}\right| = 0.653\ \text{pu}$$

$$\text{dip} = \boxed{1 - 0.653\ \text{pu}\ (35\%)}$$

10. The impedance for wye starting is three times that for delta starting, so the current will be only one-third the delta current.

$$I_{\text{starting}} = \frac{367\ \text{A}}{3} = \boxed{122.3\ \text{A}}$$

41 Lightning Protection and Grounding

PRACTICE PROBLEMS

1. What range of voltage does lightning span?

(A) 1 to 2 MV
(B) 2 to 5 MV
(C) 5 to 10 MV
(D) 10 to 1000 MV

2. What range of current does lightning span?

(A) 100 to 1000 A
(B) 1000 to 10,000 A
(C) 1000 to 100,000 A
(D) 1000 to 200,000 A

3. What is the name for an insulation level specified in terms of the crest value of the standard lightning impulse?

(A) BIL
(B) lightning impulse level
(C) TIL
(D) withstand voltage

4. A power system designed for what voltage is considered protected from direct lightning strikes by its own insulation?

(A) 15 kV
(B) 230 kV
(C) 330 kV
(D) none of the above

5. An arrester provides a protection level of 38 kV. If the minimum protective margin allowed by standards is 20%, is equipment with a 95 kV withstand voltage adequately protected?

SOLUTIONS

1. Lightning voltage varies widely but is in the range of 10 to 1000 MV.

The answer is (D).

2. The current ranges from 1 to 200 kA.

The answer is (D).

3. The voltage level that insulation is able to withstand, specified in terms of a *standard* lightning impulse, is called the *basic lightning impulse level* (BIL).

The answer is (A).

4. Systems of 230 kV or less cannot be insulated against the overvoltage caused by direct lightning strikes and must use a combination of shielding and grounding.

The answer is (C).

5. The protective margin is given by

$$\begin{aligned} \text{PM} &= \frac{V_w - V_p}{V_p} \\ &= \frac{95 \text{ kV} - 38 \text{ kV}}{38 \text{ kV}} \\ &= 1.5 \ (150\%) > 20\% \end{aligned}$$

The equipment is adequately protected.

Power–Lightning

42 Measurement and Instrumentation

PRACTICE PROBLEMS

1. What instrument term can be defined as a measure of the reproducibility of observations?

(A) accuracy
(B) error
(C) precision
(D) resolution

2. A digital voltmeter is designed for a maximum ambient temperature of 65°C. On the 10 kHz scale, what is the expected noise voltage in a 100 kΩ series resistor?

(A) 4.3 nV
(B) 0.14 μV
(C) 1.9 μV
(D) 4.3 μV

3. What is the average value of the voltage signal $v(t) = 5 \sin 377t$?

(A) 0.0 V
(B) 3.2 V
(C) 3.5 V
(D) 5.6 V

4. The voltage signal in Prob. 3 is rectified and measured by a d'Arsonval meter movement. What is the voltage reading?

(A) 0.0 V
(B) 3.2 V
(C) 3.5 V
(D) 5.6 V

5. The manufacturer's data sheet lists the sensitivity of a DC voltmeter as 1000 Ω/V. What current is required for full-scale meter deflection?

(A) 1.0 nA
(B) 1.0 μA
(C) 1.0 mA
(D) 1.0 A

6. A DC voltmeter with a coil resistance of 150 Ω is to be designed with a sensitivity of 10,000 Ω/V and a voltage range of 0 to 100 V. What type and what magnitude of resistance is required?

(A) parallel; 10 kΩ
(B) parallel; 1 MΩ
(C) series; 10 kΩ
(D) series; 1 MΩ

7. A d'Arsonval movement with a coil resistance of 150 Ω is to be designed for use as an ammeter. What is the full-scale deflection current?

(A) 0.3 mA
(B) 0.3 A
(C) 3.0 A
(D) none of the above

8. A DC voltmeter with 20,000 Ω of internal resistance measures the voltage across a resistance as 10.0 V. The current through the parallel combination of the resistor and voltmeter is measured as 1.00 mA by a millimeter with 100 Ω of internal resistance. The resistor is fed from a source with a Thevenin resistance of 600 Ω. Determine the (a) resistance of the resistor and (b) current that will flow in it if the meters are removed from the circuit.

9. A voltmeter is designed to measure voltages in the full-scale ranges of 3, 10, 30, and 100 V DC. The meter movement has an internal resistance of 50 Ω and a full-scale current of 1 mA. Using a four-contact multiposition switch, design the voltmeter and specify the four resistor values used.

10. A digital meter uses the number system base b and has n digits. The noise is one-half the place value of the least significant digit. (a) Determine the signal-to-noise ratio for the case where the reading is half the full-scale value. (b) Obtain the signal-to-noise ratios for the following cases.

n	b	S/N (dB)
3	10	20 log (S/N)
4	8	
8	2	
3	16	

SOLUTIONS

1. *Precision* is the range of values of a set of measurements and thus a measure of the reproducibility of observations.

The answer is (C).

2. The thermal noise voltage is given by

$$\begin{aligned} V_{\text{noise}} &= \sqrt{4\kappa TR(\text{BW})} \\ &= \sqrt{\begin{array}{l}(4)\left(1.3805\times 10^{-23}\ \frac{\text{J}}{\text{K}}\right)(338\ \text{K}) \\ \quad\times (100\times 10^{3}\ \Omega)\left(10\times 10^{3}\ \frac{1}{\text{s}}\right)\end{array}} \\ &= \boxed{4.3\ \mu\text{V}} \end{aligned}$$

The answer is (D).

3. The average value of a sinusoid is zero.

The answer is (A).

4. A d'Arsonval meter responds to average values. The average value of a rectified sinusoid is given by

$$\begin{aligned} V_{\text{ave}} &= \left(\frac{2}{\pi}\right)V_p \\ &= \left(\frac{2}{\pi}\right)(5\ \text{V}) \\ &= \boxed{3.18\ \text{V}} \end{aligned}$$

The answer is (B).

5. The sensitivity is defined as

$$\text{sensitivity} = \frac{1}{I_{\text{fs}}}$$

$$\begin{aligned} I_{\text{fs}} &= \frac{1}{\text{sensitivity}} = \frac{1}{1000\ \frac{\Omega}{\text{V}}} \\ &= \boxed{10^{-3}\ \text{A}} \end{aligned}$$

The answer is (C).

6. The voltmeter requires an external $\boxed{\text{series}}$ resistance given by

$$\frac{1}{I_{\text{fs}}} = \frac{R_{\text{ext}} + R_{\text{coil}}}{V_{\text{fs}}}$$

$$\begin{aligned} R_{\text{ext}} &= V_{\text{fs}}\left(\frac{1}{I_{\text{fs}}}\right) - R_{\text{coil}} \\ &= (100\ \text{V})\left(10{,}000\ \frac{\Omega}{\text{V}}\right) - 150\ \Omega \\ &= \boxed{0.999\ \text{M}\Omega} \end{aligned}$$

The answer is (D).

7. For ammeters, the standard design voltage is 50 mV. The full-scale deflection current is

$$\begin{aligned} I_{\text{fs}} &= \frac{V}{R_{\text{coil}}} = \frac{50\times 10^{-3}\ \text{V}}{150\ \Omega} \\ &= 3.33\times 10^{-4}\ \text{A} \\ &= \boxed{0.33\ \text{mA}} \end{aligned}$$

The answer is (A).

8.

1 mA
+
600 Ω
100 Ω
V_{Th}
R
10 V
20 kΩ
−

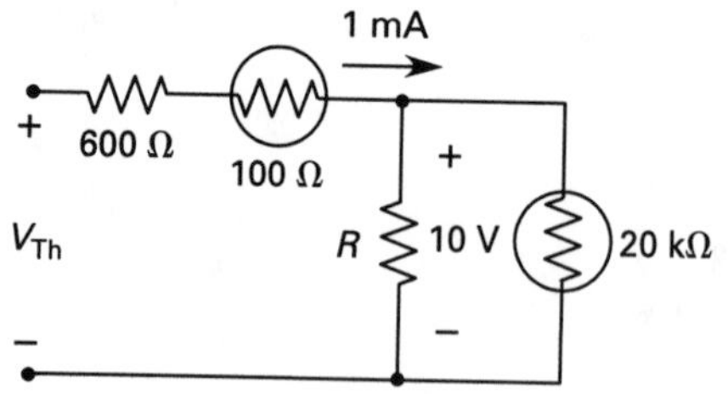

(a) Assuming the meters are perfectly accurate,

$$V_{\text{Th}} = 10\ \text{V} + (0.001\ \text{A})(700\ \Omega) = 10.7\ \text{V}$$

With 1 mA flowing into 10 V, the equivalent load resistance is 10 V/1 mA = 10 kΩ, so

$$\begin{aligned} 10\ \text{k}\Omega &= \frac{(20\ \text{k}\Omega)R}{20\ \text{k}\Omega + R} \\ (20\ \text{k}\Omega + R)(10\ \text{k}\Omega) &= (20\ \text{k}\Omega)R \\ 200\ \text{M}\Omega + (10\ \text{k}\Omega)R &= (20\ \text{k}\Omega)R \\ 200\ \text{M}\Omega &= (10\ \text{k}\Omega)R \\ R &= \boxed{20\ \text{k}\Omega} \end{aligned}$$

(b) Removing the meters, the current flowing is

$$I = \frac{10.7\ \text{V}}{20{,}600\ \Omega} = \boxed{0.52\ \text{mA}}$$

9. The meter configuration is as shown.

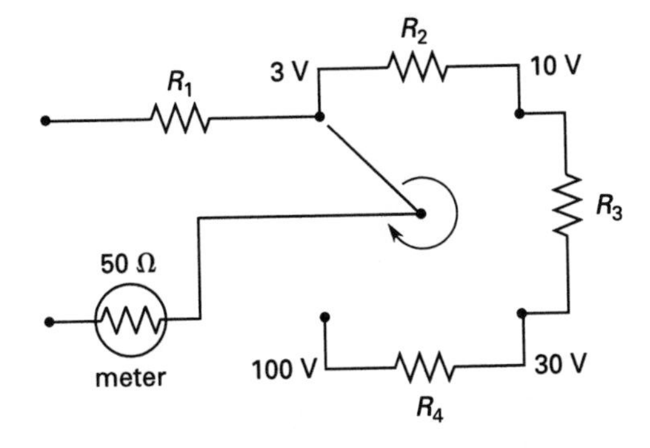

With the 1 mA full-scale current,

$$\text{3 V:}\quad 50 + R_1 = \frac{3\ \text{V}}{1\ \text{mA}} = 3000\ \Omega$$

$$R_1 = \boxed{2950\ \Omega}$$

$$\text{10 V:}\qquad R_2 = \frac{10\ \text{V} - 3\ \text{V}}{1\ \text{mA}} = \frac{7\ \text{V}}{1\ \text{mA}} = \boxed{7000\ \Omega}$$

$$\text{30 V:}\qquad R_3 = \frac{30\ \text{V} - 10\ \text{V}}{1\ \text{mA}} = \frac{20\ \text{V}}{1\ \text{mA}} = \boxed{20{,}000\ \Omega}$$

$$\text{100 V:}\qquad R_4 = \frac{100\ \text{V} - 30\ \text{V}}{1\ \text{mA}} = \frac{70\ \text{V}}{1\ \text{mA}} = \boxed{70{,}000\ \Omega}$$

10. (a) Full scale is 999_{10} for the first case. The noise is 0.5. At half-scale the signal is 500_{10}, so the signal-to-noise ratio is $500/0.5 = 10^3$ or

$$\text{S/N}|_{\text{dB}} = 20 \log 10^3 = \boxed{60\ \text{dB}}$$

(b) For the octal, full-scale in base 10 is

$$(7)(8)^3 + (7)(8)^2 + (7)(8)^1 + 7 = 4095$$

Half scale is 2048.

$$\text{S/N} = 20 \log \left(\frac{2048}{0.5}\right) = \boxed{72.2\ \text{dB}}$$

For the binary, full-scale in base 10 is

$$(2)^7 + (2)^6 + (2)^5 + (2)^4 + (2)^3 + (2)^2 + (2)^1 + 1 = 225$$

Half-scale is 128.

$$\text{S/N} = 20 \log \left(\frac{128}{0.5}\right) = \boxed{48.2\ \text{dB}}$$

For the hexadecimal, full-scale is

$$(15)(16)^2 + (15)(16)^1 + 15 = 4095$$

For half-scale, the S/N ratio is the same as the octal half-scale.

$$\text{S/N} = 20 \log \left(\frac{2048}{0.5}\right) = \boxed{72.2\ \text{dB}}$$

43 Electronic Components

PRACTICE PROBLEMS

1. Intrinsic germanium has a concentration of electrons of 2.50×10^{13}. Determine the number of holes.

(A) 1.25×10^{13} cm^{-3}
(B) 1.50×10^{13} cm^{-3}
(C) 2.00×10^{13} cm^{-3}
(D) 2.50×10^{13} cm^{-3}

2. What type of semiconductor material results if phosphor is the only impurity?

(A) acceptor
(B) donor
(C) n-type
(D) p-type

3. A silicon semiconductor is doped with 10^{15} atoms/cm^3 of phosphor. What is the concentration of electrons?

(A) 1.6×10^{-5} cm^{-3}
(B) 2.6×10^{5} cm^{-3}
(C) 0.5×10^{15} cm^{-3}
(D) 1.0×10^{15} cm^{-3}

4. A silicon semiconductor is doped with 10^{15} atoms/cm^3 of phosphor. What is the concentration of holes?

(A) 1.6×10^{-5} cm^{-3}
(B) 2.6×10^{5} cm^{-3}
(C) 0.5×10^{15} cm^{-3}
(D) 1.0×10^{15} cm^{-3}

5. A germanium semiconductor is doped with an impurity having three valence electrons. What are the majority carriers?

(A) conduction band electrons
(B) conduction band holes
(C) valence band electrons
(D) valence band holes

6. Consider the common base (CB) circuit shown.

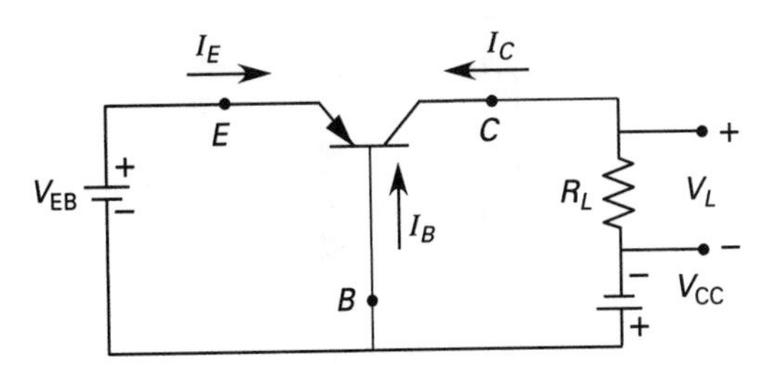

If the collector supply voltage has a magnitude of 1 V and the load is 50 Ω, determine and draw the load line.

7. A discrete silicon semiconductor diode is forward biased at 0.75 V. What is the approximate current?

(A) 2.0 mA
(B) 3.0 nA
(C) 1.5 A
(D) 3.0 MA

8. The compound GaP, with an energy gap of 2.26 eV, is to be used in an LED. What is the expected wavelength of the emitted photons?

(A) 8.79×10^{-26} m
(B) 5.49×10^{-7} m
(C) 3.42×10^{12} m
(D) 5.46×10^{14} m

9. At 20°C, a germanium diode passes 70 μA when the reverse bias is −1.5 V. (a) What is the saturation current? (b) What is the current that flows when a forward bias of +0.2 V is applied at 20°C, and (c) at 40°C?

10. A voltage $V = 20 + 5\sqrt{2}\sin 60t$ V is applied to the circuit shown. The diode characteristics are static forward resistance $r_f = 120$ Ω and dynamic resistance $r_p = 100$ Ω. What is the voltage across the inductance?

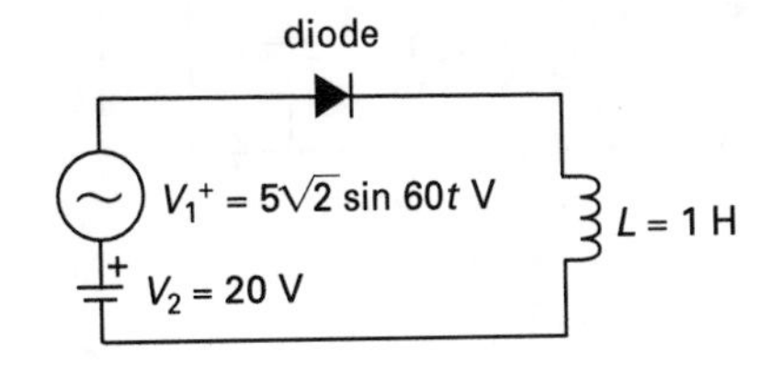

11. At 25°C, a germanium diode shows a saturation current of 100 μA. (a) What current is expected at 100°C, when the diode becomes "useless"? (b) What current is expected at 0°C? (c) At 25°C, what current is predicted for a voltage of −0.5 V?

12. The forward resistance of the diode shown is R_f = 5 kΩ. (a) What current is expected in the load resistor R_L = 5 kΩ if $V = 100 \sin \omega t$ V? (b) Plot the current through and voltage across the external load resistance versus ωt for $0 < \omega t < 2\pi$.

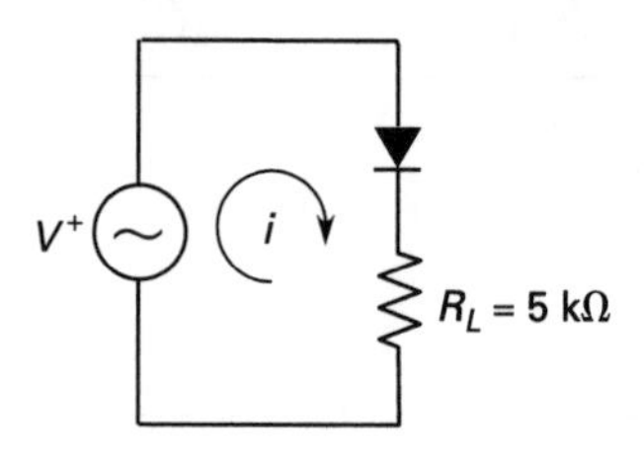

13. A silicon *npn* transistor is used in the circuit shown. The transistor parameters are $I_{CBO} = 10^{-6}$A and $\alpha_0 = 0.99$. The base-emitter voltage is 0.6 V. What is the quiescent collector current, I_C?

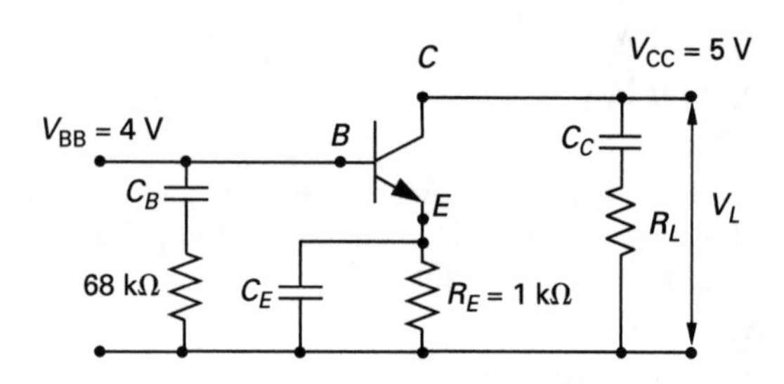

14. A silicon *npn* transistor is shown. Use $\alpha = 0.95$, $I_{CO} = 10^{-6}$ A, and $V_{BE} = 0.6$ V. What current, I_C, will result from this arrangement?

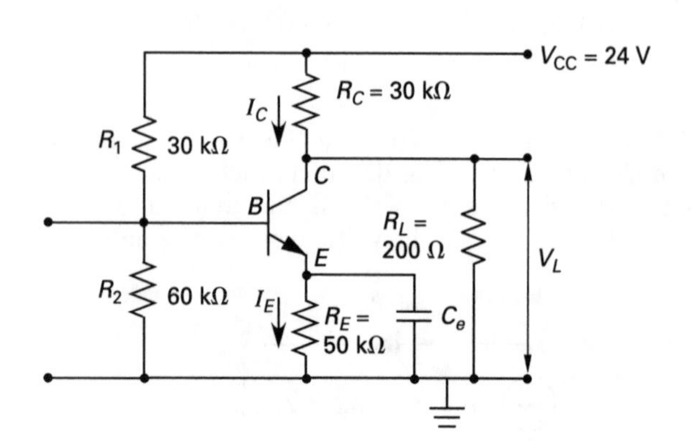

15. The equivalent circuit *h*-parameters for a bipolar junction transistor operating near the center of its active region in a common emitter configuration are h_{ie} = 500 Ω and h_{fe} = 100. Determine the values of (a) h_{ib}, (b) h_{fc}, (c) h_{ic}, and (d) h_{fb}.

16. The equivalent circuit common base *h*-parameters for a bipolar junction transistor are h_{ib} = 21.5 Ω, $h_{fb} = -0.95$, $h_{rb} = 0.0031$, and $h_{ob} = 4.7 \times 10^{-7}$ 1/Ω. What are this transistor's four common emitter *h*-parameters?

17. The *h*-parameters in the transistor shown are h_{ie} = 2200 Ω, $h_{re} = 3.6 \times 10^{-4}$, $h_{fe} = 55$, and $h_{oe} = 12.5$ μS. (a) What load resistance is needed for a signal output voltage of 0.1 V with an input signal current of 0.5 μA? (b) What input voltage is required? (c) What is the voltage gain (V_L/V_S)?

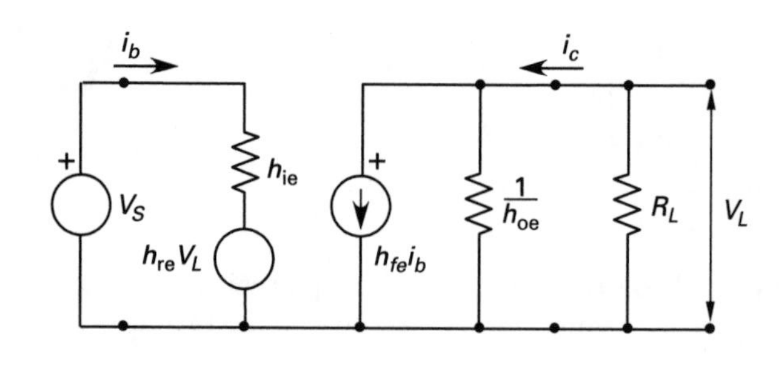

18. An AC-coupled FET amplifier circuit is shown.

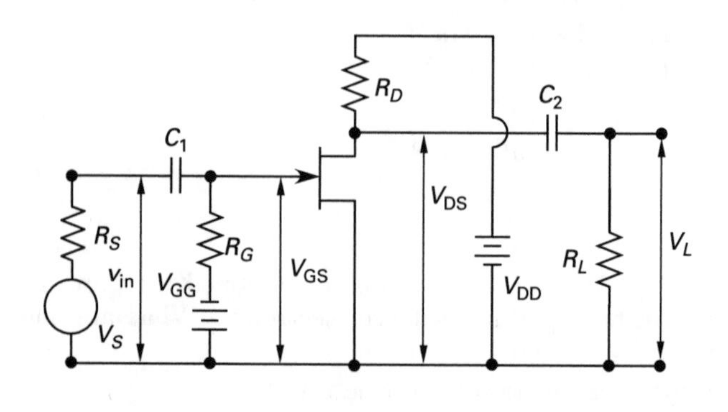

Assume the drain bias voltage is $V_{DD} = 20$ V. If I_D at $V_{DS} = 0$ is 10 mA, draw the load line.

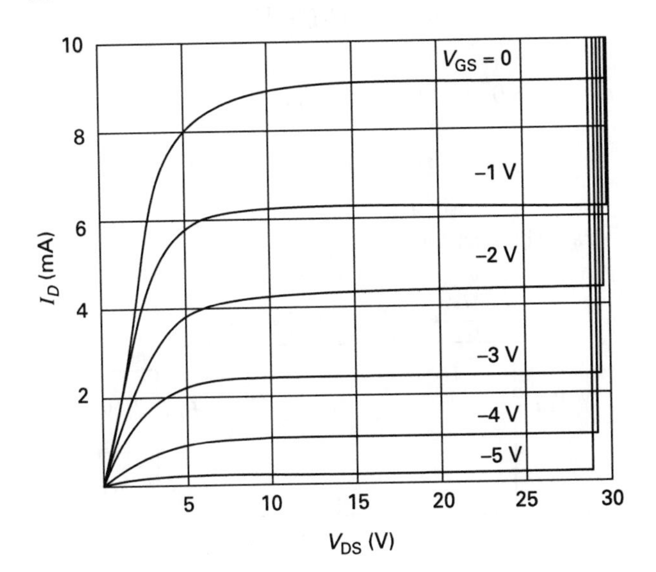

19. A FET with the characteristics shown is to be used in the circuit in Prob. 18. $R_D = 1.5$ kΩ, $V_{DD} = 20$ V, and $V_{GSQ} = -2$ V. Draw the DC load line and locate the Q-point.

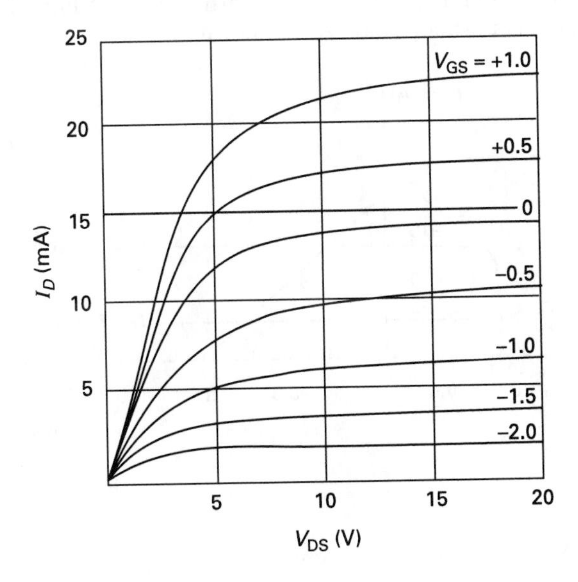

20. The manufacturer of a MOSFET has provided the following information: $V_{DS_{max}} = 20$ V, $V_{GS_{min}} = -8$ V, $I_{D_{max}} = 25$ mA, and $P_{max} = 330$ mW. The FET is to be used in the amplifier circuit shown, with $R_D = 800\ \Omega$ and $R_L = 800\ \Omega$. Draw the DC load line and a Q-point to ensure linear amplification. Use the characteristic curves from Prob. 19.

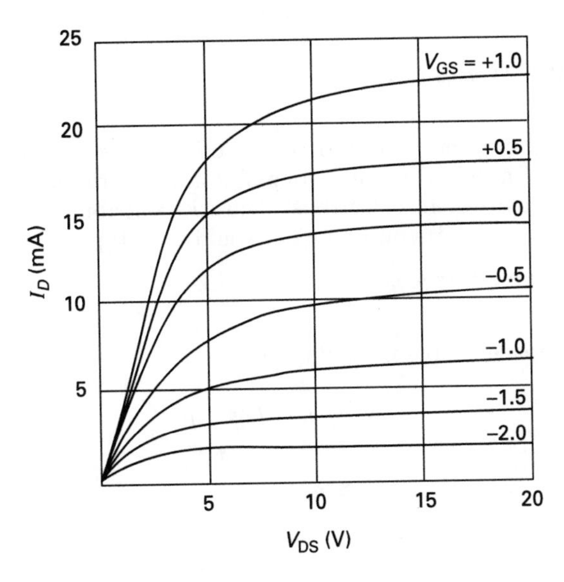

SOLUTIONS

1. In intrinsic semiconductors, $n = p$.

Thus,

$$n_i^2 = np = n^2$$

$$n_i = \boxed{2.5 \times 10^{13} \text{ cm}^{-3}}$$

The answer is (D).

2. "Acceptor" and "donor" describe types of impurities. Phosphor is from group VA on the periodic table, with five valence electrons. Thus, it contributes a free electron. The semiconductor would be n-type.

The answer is (C).

3. Space-charge neutrality gives

$$N_A + n = N_D + p$$

Since $N_A = 0$ and $n \gg p$,

$$\boxed{n \approx N_D = 10^{15} \text{ cm}^{-3}}$$

The answer is (D).

4. From Prob. 3, $N_D \approx n$. The law of mass action becomes

$$n_i^2 = np = N_D p$$

$$p = \frac{n_i^2}{N_D} = \frac{(1.6 \times 10^{10} \text{ cm}^{-3})^2}{10^{15} \text{ cm}^{-3}}$$

$$= \boxed{2.56 \times 10^5 \text{ cm}^{-3}}$$

The answer is (B).

5. Germanium has four valence electrons. The impurity has three valence electrons and thus adds a hole. The majority carriers will be holes.

The answer is (D).

6. Redraw the circuit as shown.

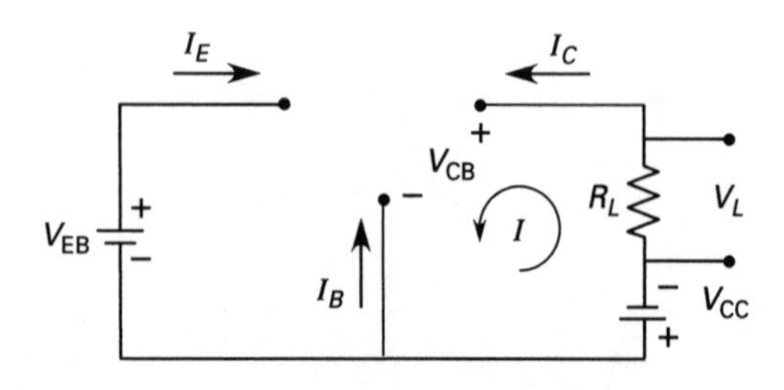

KVL around the indicated loop gives

$$-V_{CC} - I_C R_L - V_{CB} = 0$$

Since $I_C = 0$ and $V_{CC} = -1$ V,

$$V_{CC} = V_{CB} = -1 \text{ V}$$

Redraw the current as shown.

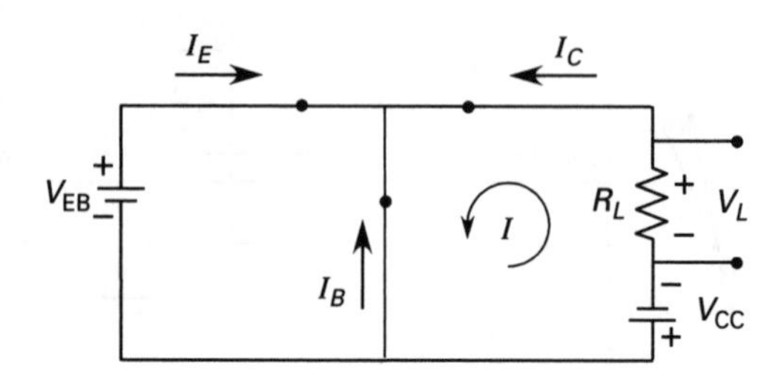

Using Ohm's law and the given information,

$$I_C = \frac{V_{CC}}{R_L} = \frac{-1 \text{ V}}{50 \ \Omega} = -20 \text{ mA}$$

Plot V_{CC} and I_C and connect with a straight line. The result is normally plotted as shown.

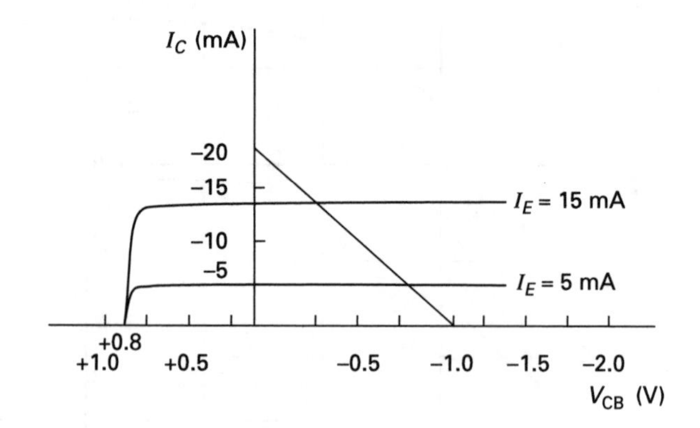

7. The diode current is given by

$$I = I_s \left(e^{\frac{V}{\eta V_T}} - 1 \right)$$

A reverse saturation current of 10^{-9} A is typical for silicon, $\eta = 2$, and V_T at room temperature is

$$V_T = \frac{\kappa T}{q} = \frac{\left(1.381 \times 10^{-23} \ \frac{\text{J}}{\text{K}}\right)(300\text{K})}{1.6 \times 10^{-19} \text{ C}}$$

$$= 0.026 \text{ V}$$

Substituting the known and calculated values gives the diode current.

$$I = \left(10^{-9}\ \text{A}\right)\left(e^{\frac{0.75\ \text{V}}{(2)(0.026\ \text{V})}} - 1\right)$$

$$= \boxed{1.84 \times 10^{-3}\ \text{A}}$$

The answer is (A).

8. The energy gap for GaP is given as 2.26 eV. The wavelength is given by

$$\lambda = \frac{hc}{E_G}$$

$$= \frac{\left(6.626 \times 10^{-34}\ \text{J·s}\right)\left(3.00 \times 10^8\ \frac{\text{m}}{\text{s}}\right)}{(2.26\ \text{eV})\left(\frac{1.602 \times 10^{-19}\ \text{J}}{1\ \text{eV}}\right)}$$

$$= \boxed{5.49 \times 10^{-7}\ \text{m}}$$

The wavelength is in the visible spectrum.

The answer is (B).

9. (a)

$$I_s = \frac{I}{e^{\frac{qV}{\eta\kappa T}} - 1}$$

For a germanium diode, $\eta \approx 1$. $T = 20°\text{C} + 273 = 293\text{K}$.

$$I_s = \frac{70\ \mu\text{A}}{e^{\left(\frac{(1.6\times10^{-19}\ \text{C})(-1.5\ \text{V})}{(1)\left(1.38\times10^{-23}\ \frac{\text{J}}{\text{K}}\right)(293\text{K})}\right)} - 1}$$

$$\approx \frac{70\ \mu\text{A}}{-1} = \boxed{-70\ \mu\text{A}}$$

The reverse saturation current is constant for reverse voltages up to the breakdown voltage.

(b)

$$I \approx I_s\left(e^{\frac{40V}{\eta}} - 1\right)$$

$$= \left(70 \times 10^{-6}\ \text{A}\right)\left(e^{\frac{\left(\frac{40\ 1}{\text{V}}\right)(0.2\ \text{V})}{1}} - 1\right)$$

$$= \boxed{0.209\ \text{A}}$$

(c)

$$\frac{I_{s,40°\text{C}}}{I_{s,20°\text{C}}} = (2)^{\frac{40°\text{C}-20°\text{C}}{10}}$$

$$(2)^2 = 4$$

$$I_{s,40°\text{C}} = 4I_{s,20°\text{C}} = (4)(70\ \mu\text{A})$$

$$= 280\ \mu\text{A}$$

$$I = I_s\left(e^{\frac{qV}{\eta\kappa T}} - 1\right)$$

$$= \left(280 \times 10^{-6}\ A\right) \times \left(e^{\left(\frac{(1.6\times10^{-19}\ \text{C})(0.2\ \text{V})}{(1)\left(\frac{1.38\times10^{-23}\ \text{J}}{\text{K}}\right)(313\text{K})}\right)} - 1\right)$$

$$= \boxed{0.4617\ \text{A}}$$

10. Since $20 > 5\sqrt{2}$, the applied voltage is never negative. Therefore, the diode does not rectify the applied voltage and serves no purpose. Model the circuit as an ideal diode with a 100 Ω resistance. (The forward resistance could also be used.)

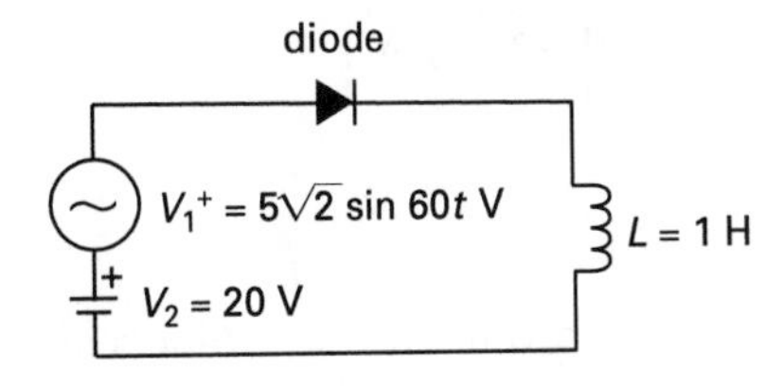

$$X_L = \omega L = \left(60\ \frac{\text{rad}}{\text{s}}\right)(1\ \text{H}) = 60\ \Omega$$

$$Z = \sqrt{R^2 + X_L^2} = \sqrt{(100\ \Omega)^2 + (60\ \Omega)^2}$$

$$= 116.6\ \Omega$$

$$\phi = \arctan\left(\frac{X_L}{R}\right) = \arctan\left(\frac{60\ \Omega}{100\ \Omega}\right) = 30.96°$$

$$\mathbf{I} = \frac{\mathbf{V}}{\mathbf{Z}} = \frac{20 + 5\sqrt{2}\angle 0°}{116.6\angle 30.96°}$$

The DC voltage across the inductor is zero. Therefore, the 20 V bias is not used.

$$\mathbf{V}_L = \mathbf{I}\mathbf{Z}_L = \left(\frac{5\sqrt{2}\angle 0°}{116.6\ \angle 30.96°}\right)(60\angle 90°)$$

$$= \boxed{3.639\angle 59.04°\ \text{V}}$$

11. The saturation current doubles for every 10°C.

(a)

$$\frac{I_{s2}}{I_{s1}} = (2)^{\frac{\Delta T}{10^\circ\text{C}}} = (2)^{\frac{100^\circ\text{C}-25^\circ\text{C}}{10^\circ\text{C}}}$$

$$= (2)^{7.5} = 181$$

$$I_{s2} = (181)(100\ \mu\text{A})$$

$$= \boxed{18{,}100\ \mu\text{A} \quad (18.1\ \text{mA})}$$

(b) At 0°C,

$$I_{s2} = (2)^{\frac{0^\circ\text{C}-25^\circ\text{C}}{10^\circ\text{C}}} I_{s1} = (0.177)(100\ \mu\text{A})$$

$$= \boxed{17.7\ \mu\text{A}}$$

(c)

$$I_{-0.5} \approx I_s\left(e^{40V} - 1\right)$$

$$= (100\ \mu\text{A})\left(e^{\left(40\ \frac{1}{\text{V}}\right)(-0.5\ \text{V})} - 1\right)$$

$$= \boxed{-100\ \mu\text{A}}$$

12. Clipping of the voltage wave will occur.

(a)

$$I = \frac{100 \sin \omega t\ \text{V}}{5000\ \Omega + 5000\ \Omega}$$

$$= \boxed{10^{-2} \sin \omega t\ \text{A}}$$

(b)

13.

$$I_C = \alpha I_E - I_{\text{CBO}}$$

$$= \alpha\left(\frac{V_{\text{BB}} - V_{\text{BE}}}{R_E}\right) - I_{\text{CBO}}$$

$$= (0.99)\left(\frac{4\ \text{V} - 0.6\ \text{V}}{1000\ \Omega}\right) - 10^{-6}\ \text{A}$$

$$= \boxed{0.003365\ \text{A}}$$

(Alternate Solution)

Writing Kirchhoff's voltage law around the base loop,

$$V_{\text{BB}} - V_{\text{BE}} - (I_C + I_B)R_E = 0$$

$$\frac{I_C}{I_B} = \frac{\alpha}{1-\alpha} = \frac{0.99}{1-0.99} = 99$$

$$4\ \text{V} - V_{\text{BE}} - \left(I_C + \frac{I_C}{99}\right)1000\ \Omega = 0$$

$$4\ \text{V} - 0.6\ \text{V} - (1010\ \Omega)I_C = 0$$

$$I_C = \boxed{3.366 \times 10^{-3}\ \text{A} \quad (3.366\ \text{mA})}$$

14. R_1 and R_2 form a voltage divider.

$$V_B = \left(\frac{R_2}{R_1 + R_2}\right)V_{\text{CC}}$$

$$= \left(\frac{60\ \text{k}\Omega}{30\ \text{k}\Omega + 60\ \text{k}\Omega}\right)(24\ \text{V}) = 16\ \text{V}$$

For the purpose of the signal, R_1 and R_2 both connect to ground and are in parallel.

$$R_{\text{in}} = \frac{R_1 R_2}{R_1 + R_2} = \frac{(30\ \text{k}\Omega)(60\ \text{k}\Omega)}{30\ \text{k}\Omega + 60\ \text{k}\Omega}$$

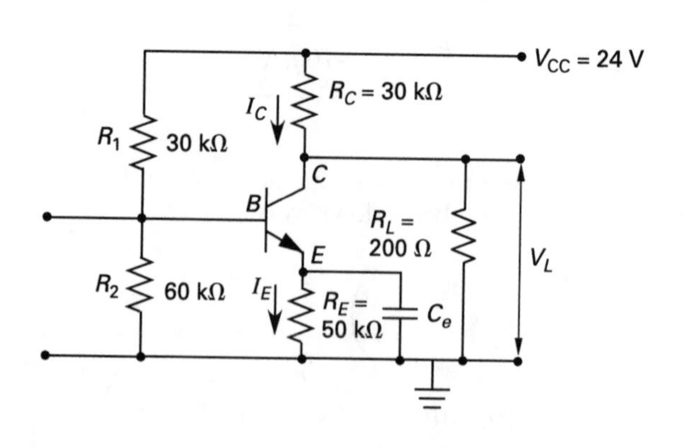

$$\beta = \frac{I_C}{I_B} = \frac{\alpha}{1-\alpha} = \frac{0.95}{1-0.95} = 19$$

So, $I_B = I_C/19$.

Writing Kirchhoff's voltage law around the base circuit,

$$V_B - V_{BE} - (I_C + I_B)R_E - I_B R_{in} = 0$$

$$16\text{ V} - 0.6\text{ V} - \left(I_C + \frac{I_C}{19}\right)(50{,}000\ \Omega) - \left(\frac{I_C}{19}\right)(20{,}000\ \Omega) = 0$$

$$I_C = 2.869 \times 10^{-4}\text{ A (0.2869 mA)}$$

I_{CO} is given, so it should be used even though its effect is small.

$$I'_C = I_C + I_{CO} = 2.869 \times 10^{-4}\text{ A} + 10^{-6}\text{ A} = \boxed{2.879 \times 10^{-4}\text{ A} \quad (0.2879\text{ mA})}$$

The exact solutions based on node-voltage analysis are

$$V_B = 15.70\text{ V}$$
$$I_B = 0.0151\text{ mA}$$
$$I_C = 0.2869\text{ mA}$$

15. (a)

$$h_{ib} = \frac{h_{ie}}{1 + h_{fe}} = \frac{500\ \Omega}{1 + 100} = \boxed{4.95\ \Omega}$$

(b)

$$h_{fc} = -1 - h_{fe} = -1 - 100 = \boxed{-101}$$

(c)

$$h_{ic} = h_{ie} = \boxed{500\ \Omega}$$

(d)

$$h_{fb} = \frac{-h_{fe}}{1 + h_{fe}} = \frac{-100}{1 + 100} = \boxed{-0.99}$$

16.

$$h_{ie} = \frac{h_{ib}}{1 + h_{fb}} = \frac{21.5\ \Omega}{1 + (-0.95)} = \boxed{430\ \Omega}$$

$$h_{re} = \frac{h_{ib}h_{ob}}{1 + h_{fb}} - h_{rb} = \frac{(21.5\ \Omega)\left(4.7 \times 10^{-7}\ \frac{1}{\Omega}\right)}{1 + (-0.95)} - 0.0031 = \boxed{-0.00290}$$

$$h_{fe} = \frac{-h_{fb}}{1 + h_{fb}} = \frac{-(-0.95)}{1 + (-0.95)} = \boxed{19}$$

$$h_{oe} = \frac{h_{ob}}{1 + h_{fb}} = \frac{4.7 \times 10^{-7}\ \frac{1}{\Omega}}{1 + (-0.95)} = \boxed{9.4 \times 10^{-6}\ \frac{1}{\Omega}}$$

17.

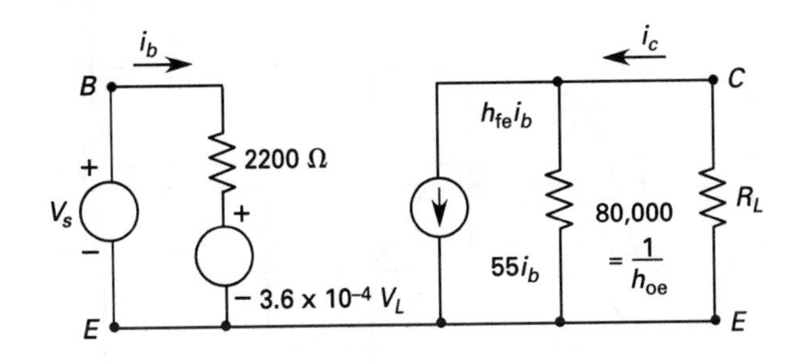

(a)

$$-i_c R_L = V_L$$
$$i_b = 0.5\ \mu\text{A}$$
$$(55 i_b)\left(\frac{80{,}000 R_L}{R_L + 80{,}000}\right) = V_L = 0.1\text{ V}$$
$$80{,}000 R_L = (R_L + 80{,}000) \times \left(\frac{0.1}{(55)(0.5 \times 10^{-6}\text{ A})}\right)$$
$$= 3636.36 R_L + (290.91 \times 10^6)$$
$$R_L = \boxed{3809\ \Omega}$$

(b)

$$\begin{aligned} V_s &= i_b(2200\ \Omega) + (3.6 \times 10^{-4})V_L \\ &= (0.5 \times 10^{-6}\ \text{A})(2200\ \Omega) \\ &\quad + (3.6 \times 10^{-4})(0.1\ \text{V}) \\ &= (1.1 \times 10^{-3}) + (36 \times 10^{-6}) \\ &= \boxed{1.136 \times 10^{-3}\ \text{V}\ (1.136\ \text{mV})} \end{aligned}$$

(c)

$$\begin{aligned} A &= \frac{V_L}{V_s} = \frac{0.1\ \text{V}}{1.136 \times 10^{-3}} \\ &= \boxed{88.03} \end{aligned}$$

18. $P_1\ (V_{DS} = 20\ \text{V},\ I_D = 0)$

$P_2\ (V_{DS} = 0,\ I_D = 10\ \text{mA})$

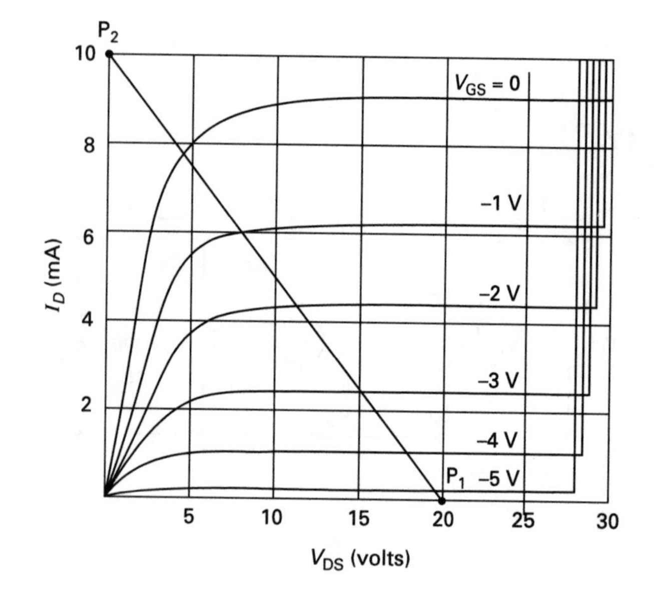

19. $P_1\ (V_{DS} = 20\ \text{V}, 0)$

$$P_2\left(0, I_D = \frac{20\ \text{V}}{1500\ \Omega} = 0.01333\ \text{A}\ (13.33\ \text{mA})\right)$$

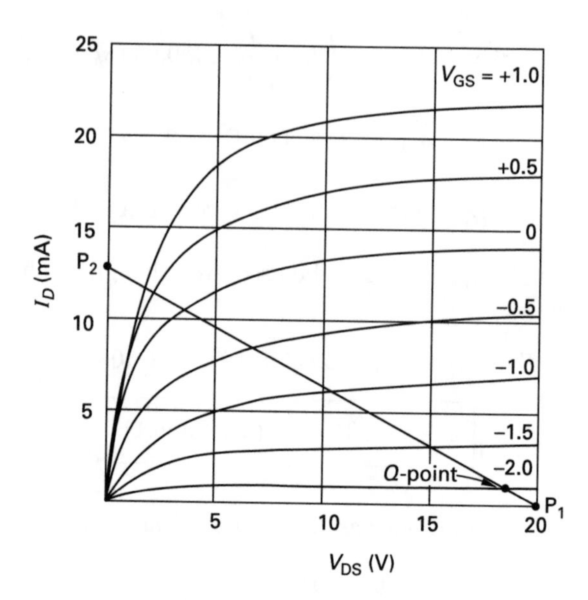

20. The highly negative V_{GS} curves are not shown and are, presumably, crowded together near the $I_D = 0$ axis. Because of this, if the Q-point were in the highly negative region, positive swings in V_{GS} would accompany large increases in I_D, but negative swings in V_{GS} would decrease I_D very little. Therefore, the quiescent operating point needs to be near the $V_{GS} = 0$ line. In this position, changes in I_D and V_{GS} are approximately proportional. Power dissipation is another consideration. It should be evaluated at the extreme points of the signal swing, not only at the Q-point.

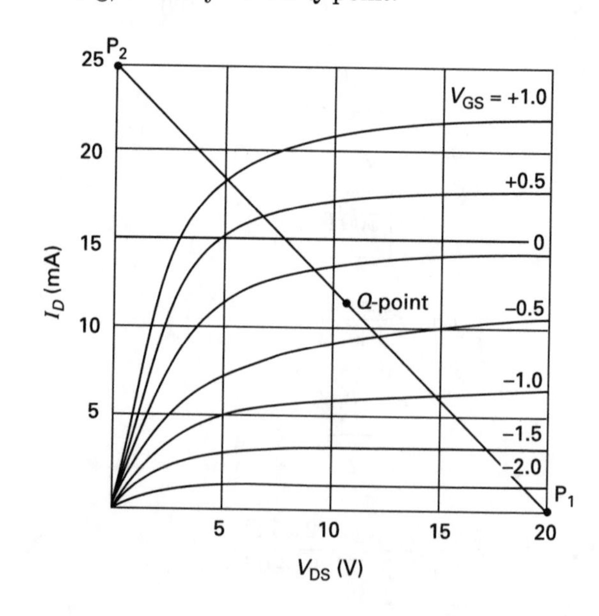

44 Amplifiers

PRACTICE PROBLEMS

1. An operational amplifier has a gain of 10^5. The frequencies at the half-power points are $f_L = 500$ Hz and $f_H = 1000$ Hz. What is the figure of merit?

(A) 1×10^6 rad/s
(B) 5×10^7 rad/s
(C) 3×10^8 rad/s
(D) 6×10^8 rad/s

2. During testing of an operational amplifier, a signal of equal magnitude is applied to the two terminals. That is, $v^+ = -v^-$. What gain is being measured?

(A) A_{cm}
(B) A_{dm}
(C) A_V
(D) G_P

3. An operational amplifier data sheet indicates a voltage gain of 10^5 and a power supply voltage of 12 V. What is the maximum voltage difference between the input terminals that will ensure linear operation?

(A) 60 μV
(B) 90 μV
(C) 120 μV
(D) 150 μV

4. Consider the figure shown. The feedback resistance is 10 kΩ, and the input resistance is 100 Ω. If the op amp gain is 10^5, what is the gain of the circuit?

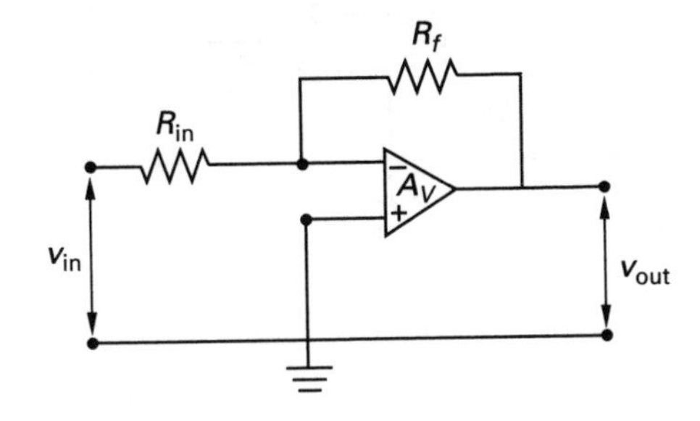

(A) −100
(B) −99.9
(C) +100
(D) 10^5

5. An operational amplifier has an input resistance of 10^5 Ω and a bandwidth of 1000 Hz, and is at a room temperature of 300K. If 10 dB above the noise is required for proper operation, what is the required signal power?

(A) 0.1 μV
(C) 2.0 μV
(C) 4.0 μV
(D) 16 μV

6. (a) If $R_f = 1$ MΩ and $R_{in} = 50$ Ω, what is the gain? (b) If $v_{in} = 1$ mV, what is v_{out}?

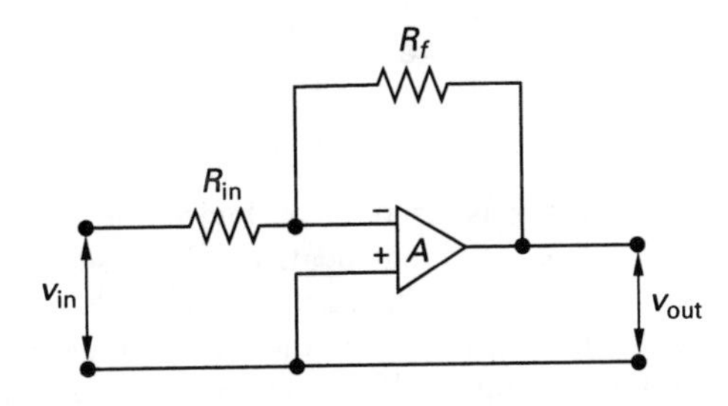

7. For the circuit shown, $v_1 = 10 \sin 200t$ and $v_2 = 15 \sin 200t$. What is v_{out}? The op amp is ideal and has infinite gain.

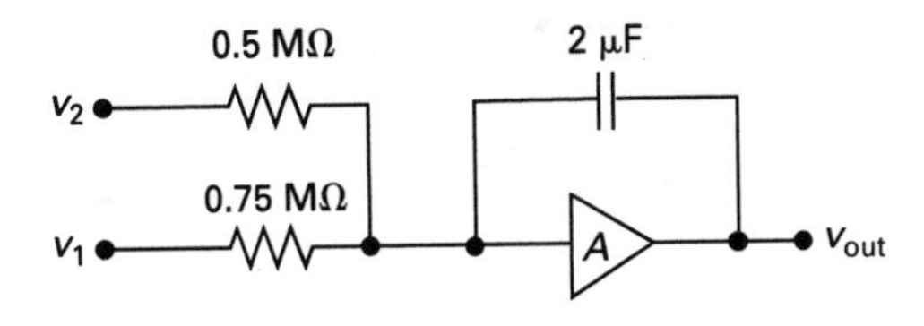

8. If $v_{in} = \sin 30t$, what is v_{out}? The op amp shown is ideal and has infinite gain.

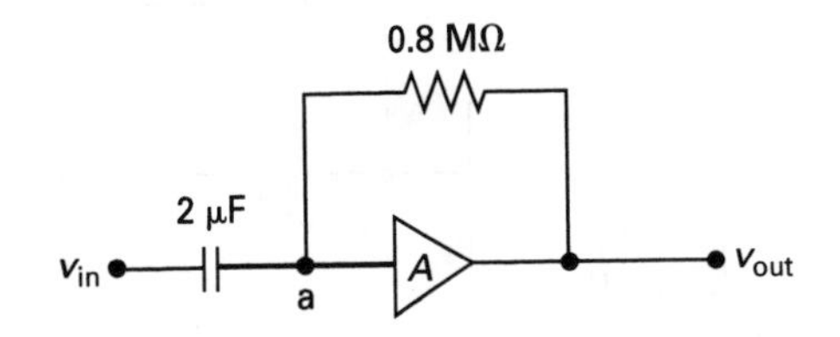

9. If $v_1 = v_2 = v_3 = 12$ V, what is v_{out}? The op amp shown is ideal and has infinite gain.

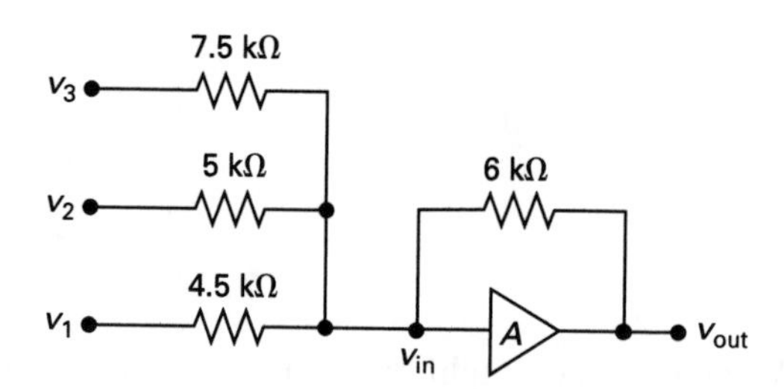

10. Given the following circuit, (a) what value of R_f will give an output of −8 V? (b) If the maximum of each input is 10 V and the output must not exceed −15 V, what value must be used for R_f? The op amp is ideal and has infinite gain.

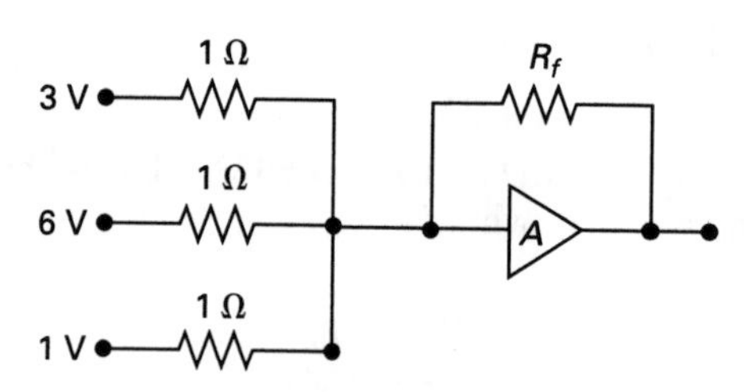

11. An integrator is used in a device for measuring displacement from a fixed location. The scale of the device is 1 V to 1 in. The switch is closed to set the output for initial displacement and then opened at $t = 0$ to initiate measurement. If the displacement is initially 12 in d_0, what is the output voltage for an input of $8 \cos 10t$ V? The op amp is ideal and has infinite gain.

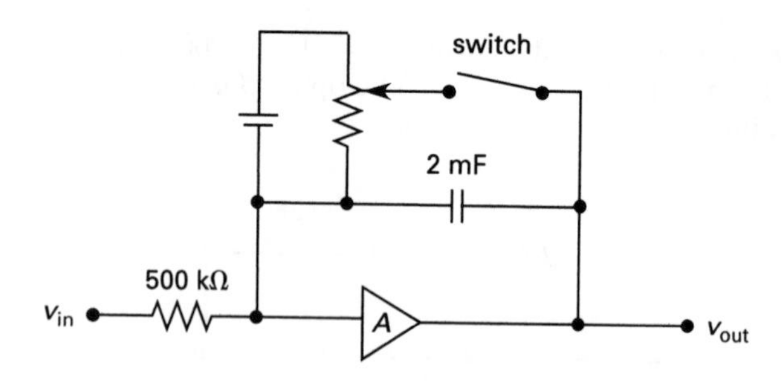

12. What is the differential equation that is solved by the following circuit? The op amp is ideal and has infinite gain.

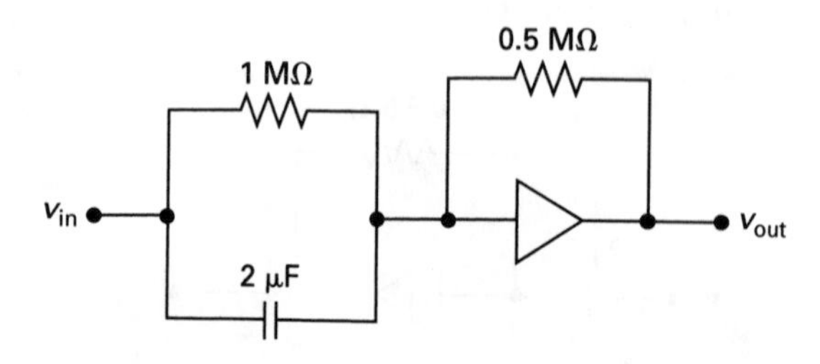

13. What is the differential equation that writes v_{out} in terms of v_{in}? The op amp is ideal and has infinite gain.

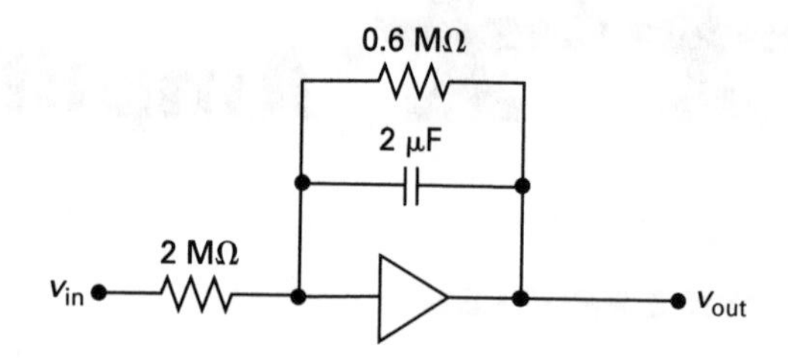

14. For the circuit shown, determine the output voltage as a function of the input voltage.

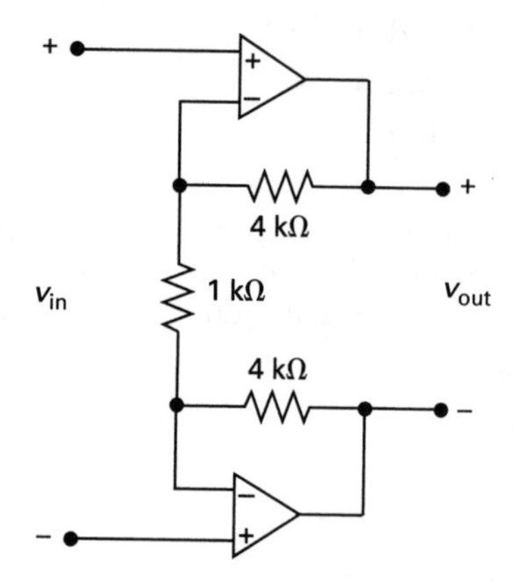

15. The op amp circuit shown is to provide a characteristic similar to a relay, so that when the output is negative and the input current increases, the output will go to the positive limit of +12 V when the input voltage exceeds +1 V, and will remain there until the input voltage becomes less than −1 V, at which time the output will go to the negative limit of −12 V. Determine an appropriate value for R.

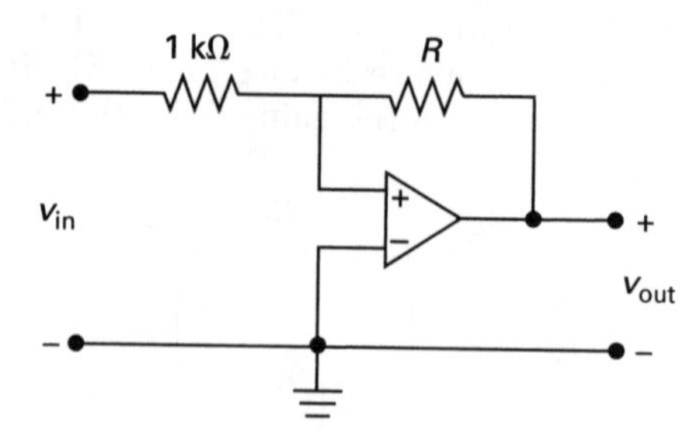

SOLUTIONS

1. The figure of merit is given by

$$\begin{aligned} F_m &= A_{\text{ref}}(\text{BW}) = A_{\text{ref}}\,(2\pi f_H - 2\pi f_L) \\ &= \left(10^5\right)(2\pi)\,(1000\text{ Hz} - 500\text{ Hz}) \\ &= \boxed{3.14 \times 10^8\text{ rad/s}} \end{aligned}$$

The answer is (C).

2. The differential-mode voltage is given by

$$v_{\text{dm}} = v^+ - v^-$$

Thus, with $v^+ = -v^-$,

$$v_{\text{dm}} = v^+ + v^+ = 2v^+$$

The common-mode voltage is given by

$$v_{\text{cm}} = \tfrac{1}{2}(v^+ + v^-) = \tfrac{1}{2}\left(v^+ + (-v^-)\right) = 0$$

Therefore, only the difference is being amplified and the gain so determined is A_{dm}.

The answer is (B).

3. The maximum voltage difference is given by

$$\begin{aligned} \left|v^+ - v^-\right| &< \frac{V_{\text{DC}} - 3\text{ V}}{A_V} \\ |\Delta V| &< \frac{12\text{ V} - 3\text{ V}}{10^5} = \frac{9\text{ V}}{10^5} = 90 \times 10^{-6}\text{ V} \\ &= \boxed{90\ \mu\text{V}} \end{aligned}$$

The answer is (B).

4. Consider the figure with current directions at the negative terminal assumed, as indicated.

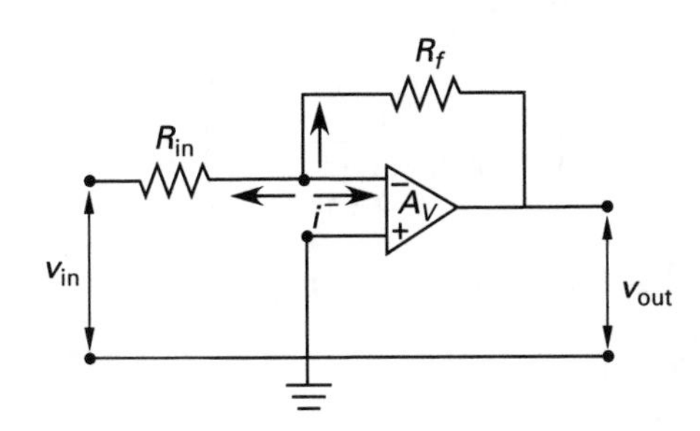

Writing KCL at the inverting terminal node gives

$$\frac{V^- - v_{\text{in}}}{R_{\text{in}}} + \frac{V^- - v_{\text{out}}}{R_f} + i^- = 0$$

Applying the ideal op amp assumptions means $i^- = 0$ and $v^- = 0$. Substituting and solving for the gain gives

$$\frac{0 - v_{\text{in}}}{R_{\text{in}}} + \frac{0 - v_{\text{out}}}{R_f} + 0 = 0$$

$$\begin{aligned} \frac{v_{\text{out}}}{v_{\text{in}}} &= \frac{-R_f}{R_{\text{in}}} = A_{V,\text{ideal}} \\ &= \frac{-10 \times 10^3\ \Omega}{100\ \Omega} = -100 \end{aligned}$$

But this op amp is real and has a finite gain. The constraint equation (that is, the fundamental op amp equation) rearranged is

$$V^- - V^+ = \Delta = \frac{-v_{\text{out}}}{A_V}$$

Since V^+ is at 0 V,

$$V^- = \frac{-v_{\text{out}}}{A_V}$$

Substituting into the original equation gives

$$\frac{\dfrac{-v_{\text{out}}}{A_V} - v_{\text{in}}}{R_{\text{in}}} + \frac{\dfrac{-v_{\text{out}}}{A_V} - v_{\text{out}}}{R_f} + 0 = 0$$

The input current is still small enough to be ignored. Rearranging gives

$$\frac{v_{\text{out}}}{v_{\text{in}}} = -A_V\left(\frac{R_f}{R_f + R_{\text{in}} + A_V R_{\text{in}}}\right)$$

Note that when the gain is large, this expression reduces to the expression for gain in the ideal case.

Substituting the given values gives the following result.

$$\begin{aligned} \frac{v_{\text{out}}}{v_{\text{in}}} &= \left(-10^5\right)\left(\frac{10 \times 10^3\ \Omega}{\begin{array}{c}10 \times 10^3\ \Omega + 100\ \Omega \\ +\ (10^5)(100\ \Omega)\end{array}}\right) \\ &= \boxed{-99.9} \end{aligned}$$

The answer is (B).

5. The signal power is found from

$$\text{SNR} = 10\log\left(\frac{P_s}{P_n}\right)$$

$$\text{antilog}\left(\frac{\text{SNR}}{10}\right) = \frac{P_s}{P_n}$$

$$P_s = P_n\,\text{antilog}\left(\frac{\text{SNR}}{10}\right)$$

The noise power is unknown. The noise power is

$$P_n = \frac{V_{\text{rms}}^2}{4R} = (\text{BW})kT$$
$$= (1000\ \text{Hz})\left(1.3805 \times 10^{-23}\ \frac{\text{J}}{\text{K}}\right)(300\text{K})$$
$$= 4.14 \times 10^{-18}\ \text{W}$$

Substituting,

$$P_s = \left(4.14 \times 10^{-18}\text{W}\right)\ \text{antilog}\left(\frac{10}{10}\right)$$
$$= 41.4 \times 10^{-18}\ \text{W}$$

The voltage is found from

$$P_s = \frac{V_{\text{rms}}^2}{4R}$$
$$V_{\text{rms}} = \sqrt{4RP_s} = \sqrt{(4)(10^5\ \Omega)(41.4 \times 10^{-18}\ \text{W})}$$
$$= \boxed{4.07 \times 10^{-6}\ \text{V}}$$

The answer is (C).

6. (a)

$$A = \frac{-R_f}{R_{\text{in}}} = \frac{-10^6\ \Omega}{50\ \Omega}$$
$$= \boxed{-20{,}000}$$

(b)

$$v_{\text{out}} = Av_{\text{in}} = (-20{,}000)(0.001\ \text{V})$$
$$= \boxed{-20\ \text{V}}$$

7. This is a summing circuit.

$$v_{\text{out}} = -\left(\frac{Z_f}{Z_1}\right)v_1 - \left(\frac{Z_f}{Z_2}\right)v_2$$
$$= -\left(\frac{1}{sC}\right)\left(\frac{v_1}{R_1} + \frac{v_2}{R_2}\right)$$
$$= \frac{-1}{(2 \times 10^{-6}\ \text{F})(0.75 \times 10^6\ \Omega)} \int 10 \sin 200t\ dt - \frac{1}{(2 \times 10^{-6}\ \text{F})(0.5 \times 10^6\ \Omega)} \int 15 \sin 200t\ dt$$
$$= \left(\frac{2}{3}\right)\left(\frac{1}{200}\right)(10)\cos 200t + \left(\frac{1}{1}\right)\left(\frac{15}{200}\right)\cos 200t + C$$
$$= \boxed{\left(\frac{1}{30}\right)\cos 200t + \left(\frac{3}{40}\right)\cos 200t + C\ \text{V}}$$

8. The circuit is a differentiator.

$$v_{\text{out}} = -RC\,\frac{dv_{\text{in}}}{dt}$$
$$= -(0.8 \times 10^6\ \Omega)(2 \times 10^{-6}\ \text{F})\left(\frac{d(\sin 30t)}{dt}\right)$$
$$= \boxed{-48 \cos 30t\ \text{V}}$$

9. This is a summing circuit.

$$v_{\text{out}} = \left(\frac{-6\ \text{k}\Omega}{4.5\ \text{k}\Omega}\right)v_1 - \left(\frac{6\ \text{k}\Omega}{5\ \text{k}\Omega}\right)v_2 - \left(\frac{6\ \text{k}\Omega}{7.5\ \text{k}\Omega}\right)v_3$$
$$= (-12\ \text{V})\left(\frac{6\ \text{k}\Omega}{4.5\ \text{k}\Omega} + \frac{6\ \text{k}\Omega}{5\ \text{k}\Omega} + \frac{6\ \text{k}\Omega}{7.5\ \text{k}\Omega}\right)$$
$$= \boxed{-40\ \text{V}}$$

10. (a) This is a summing circuit.

$$v_{\text{out}} = -\left(\left(\frac{R_f}{1\ \Omega}\right)(1\ \text{V}) + \left(\frac{R_f}{1\ \Omega}\right)(6\ \text{V}) + \left(\frac{R_f}{1\ \Omega}\right)(3\ \text{V})\right)$$
$$= -8\ \text{V}$$
$$R_f = \boxed{0.8\ \Omega}$$

(b)

$$v_{\text{out}} = -\left(\left(\frac{R_f}{1\ \Omega}\right)(10\ \text{V}) + \left(\frac{R_f}{1\ \Omega}\right)(10\ \text{V}) + \left(\frac{R_f}{1\ \Omega}\right)(10\ \text{V})\right)$$
$$= -15\ \text{V}$$
$$R_f = \boxed{0.5\ \Omega}$$

11.

$$v_{\text{out}} = -\frac{1}{RC}\int v_{\text{in}}dt$$
$$= -\left(\frac{1}{(500{,}000\ \Omega) \times (2 \times 10^{-3}\ \text{F})}\right)\int 8 \cos 10t\ dt$$
$$= (-8 \times 10^{-4}) \sin 10t + V_0$$

At $T = 0, d_o = 12$ in.

$$V_0 = (12\ \text{in})\left(\frac{1\ \text{V}}{1\ \text{in}}\right) = 12\ \text{V}$$
$$v_{\text{out}} = \boxed{(-8 \times 10^{-4}) \sin 10t + 12\ \text{V}}$$

12.

$$\frac{v_{\text{out}}}{v_{\text{in}}} = -\frac{Z_f}{Z_{\text{in}}} = \frac{-500{,}000\ \Omega}{\frac{(10^6)\left(\frac{1}{s(2\times 10^{-6})}\right)}{10^6 + \frac{1}{s(2\times 10^{-6})}}}$$

$$= -\frac{\frac{1}{2}}{\frac{1}{2s+1}} = -\left(s+\frac{1}{2}\right)$$

$$v_{\text{out}} = -\left(s+\frac{1}{2}\right) v_{\text{in}}$$

$$= \boxed{-\frac{dv_{\text{in}}}{dt} - \frac{v_{\text{in}}}{2}}$$

13.

$$\frac{v_{\text{out}}}{v_{\text{in}}} = -\frac{Z_f}{Z_{\text{in}}}$$

$$= -\frac{\frac{(600{,}000)\left(\frac{1}{s(2\times 10^{-6})}\right)}{600{,}000 + \left(\frac{1}{s(2\times 10^{-6})}\right)}}{2{,}000{,}000}$$

$$= -\frac{0.3}{1.2s+1}$$

$$v_{\text{out}}(1.2s+1) = -0.3v_{\text{in}}$$

$$(1.2)\left(\frac{dv_{\text{out}}}{dt}\right) + v_{\text{out}} = -0.3v_{\text{in}}$$

$$\boxed{(1.2)\left(\frac{dv_{\text{out}}}{dt}\right) + v_{\text{out}} + 0.3v_{\text{in}} = 0}$$

14.

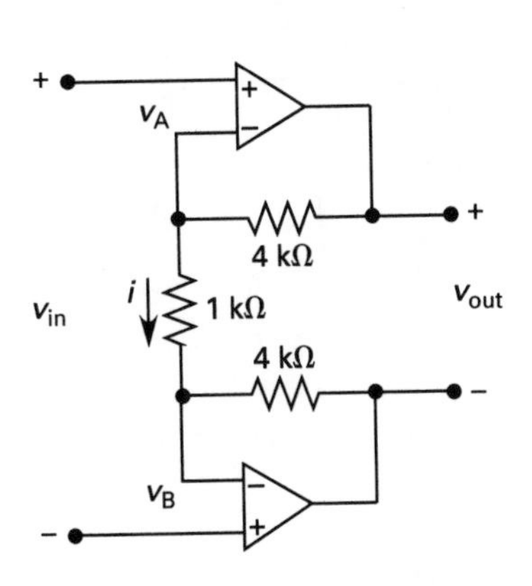

$$v_A = 0\ \text{V}$$

$$v_B = 0\ \text{V}$$

$$v_{\text{in}} = i(1\ \text{k}\Omega)$$

$$i = \frac{v_{\text{in}}}{1\ \text{k}\Omega}$$

$$v_{\text{out}} = (4\ \text{k}\Omega + 1\ \text{k}\Omega + 4\ \text{k}\Omega)i$$

$$= \left(\frac{9\ \text{k}\Omega}{1\ \text{k}\Omega}\right) v_{\text{in}}$$

$$= \boxed{9v_{\text{in}}}$$

15. Except while slewing from one saturation level to the other, the op amp is not operating in its linear region.

The current flowing in the resistors is

$$i_{\text{in}} = \frac{v_{\text{in}} - v_{\text{out}}}{1\ \text{k}\Omega + R}$$

The voltage of the positive terminal is

$$v_{\text{in}} - i_{\text{in}}(1\ \text{k}\Omega) = v_{\text{in}} - \frac{(1\ \text{k}\Omega)v_{\text{in}} - (1\ \text{k}\Omega)v_{\text{out}}}{1\ \text{k}\Omega + R}$$

$$v^+ = \frac{Rv_{\text{in}} + (1\ \text{k}\Omega)v_{\text{out}}}{R + 1\ \text{k}\Omega}$$

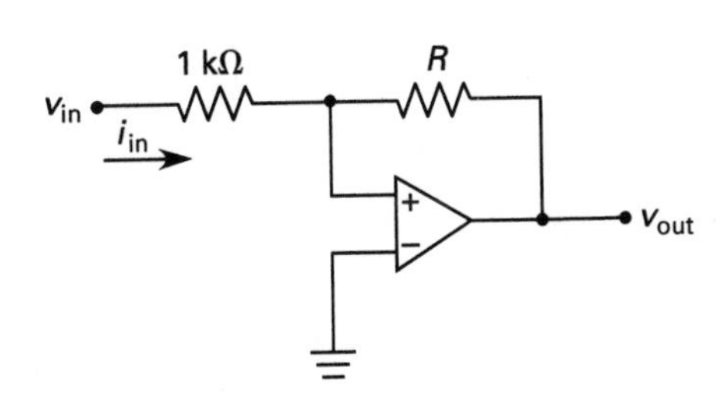

For v^+ positive, v_{out} is +12 V. v_{out} will reverse when v^+ becomes negative.

$$v_{\text{in}} = (-1-\delta)\ \text{V}$$

$$-R(1\ \text{V}+\delta) + (1\ \text{k}\Omega)(12\ \text{V}) < 0$$

$$-R + 12\ \text{k}\Omega - \delta R < 0$$

Let δ approach zero. Then,

$$R = \boxed{12\ \text{k}\Omega}$$

Electronics

45 Pulse Circuits: Waveform Shaping and Logic

PRACTICE PROBLEMS

1. A germanium diode with an offset voltage of 0.2 V is used in a precision diode circuit. The power supply voltages are ±15 V. The amplifier voltage gain is 10^7. What is the threshold voltage of the precision diode?

(A) 20 nV
(B) 100 nV
(C) 1.5 μV
(D) 9.0 μV

The following circuit is applicable to Probs. 2 through 5. The output is to be maintained at 24 V, while the input varies from 120 to 126 V with a load current ranging from 0 to 5 A. Assume ideal conditions, that is, the keep-alive current is zero and the zener resistance, R_Z, is negligible.

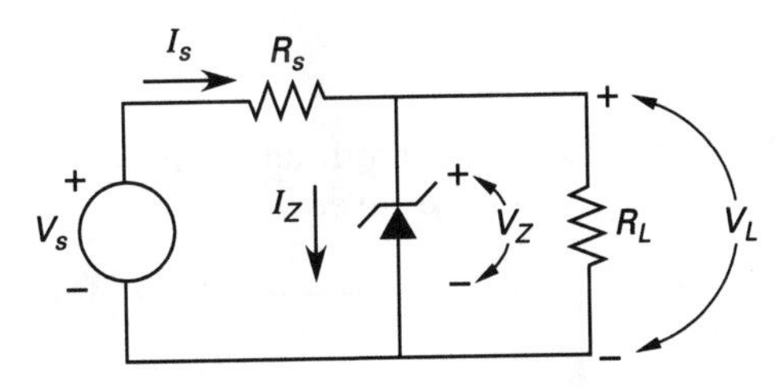

2. What is the required supply resistance to ensure proper operation of the zener regulator circuit?

(A) 19 Ω
(B) 20 Ω
(C) 24 Ω
(D) 25 Ω

3. What is the maximum zener current?

(A) 4.1 A
(B) 4.3 A
(C) 5.1 A
(D) 5.3 A

4. What is the minimum diode power rating that can be used?

(A) 100 W
(B) 150 W
(C) 300 W
(D) 600 W

5. What minimum supply resistor power rating is required?

(A) 400 W
(B) 450 W
(C) 500 W
(D) 550 W

The following circuit is applicable to Probs. 6 through 9. The circuit is to be operated at low frequencies. The transistor has the following h-parameter values: $h_{ie} = 4 \times 10^3$ Ω, $h_{fe} = 200$, $h_{oe} = 100 \times 10^{-6}$ S, and $h_{re} = 4 \times 10^3$.

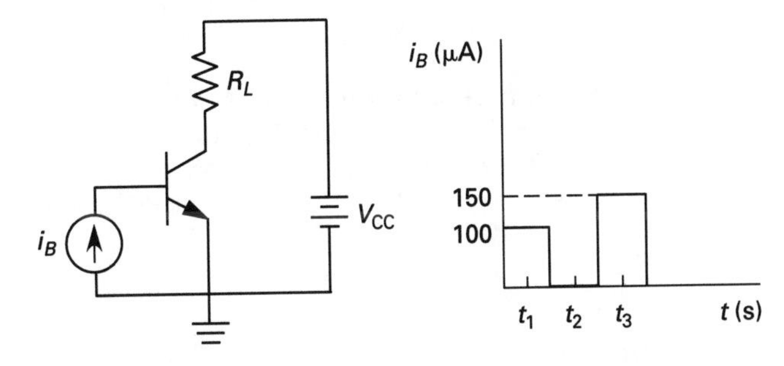

6. If the load resistance is 400 Ω, what collector supply voltage is required for the transistor to be on at time t_1?

(A) 4 V
(B) 8 V
(C) 12 V
(D) 16 V

7. If the collector supply voltage is 12 V and the load resistor is 400 Ω, what is the condition of the transistor at time t_2?

(A) active
(B) off
(C) on
(D) saturated

8. The collector supply voltage is selected as 12 V. The transistor must be on for pulses of 150 μA or greater only. What is the required value of the load resistance?

(A) 200 Ω
(B) 300 Ω
(C) 400 Ω
(D) 600 Ω

9. For the circuit parameters in Prob. 8, what is the actual saturation current for a silicon transistor?

(A) 20 mA
(B) 25 mA
(C) 29 mA
(D) 30 mA

10. Consider the inverter, or NOT gate, circuit shown.

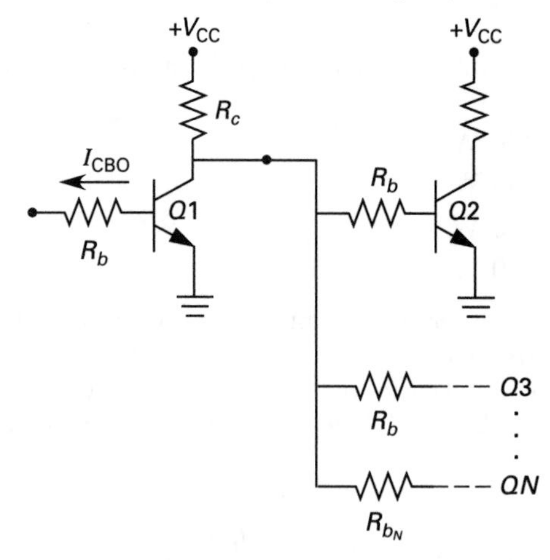

Parameters taken from a manufacturer's data sheet indicate that $V_{CC} = 5.0$ V, $I_{B,\text{sat}} > 0.5$ mA, $V_{BE,\text{sat}} < 1.5$ V, $R_c = 500\ \Omega$, and $R_b = 1000\ \Omega$. How much does a reverse-saturation current of 1 mA lower the fan-out of $Q1$?

(A) 2
(B) 4
(C) 10
(D) 12

11. For the transistor circuit shown, determine (a) the necessary base current for this gate to sink five identical gates connected to its output, and (b) the minimum size of R_b to allow this gate to source five identical gates.

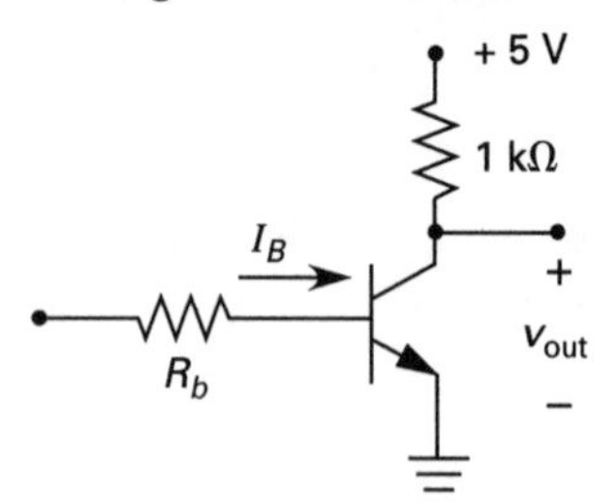

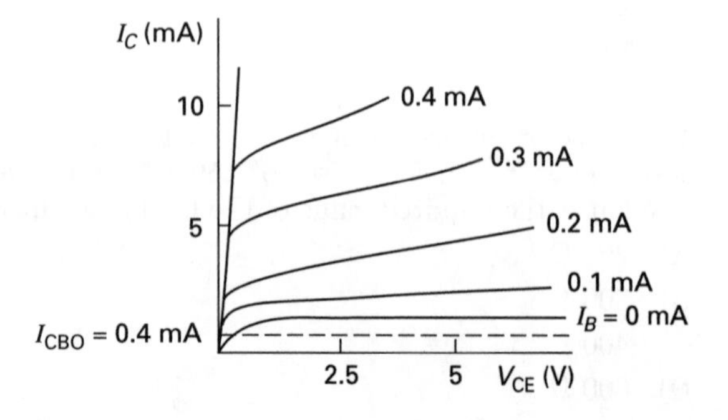

12. A particular logic gate has the following parameters: $t_d = 5$ ns, $t_r = 50$ ns, $t_s = 30$ ns, and $t_f = 30$ ns. Determine the maximum frequency at which this gate can operate.

13. Use an operational amplifier to convert a rectangular wave voltage of ±5 V and a period of 10 ms to a triangular wave with a peak output of ±5 V with the same period.

14. For the circuit shown, determine the transfer characteristic of the output voltage versus the input voltage. The diodes are ideal.

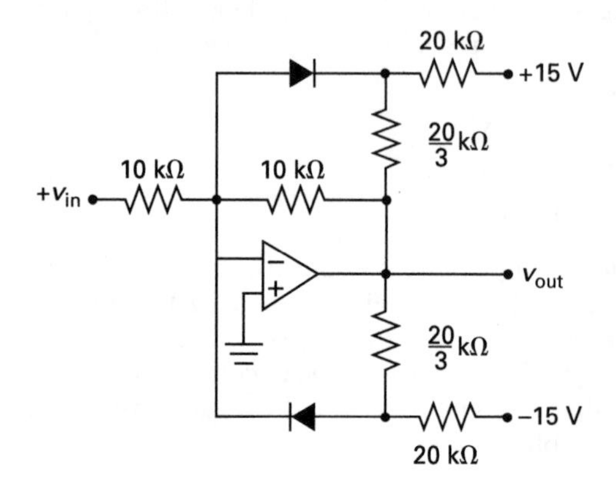

15. Determine the power requirement for the DTL circuit shown for a fan-out of five.

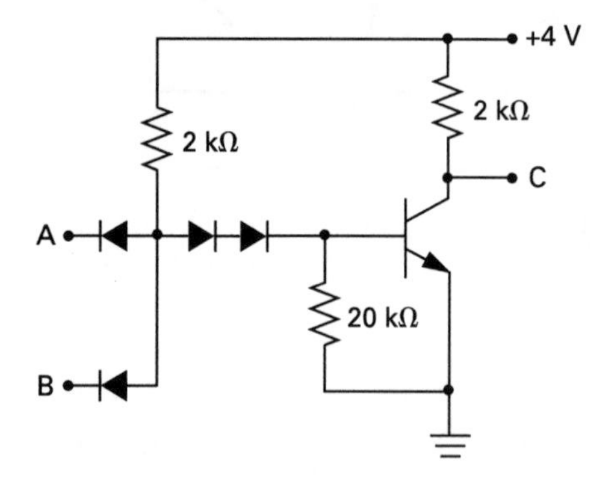

SOLUTIONS

1. The threshold voltage is reduced by an amount equal to the second term in Eq. 45.5, which is the transfer equation for a precision diode.

$$V_{\text{out}} = V_{\text{in}} - \frac{V_F}{A_v}$$

Substituting the given information gives

$$\frac{V_F}{A_v} = \frac{0.2 \text{ V}}{10^7} = 20 \times 10^{-9} \text{ V} = \boxed{20 \text{ nV}}$$

The answer is (A).

2. The supply resistance is given by Eq. 45.8.

$$\begin{aligned} R_s &= \frac{V_{\text{in,min}} - V_{\text{ZM}}}{I_{L,\text{max}}} \\ &= \frac{120 \text{ V} - 24 \text{ V}}{5 \text{ A}} = \boxed{19.2 \ \Omega} \end{aligned}$$

The answer is (A).

3. The maximum zener current is determined from Eq. 45.9.

$$\begin{aligned} I_{Z,\text{max}} &= \left(\frac{V_{\text{in,max}} - V_{\text{ZM}}}{V_{\text{in,min}} - V_{\text{ZM}}} \right) I_{L,\text{max}} \\ &= \left(\frac{126 \text{ V} - 24 \text{ V}}{120 \text{ V} - 24 \text{ V}} \right) (5 \text{ A}) \\ &= \boxed{5.3 \text{ A}} \end{aligned}$$

The answer is (D).

4. The diode power is given by Eq. 45.10.

$$\begin{aligned} P_D &= I_{Z,\text{max}} V_{\text{ZM}} \\ &= (5.3 \text{ A})(24 \text{ V}) \\ &= \boxed{127.2 \text{ W}} \end{aligned}$$

The answer is (B).

5. The minimum power rating of the supply resistor is given by Eq. 45.11.

$$\begin{aligned} P_{R_s} &= I^2_{Z,\text{max}} R_s \\ &= (5.3 \text{ A})^2 (19.2 \ \Omega) \\ &= \boxed{539.3 \text{ W}} \end{aligned}$$

The answer is (D).

6. At low frequencies $h_{\text{fe}} \approx h_{\text{FE}}$, so the small-signal parameter given can be used. The base current necessary for saturation is given by Eq. 45.25.

$$I_B \geq \frac{V_{\text{CC}}}{h_{\text{FE}} R_L}$$

For a load resistance of 400 Ω, the collector supply voltage is

$$\begin{aligned} V_{\text{CC}} &\leq I_{B_{t1}} h_{\text{FE}} R_L \\ &\leq (100 \times 10^{-6} \text{ A})(200)(400 \ \Omega) \\ &\leq \boxed{8 \text{ V}} \end{aligned}$$

The answer is (B).

7. If the base current is zero, the transistor is off.

The answer is (B).

8. The minimum base current, which will be 150×10^{-6} A to ensure that the transistor is not on at t_1, is given by Eq. 45.25.

$$I_B \geq \frac{V_{\text{CC}}}{h_{\text{FE}} R_L} \approx \frac{V_{\text{CC}}}{h_{\text{fe}} R_L}$$

Rearranging gives

$$\begin{aligned} R_L &\geq \frac{V_{\text{CC}}}{h_{\text{fe}} I_B} \\ &\geq \frac{12 \text{ V}}{(200)(150 \times 10^{-6} \text{ A})} \\ &\geq \boxed{400 \ \Omega} \end{aligned}$$

The answer is (C).

9. The actual saturation current is given by Eq. 45.24.

$$\begin{aligned} I_{C,\text{sat}} &= \frac{V_{\text{CC}} - V_{\text{CE,sat}}}{R_L} \\ &= \frac{12 \text{ V} - 0.2 \text{ V}}{400 \ \Omega} \\ &= \boxed{29.5 \times 10^{-3} \text{ A}} \end{aligned}$$

Note that if the collector-emitter voltage drop is ignored, the saturation current will be 30×10^{-3} A. Thus, except for exacting calculations, the ON transistor can be modeled as a short circuit.

The answer is (C).

10. A transistor inverter is a common-emitter configured transistor as shown.

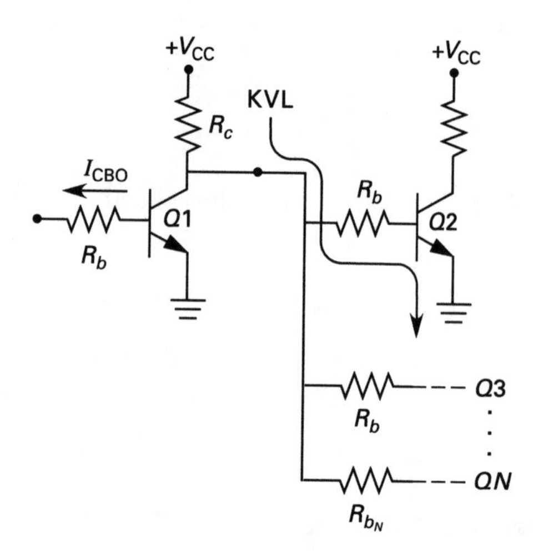

Initially ignoring the reverse saturation current, I_{CBO}, means $Q2$ can be modeled as an open circuit. Then, writing KVL in the loop shown gives

$$V_{\text{CC}} - NI_{B,\text{min}}R_c - I_{B,\text{min}}R_b - V_{\text{BE2}} = 0$$

The minimum base current is the current necessary to cause $Q2$ and other load transistors to operate in the saturation region. This is given as $I_{B,\text{sat}} > 0.5$ mA. The base-emitter voltage is the voltage of the transistor while saturated, given as $V_{\text{BE,sat}} < 1.5$ V. Substituting and solving for the fan-out gives

$$N = \frac{V_{\text{CC}} - V_{\text{BE,sat}} - I_{B,\text{min}}R_b}{I_{B,\text{min}}R_c}$$

$$= \frac{5\text{ V} - 1.5\text{ V} - (0.5 \times 10^{-3}\text{ A})(1000\ \Omega)}{(0.5 \times 10^{-3}\text{ A})(500\ \Omega)}$$

$$= 12$$

To account for I_{CBO}, add the additional voltage drop $I_{\text{CBO}}R_c$ to the original KVL equation, or use Eq. 45.33 directly.

$$V_{\text{CC}} - NI_{B,\text{min}}R_c - I_{\text{CBO}}R_c - I_{B,\text{min}}R_b - V_{\text{BE2}} = 0$$

$$N + \frac{R_b}{R_c} = \frac{V_{\text{CC}} - V_{\text{BE2}} - I_{\text{CBO}}R_c}{I_{B,\text{min}}R_c}$$

Substituting and solving gives

$$N = 10$$

Thus, the reverse saturation current lowers the fan-out by

$$12 - 10 = \boxed{2}$$

The answer is (A).

11. The load line with no loads has $I_C \approx 4.9$ mA, which requires I_B to be slightly more than 0.3 mA.

To sink five additional loads in the off condition, as $I_{\text{CBO}} = 0.4$ mA, I_C must increase by $(5)(0.4\text{ mA}) = 2$ mA to ~ 6.9 mA. This will require $I_B \approx 0.4$ mA.

(a)

$$I_{B,\text{low output}} = \boxed{0.4\text{ mA}}$$

(b) When this transistor is off, the five loads draw 0.4 mA each, while the bases have 0.7 V.

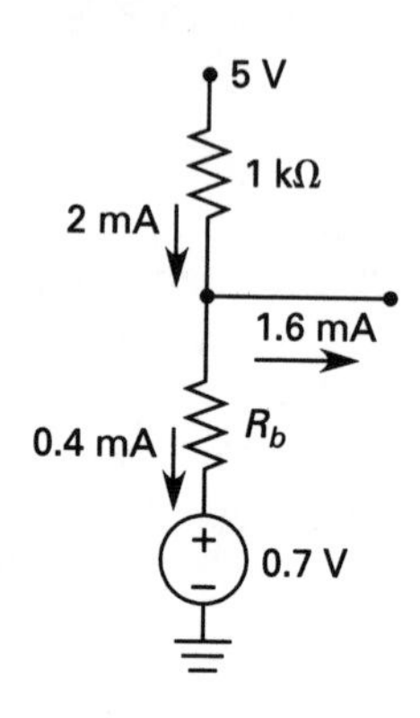

$$5\text{ V} = (2\text{ mA})(1\text{ k}\Omega) + (0.4\text{ mA})R_b + 0.7\text{ V}$$

$$R_b = \frac{5\text{ V} - 0.7\text{ V} - 2\text{ V}}{0.4\text{ mA}} = \boxed{5.75\text{ k}\Omega}$$

12.

$$\tau_{\text{min}} = \tau_d + \tau_r + \tau_s + \tau_f = 115\text{ ns}$$

$$= 5\text{ ns} + 50\text{ ns} + 30\text{ ns} + 30\text{ ns}$$

$$f_{\text{max}} = \frac{1}{\tau_{\text{min}}} = \frac{10^9\text{ s/ns}}{115\text{ ns}}$$

$$= \boxed{8.7\text{ MHz}}$$

13.

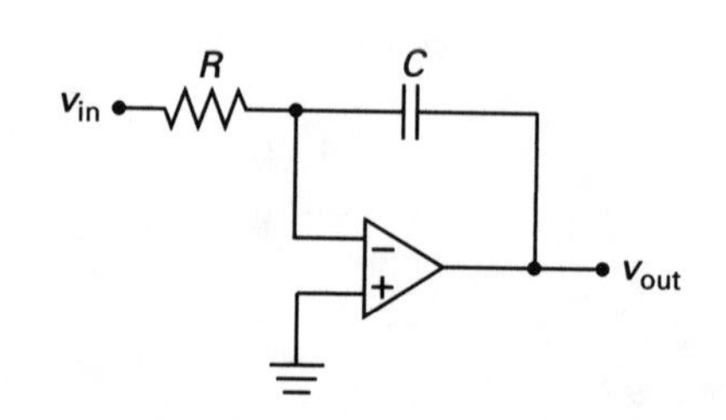

$$v_{\text{out}} = \frac{-1}{RC}\int_0^{0.005} 5dt + 5\text{ V} = -5\text{ V}$$

$$\frac{(0.005\text{ s})(5\text{ V})}{RC} = 10\text{ V}$$

$$RC = \boxed{2.5 \times 10^{-3}\text{ s}}$$

14. With both diodes open,

$$\frac{v_{\text{out}}}{v_{\text{in}}} = -1$$

With the upper diode conducting,

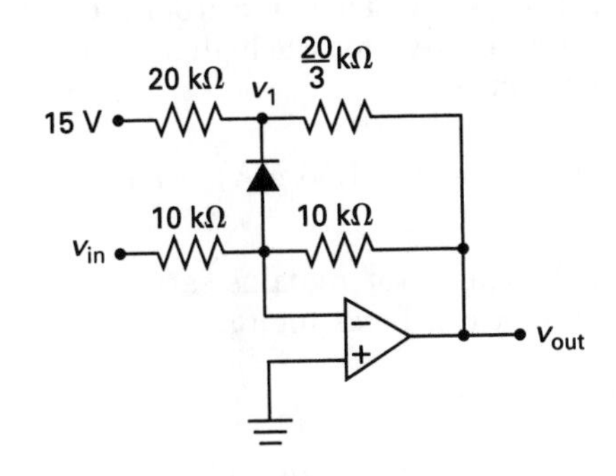

$$\frac{\left(\frac{20}{3}\text{ k}\Omega\right)(10\text{ k}\Omega)}{\left(\frac{20}{3}\text{ k}\Omega\right) + 10\text{ k}\Omega} = 4\text{ k}\Omega$$

$$\frac{15\text{ V}}{20\text{ k}\Omega} + \frac{v_{\text{in}}}{10\text{ k}\Omega} + \frac{v_{\text{out}}}{4\text{ k}\Omega} = 0$$

$$v_{\text{out}} = -0.4v_{\text{in}} - 3\text{ V}$$

For the lower diode conducting,

$$v_{\text{out}} = -0.4v_{\text{in}} + 3\text{ V}$$

For both diodes open,

$$v_1 = (15\text{ V})\left(\frac{20\text{ k}\Omega}{\frac{20}{3}\text{ k}\Omega + 20\text{ k}\Omega}\right) + v_{\text{out}}\left(\frac{20\text{ k}\Omega}{\frac{20}{3}\text{ k}\Omega + 20\text{ k}\Omega}\right)$$

$$= \frac{(15\text{ V})\left(\frac{20}{3}\text{ k}\Omega\right) + v_{\text{out}}(20\text{ k}\Omega)}{\frac{20}{3}\text{ k}\Omega + 20\text{ k}\Omega}$$

$$= \frac{15\text{ V} + 3v_{\text{out}}}{4} = \frac{15\text{ V} - 3v_{\text{in}}}{4}$$

Since the negative terminal is at virtual ground, for the upper diode to be off, $v_1 > 0$ V, so

$$15\text{ V} - 3v_{\text{in}} > 0$$

$$v_{\text{in}} < 5\text{ V}$$

For the lower diode,

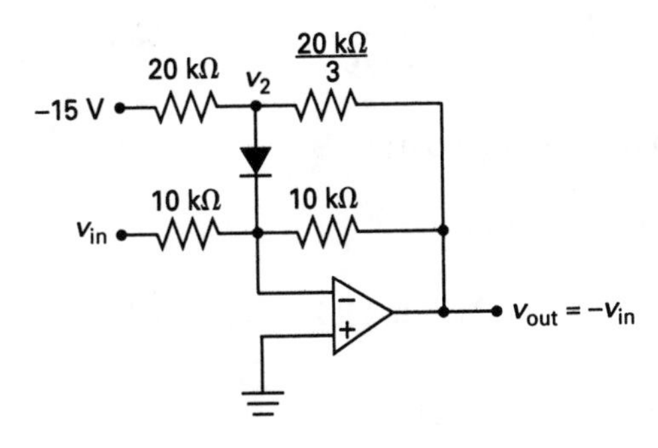

For both diodes open,

$$v_2 = (-15\text{ V})\left(\frac{\frac{20}{3}\text{ k}\Omega}{\frac{20}{3}\text{ k}\Omega + 20\text{ k}\Omega}\right) - v_{\text{in}}\left(\frac{\frac{20}{3}\text{ k}\Omega}{\frac{20}{3}\text{ k}\Omega + 20\text{ k}\Omega}\right)$$

$$= \frac{(-15\text{ V})\left(\frac{20}{3}\text{ k}\Omega\right) - v_{\text{in}}(20\text{ k}\Omega)}{\frac{20}{3}\text{ k}\Omega + 20\text{ k}\Omega}$$

$$= \frac{-15\text{ V} - 3v_{\text{in}}}{4} < 0$$

$$v_{\text{in}} > -5\text{ V}$$

For $-5\text{ V} < v_{\text{out}} < 5\text{ V}, v_{\text{out}} = -v_{\text{in}}$

$$v_{\text{in}} < -5\text{ V}, v_{\text{out}} = -0.4v_{\text{in}} - 3\text{ V}$$

$$v_{\text{in}} > 5\text{ V}, v_{\text{out}} = -0.4v_{\text{in}} + 3\text{ V}$$

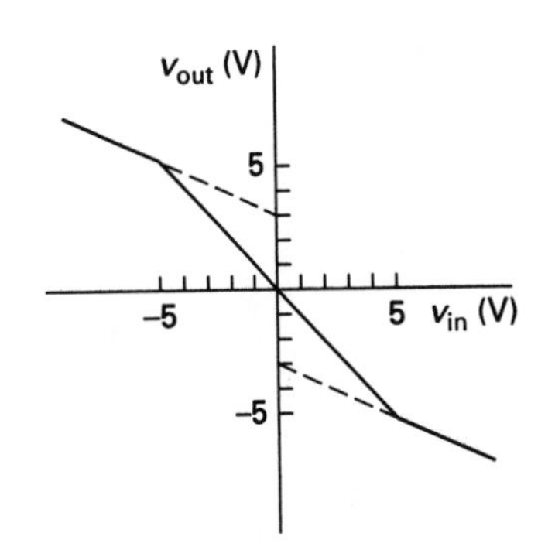

15. With at least one input low, the collector is high and no current flows in the collector circuit.

One or more low inputs of 0.2 V, when combined with a diode drop of 0.7 V, results in a current through the left resistor of

$$I = \frac{\Delta V}{R}$$

$$I = \frac{4\ \text{V} - 0.7\ \text{V} - 0.2\ \text{V}}{2\ \text{k}\Omega} = 1.55\ \text{mA}$$

The power for high output is

$$P = IV$$

$$= (1.55\ \text{mA})(4\ \text{V}) = 6.2\ \text{mW}$$

With both inputs high,

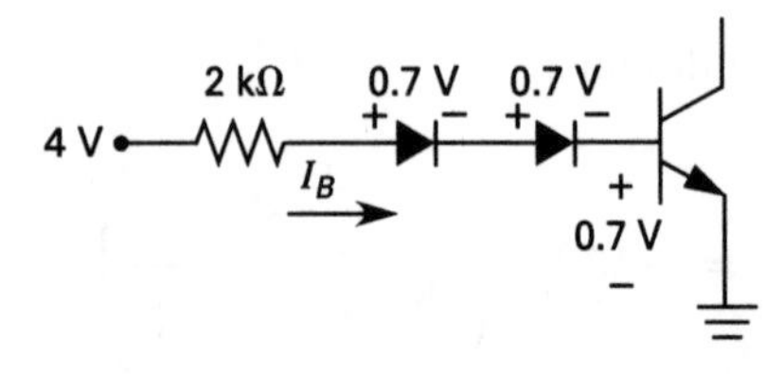

$$I = \frac{\Delta V}{R}$$

$$I_B = \frac{4\ \text{V} - (3)(0.7\ \text{V})}{2\ \text{k}\Omega} = 0.95\ \text{mA}$$

This saturates the transistor, so

$$I = \frac{\Delta V}{R}$$

$$I_C = \frac{4\ \text{V} - 0.2\ \text{V}}{2\ \text{k}\Omega} = 1.9\ \text{mA}$$

The power supply provides

$$P_{\text{supply}} = IV$$

$$(1.9\ \text{mA} + 0.95\ \text{mA})(4\ \text{V}) = 11.4\ \text{mW}$$

However, the transistor must absorb power from the loads, which must be calculated from a current of five times the input current for the high output. This excess collector current is

$$I_c = 5I = (5)(1.55\ \text{mA}) = 7.75\ \text{mA}$$

Assuming the transistor remains saturated, the excess transistor power can be as high as

$$P_{\text{excess}} = IV$$

$$= (7.75\ \text{mA})(0.2\ \text{V}) = 1.55\ \text{mW}$$

The maximum total power is

$$P_{\text{total}} = P_{\text{supply}} + P_{\text{excess}}$$

$$= 1.55\ \text{mW} + 11.4\ \text{mW} = \boxed{12.95\ \text{mW}}$$

46 Computer Hardware Fundamentals

PRACTICE PROBLEMS

1. A disk has 1000 tracks with 10,000 bits per track. The disk diameter is 15.24 cm. The average seek time is 20 ms, and the rotational speed is 2000 rpm. (a) What is the areal density of the disk? (b) Estimate the average access time.

2. What length of nine-track tape with density 1600 Bpi is required to store 1 MB of data?

3. The entire contents of a full 20 MB disk are sent in an asynchronous transmission at a 2400 baud rate. What is the time required for transmission?

SOLUTIONS

1. (a) areal density

$$\begin{aligned}&= (\text{no. bits per track})\\&\quad\times(\text{no. tracks per radial centimeter})\\&= \left(10{,}000\ \frac{\text{bits}}{\text{track}}\right)\left(\frac{1000\ \text{tracks}}{\frac{D}{2}}\right)\\&= \left(10{,}000\ \frac{\text{bits}}{\text{track}}\right)\left(\frac{2000\ \text{tracks}}{15.24\ \text{cm}}\right)\\&= \boxed{1.31\times10^6\ \text{bits/radial cm}}\end{aligned}$$

(b) access time = average latency + average seek time

The average latency can be approximated as one-half the time for the full revolution.

$$\left(\frac{1}{2}\ \text{rev}\right)\left(\frac{1\ \text{min}}{2000\ \text{rev}}\right)\left(60\ \frac{\text{sec}}{\text{min}}\right)\left(1000\ \frac{\text{ms}}{\text{sec}}\right) = 15\ \text{ms}$$

$$\begin{aligned}\text{average access time} &= 15\ \text{ms} + 20\ \text{ms}\\&= \boxed{35\ \text{ms}}\end{aligned}$$

2. Since eight tracks are used for data recording, 1 byte of data is stored across the width of the tape. Therefore, 1600 bytes of data are stored per 2.54 cm of length. The length required for 1 MB is

$$\frac{1{,}048{,}576\ \text{bytes}}{\left(\frac{1600\ \text{bytes}}{2.54\ \text{cm}}\right)\left(100\frac{\text{cm}}{\text{m}}\right)} = \boxed{16.6\ \text{m}}$$

This neglects any interblock gaps.

3. The time required is

$$\begin{aligned}t &= \frac{(20\ \text{MB})\left(1{,}048{,}576\ \frac{\text{bytes}}{\text{MB}}\right)\left(10\ \frac{\text{bits}}{\text{byte}}\right)}{\left(2400\ \frac{\text{bits}}{\text{sec}}\right)\left(60\ \frac{\text{sec}}{\text{min}}\right)\left(60\ \frac{\text{min}}{\text{hr}}\right)}\\&= \boxed{24.3\ \text{hr}}\end{aligned}$$

47 Computer Software Fundamentals

PRACTICE PROBLEMS

1. Flowchart the following procedure: If A is greater than B and A is greater than C, then add B to C and go to PLACE1. If B is greater than A and B is greater than C, then add A to C and go to PLACE2. If B is greater than A but less than C, then subtract A from B and go to PLACE1.

2. Flowchart the following procedure: If A is greater than 10 and if A is less than 14, subtract X from A. If A is less than 10 or if A is greater than 14, exit the program.

3. Draw a flowchart representing an algorithm that finds the real roots of a second-degree equation with real coefficients. The coefficients are to be read from an external peripheral, and the results are to be printed.

4. Draw a flowchart representing the bubble sort algorithm. (Sort into ascending order.)

5. For the binary search procedure, derive the number of required comparisons in a list of n elements.

SOLUTIONS

1.

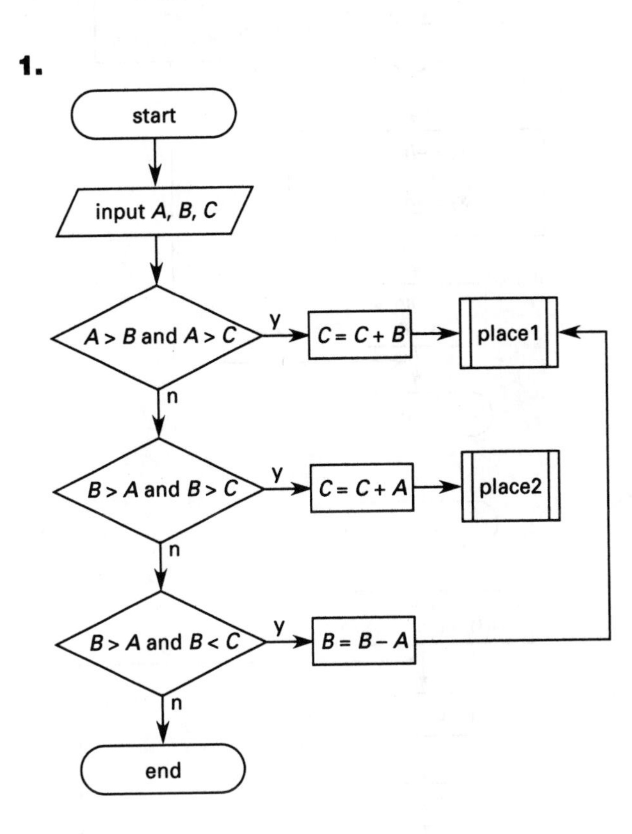

2.

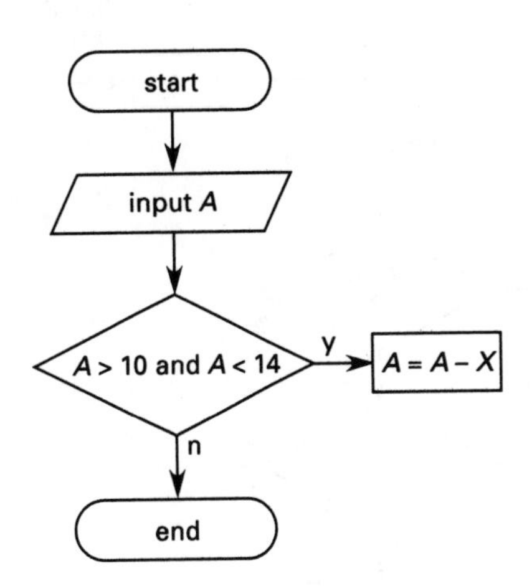

3.

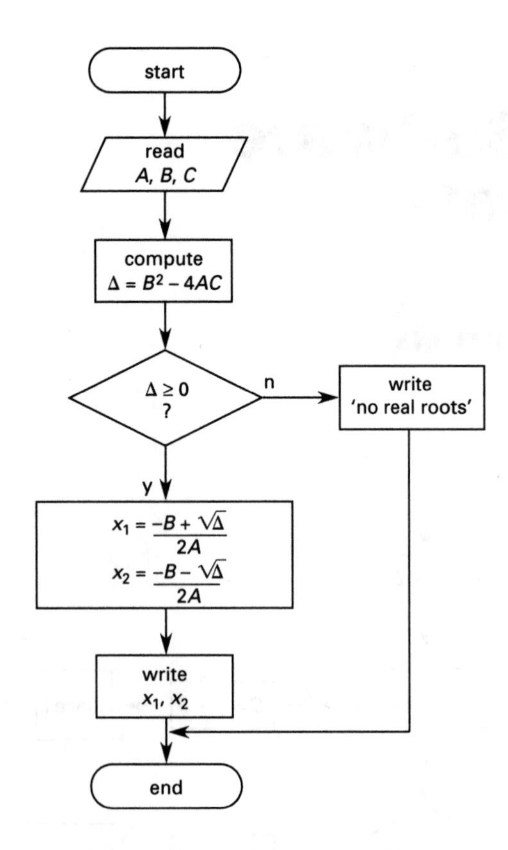

4.

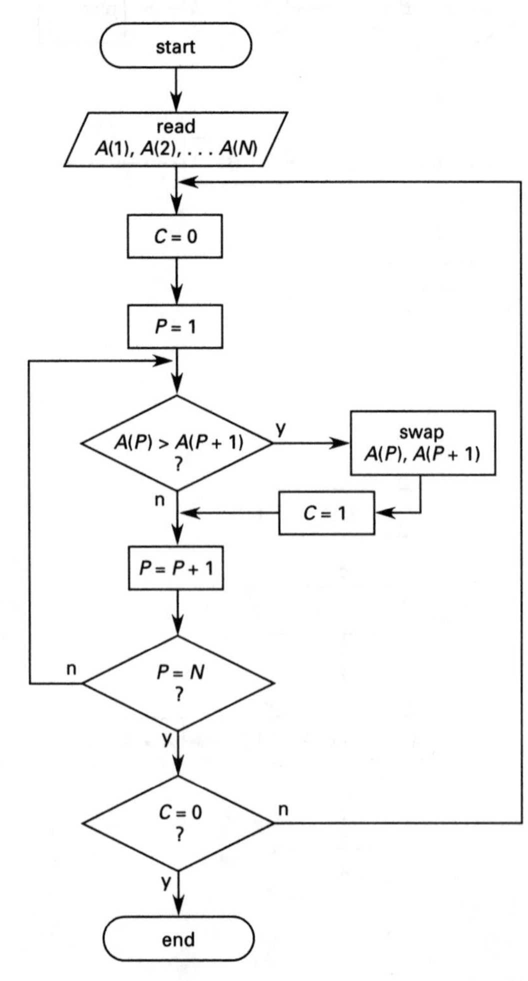

5. To simplify, assume that the number of elements in the list is some integer power of two.

$$n = 2^p$$

Assume that the sought-for element is the first element in the list. This will produce a worst-case search. Let $a(n)$ be the number of comparisons needed to find the element in the list. Since the binary search procedure discards (disregards) half of the list after each comparison,

$$a(n) = 1 + a\left(\frac{n}{2}\right) \quad \text{[Eq. I]}$$

But, $n = 2^p$.

$$a(n) = a2^p = 1 + a\left(\frac{2^p}{2}\right) = 1 + a2^{p-1} \quad \text{[Eq. II]}$$

Equation I can be applied to the second term of Eq. II.

$$a2^{p-1} = 1 + a\left(\frac{2^{p-1}}{2}\right) = 1 + a2^{p-2} \quad \text{[Eq. III]}$$

Combining Eqs. II and III,

$$a(n) = a2^p = 1 + 1 + a2^{p-2}$$

Equation I can continue to be applied to the last term until the list contains only two elements, after which the search becomes trivial. This results in a string of p ones.

$$a(n) = 1 + 1 + 1 + \cdots + 1 = p$$

An argument for $a(n) = p+1$ can be made depending on how the search logic handles a miss with two elements in the list. If it merely selects the remaining element without comparing it, then $a(n) = p$. If a singular element is investigated, then $a(n) = p + 1$.

Now,

$$n = 2^p$$

$$\log_2 n = \log_2 2^p = p$$

But $p = a(n) = \log_2 n$.

Using logarithm identities,

$$\boxed{p = \log_2 n = \frac{\log_{10} n}{\log_{10}(2)}}$$

48 Computer Architecture

PRACTICE PROBLEMS

1. Which of the following is not a type of information?

(A) character
(B) data
(C) logical
(D) numerical

2. What type of instruction set is used for a read/write (R/W) command?

(A) arithmetic
(B) logic
(C) memory
(D) control

3. What classification is used to describe a computer that handles instruction sets using a single high-level language instruction?

(A) CISC
(B) DISC
(C) PC
(D) RISC

4. A control line uses four bits of information. What is the maximum number of operations possible?

(A) 4
(B) 8
(C) 15
(D) 16

5. A register is accessed using double-word granularity. How many bits of information are being accessed at one time?

(A) 4
(B) 8
(C) 16
(D) 32

SOLUTIONS

1. "Data" is a general term for the quantities that input to a computer and is not a type of information.

The answer is (B).

2. A read/write (R/W) command involves the control and transfer of information.

The answer is (D).

3. Use of a single high-level language instruction is meant to simplify the programming process for the designer but involves more complex computer-level instructions. Thus, a computer that uses a single high-level language instruction is classified is as a complex instruction-set computer (CISC).

The answer is (A).

4. A 4-bit line can contain $2^4 = 16$ different signals.

The answer is (D).

5. Two bytes (16 bits) are considered to be a single word. Double-word granularity, then, is 32 bits.

The answer is (D).

49 Digital Logic

PRACTICE PROBLEMS

1. What is the state of a TTL cell at 4.0 V?

(A) 0
(B) indeterminate
(C) 1
(D) saturated

2. A computer accesses a register in 4-bit packets, sometimes called a *nibble*. What is the total number of values that can be represented in the nibble?

(A) 4
(B) 8
(C) 15
(D) 16

3. A computer data sheet indicates a 64K-byte RAM. How many bytes of memory are available?

(A) 8
(B) 1024
(C) 64,000
(D) 65,536

4. How many bits of information are contained within a 64K-byte RAM?

(A) 64
(B) 8192
(C) 512,000
(D) 524,288

5. If $x_1 = 0$, $x_2 = 1$, and $x_3 = 0$, what is the value of the following expression?

$$f(x_1, x_2, x_3) = \overline{x}_1 x_2 x_3 + x_1 \overline{x}_2 x_3 \cdot x_1 x_2 \overline{x}_3$$

6. A four-variable function has a minterm in row five. That minterm is given by which of the following expressions?

(A) $\overline{A}B\overline{C}$
(B) $A\overline{B}C$
(C) $\overline{A}B\overline{C}D$
(D) $A\overline{B}C\overline{D}$

The following truth table is applicable to Probs. 7 and 8.

A	B	C	$F(A,B,C)$
0	0	0	0
0	0	1	0
0	1	0	0
0	1	1	1
1	0	0	0
1	0	1	0
1	1	0	0
1	1	1	1

7. What is the canonical sum-of-products (SOP) expression for the function in the truth table shown?

(A) $\overline{A}\,\overline{B}\,\overline{C} + \overline{A}\,\overline{B}C + \overline{A}B\overline{C}$
(B) $A\overline{B}\,\overline{C} + A\overline{B}\,\overline{C} + AB\overline{C}$
(C) (A) and (B) combined
(D) $\overline{A}BC + ABC$

8. What maxterms are necessary to create the canonical product-of-sums (POS) expression for the function in the truth table shown?

(A) $M_0M_1M_2$
(B) $M_4M_5M_6$
(C) (A) and (B) combined
(D) M_3M_7

The following truth table is applicable to Probs. 9 and 10.

A	B	$F(A,B)$
−5 V	−5 V	−5 V
−5 V	0	0
0	−5 V	0
0	0	−5 V

9. If positive logic is applied to the truth table, what logic operation is obtained?

(A) coincidence
(B) NOR
(C) XOR
(D) XNOR

10. If negative logic is applied to the truth table, what logic operation is obtained?

(A) coincidence
(B) NOR
(C) OR
(D) XOR

SOLUTIONS

1. The state of a cell is its binary value. For TTL, any voltage between 2.0 and 5.0 V is state 1.

The answer is (C).

2. The total number of values in any n-tuple is given by Eq. 49.1.

$$N = 2^n = 2^4 = \boxed{16}$$

The answer is (D).

3. The K indicates $2^{10} = 1024$. Thus,

$$64\text{K} = (64)(2^{10}) = \boxed{65{,}536 \text{ bytes}}$$

The answer is (D).

4. One byte is 8 bits. Thus,

$$(65{,}536 \text{ bytes})\left(\frac{8 \text{ bits}}{1 \text{ byte}}\right) = \boxed{524{,}288 \text{ bits}}$$

The answer is (D).

5. The function is

$$f(x_1, x_2, x_3) = \overline{x}_1 x_2 x_3 + x_1 \overline{x}_2 x_3 \cdot x_1 x_2 \overline{x}_3$$

Substituting the given values and using the order of precedence rules gives

$$\begin{aligned} f(0,1,0) &= (\overline{0} \cdot 1 \cdot 0) + (0 \cdot \overline{1} \cdot 0) \cdot (0 \cdot 1 \cdot \overline{0}) \\ &= (1 \cdot 1 \cdot 0) + (0 \cdot 0 \cdot 0) \cdot (0 \cdot 1 \cdot 1) \\ &= 0 + 0 \cdot 0 \\ &= 0 + 0 \\ &= \boxed{0} \end{aligned}$$

6. A function with four variables, that is, with $n = 4$, with a value of one in row five, has a minterm in that row. Changing row five from decimal to binary gives

$$5_{10} = 0101_2$$

The minterm is deduced from its binary representation as

$$\boxed{\overline{A}B\overline{C}D = m_5}$$

The answer is (C).

7. The minterms are in rows three and seven—that is, where the function equals one. The canonical sum-of-product form is

$$F(A,B,C) = m_3 + m_7 = \boxed{\overline{A}BC + ABC}$$

The answer is (D).

8. The maxterms are in the rows where the function has a value of zero. Thus,

$$F(A,B,C) = \boxed{\Pi M(0,1,2,4,5,6)}$$

The answer is (C).

9. If positive logic is applied, the high voltage has a value of one. Rewriting the truth table with −5 V as logic 0 and 0 V as logic 1 gives

A	B	$F(A,B)$
0	0	0
0	1	1
1	0	1
1	1	0

This is the exclusive OR (XOR) operation.

The answer is (C).

10. If negative logic is applied, the low voltage has a value of one. Rewriting the truth table with −5 V as logic 1 and 0 V as logic 0 gives

A	B	$F(A,B)$
1	1	1
1	0	0
0	1	0
0	0	1

Reversing the order to place the table in standard format gives

A	B	$F(A,B)$
0	0	1
0	1	0
1	0	0
1	1	1

This is the XNOR, or coincidence, logic.

The answer is (A).

50 Logic Network Design

PRACTICE PROBLEMS

1. A 3-tuple logic expression has maxterms M_1, M_2, and M_3. Which of these maxterms are adjacent?

(A) M_1 and M_2
(B) M_1 and M_3
(C) M_2 and M_3
(D) (B) and (C)

2. What is the POS expression for the truth table shown?

F: C \ AB	00	01	11	10
0		1	1	1
1				

(A) $\overline{C}\cdot B+\overline{C}\cdot A$
(B) $(\overline{C}+B)\cdot(\overline{C}+A)$
(C) $(A+B)\cdot(A+\overline{C})\cdot(\overline{A}+\overline{C})$
(D) $(A+B)\cdot(\overline{C})$

3. What type of logic directly realizes the maxterm formulation of the truth table in Prob. 2?

(A) AND
(B) AND-OR
(C) NAND-NAND
(D) NOR-NOR

4. What is the minimal expression that can be obtained using the given Karnaugh map only?

F: CD \ AB	00	01	11	10
00		1		1
01	1	1	1	1
11		1	1	
10	1			

(A) $ABCD+A\overline{B}C+\overline{A}BC+C\overline{D}+\overline{B}\,\overline{D}$
(B) $\overline{A}\,\overline{B}\,\overline{C}\,\overline{D}+A\overline{B}C+\overline{A}BC+C\overline{D}+\overline{B}\,\overline{D}$
(C) $\overline{A}\,\overline{B}C\overline{D}+\overline{A}B\overline{C}+A\overline{B}\,\overline{C}+\overline{C}D+BD$
(D) none of the above

5. The design of a coin-changer results in the following truth table. What is the minimal logic expression that results in the truth table output?

F: C \ AB	00	01	11	10
0	1	1	1	1
1		d	d	1

(A) $A+\overline{C}$
(B) $\overline{A}+\overline{C}$
(C) $B+\overline{C}$
(D) $\overline{B}+C$

6. What is the NOR only expression for

$$F(A,B,C)=\prod M(0,1,2,5)$$

(A) $\overline{\overline{(A+C)}+\overline{(B+\overline{C})}}$
(B) $\overline{(A+B)(A+C)}\;\overline{(\overline{A}+B+C)}$
(C) $\overline{\overline{(A+B)}+\overline{(A+C)}+\overline{(\overline{A}+B+C)}}$
(D) $\overline{\overline{(A+B)}+\overline{(A+C)}+\overline{(\overline{A}+B+\overline{C})}}$

7. For the truth table given, obtain the Karnaugh map and Veitch diagram for the outputs D, E, and F. An X entry in the truth table indicates a don't care, so zero or one can be used for maximum convenience.

inputs			outputs		
A	B	C	D	E	F
0	0	0	0	1	1
0	0	1	0	X	1
0	1	0	X	1	0
0	1	1	0	0	0
1	0	0	1	1	X
1	0	1	1	1	X
1	1	0	0	X	X
1	1	1	X	1	X

8. For the outputs D, E, and F in Prob. 7, obtain the minimum gate realization between the SOP (NAND-NAND) and the POS (NOR-NOR) implementations. (Hint: Determine which canonical form will result in the minimum number of gates after minimization.)

9. A microprocessor has eight address lines—$A7$, $A6$, $A5$, $A4$, $A3$, $A2$, $A1$, and $A0$—and is required to address up to 16 input/output (I/O) ports. A decoder (4:16 with two active low enables), as shown, is to be used to provide the 16 active low outputs to select the input/output ports based on the least significant address bits $A3$, $A2$, $A1$, $A0$ when the address bits $A7$, $A6$, $A5$, $A4$ = HHHH = 1111. Using a minimum number of two-input NAND gates, design the logic to accomplish this.

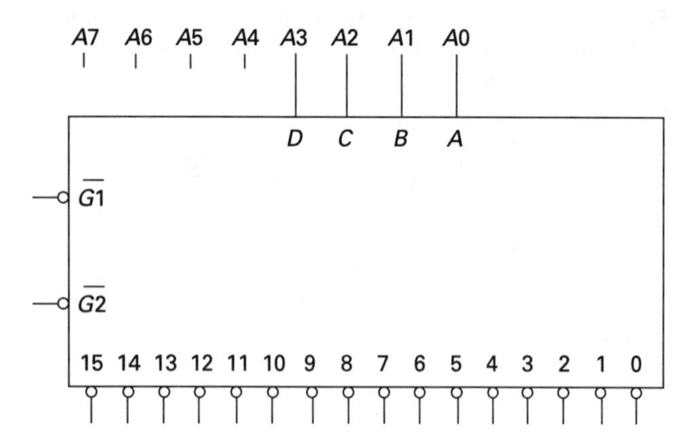

10. An input/output port selector is to use a 4:16 decoder (with two active low enables) with the most significant bits of an address bus as shown.

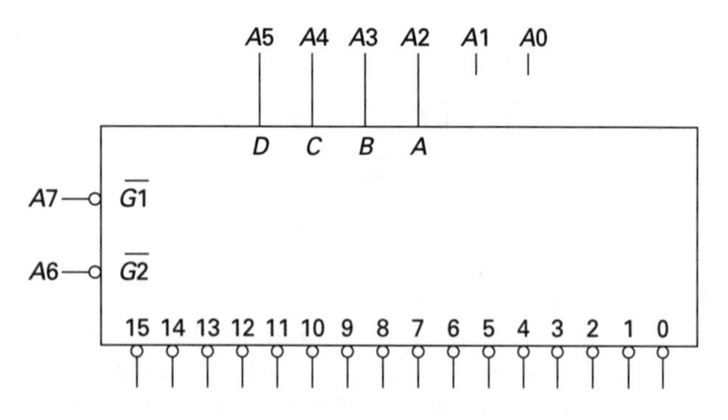

In addition, the selection of a port is to take place in coincidence with a timing strobe (asserted low) $\overline{X}$. Using only a quad two-input NAND gate package and a package with six NOT gates, design the circuit to select the address 00101101. Indicate which output from the 4:16 decoder ($\overline{Y}$) is to be used. Inputs to the logic are to be the correct output from the decoder, address lines $A1$ and $A0$, and the strobe $\overline{X}$.

SOLUTIONS

1. Maxterms M_1, M_2, and M_3 in binary form are given by

$$M_1 = 001$$
$$M_2 = 010$$
$$M_3 = 011$$

Adjacency exists when the terms differ by one value only. Thus, M_1 is adjacent to M_3, but not to M_2. M_2 is adjacent to M_3, but not to M_1. (This is the reason for the order of the binary numbers on the Karnaugh map, that is, 00, 01, 11, 10.)

The answer is (D).

2. Maxterms have an output value of zero. Therefore, the truth table is

(F) C \ AB	00	01	11	10
0	0			
1	0	0	0	0

The procedure for finding the minimal Karnaugh map results in the groupings shown. The maxterms M_0 and M_1 combine to give $A+B$. The bottom row combination of four variables eliminates two, specifically A and B, leaving $\overline{C}$. The resulting POS expression is

$$F(A,B,C) = \boxed{(A+B)\cdot\overline{C}}$$

The answer is (D).

3. Maxterm POS realization occurs with NOR-NOR logic as shown.

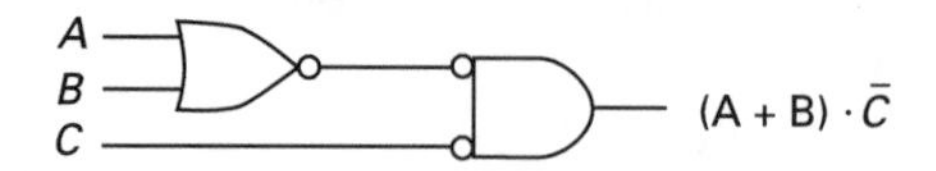

The answer is (D).

4. The minimal Karnaugh map is found using the following procedure.

step 1: Form all unique one-squares.

(F) CD \ AB	00	01	11	10
00		1		1
01	1	1	1	1
11		1	1	
10	1			

step 2: Form unique two-squares.

F: CD \ AB	00	01	11	10
00		1		1
01	1	1	1	1
11		1	1	
10	1			

step 3: Form unique four-squares.

F: CD \ AB	00	01	11	10
00		1		1
01	1	1	1	1
11		1	1	
10	1			

step 4: Form unique eight-squares.

step 5: Assign ungrouped one-squares into the largest group possible using d squares, if available. The resulting expression, taken from step 3 using the one-, two-, and four-minterm groupings is

$$F(A,B,C,D) = \boxed{\overline{A}\,\overline{B}C\overline{D} + \overline{A}B\overline{C} + A\overline{B}\,\overline{C} + \overline{C}D + BD}$$

The answer is (C).

5.

F: C \ AB	00	01	11	10
0	1	1	1	1
1		d	d	1

Using the procedure for minimization of a Karnaugh map, the two groups shown are formed. Grouping the top row of minterms eliminates A and B. Grouping columns 11 and 10 eliminates B and C. The don't care condition, d, in position 111 is given the arbitrary value of one. The resulting expression is

$$F(A,B,C) = \boxed{\overline{C} + A}$$

The answer is (A).

6. The K-V map with grouping is

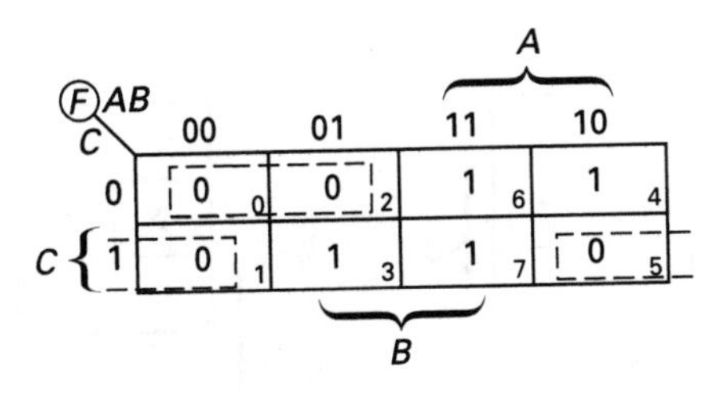

M_0 and M_2 combine to give $A+C$. The combination of corner maxterms M_1 and M_5 gives $B+\overline{C}$. The resulting expression is

$$F(A,B,C) = (A+C)(B+\overline{C})$$

Complementing the result gives

$$\overline{F(A,B,C)} = \overline{(A+C)(B+\overline{C})}$$

Applying De Morgan's law results in

$$\overline{F(A,B,C)} = \overline{(A+C)} + \overline{(B+\overline{C})}$$

Complementing the result puts the expression in NOR only form.

$$F(A,B,C) = \boxed{\overline{\overline{\overline{(A+C)} + \overline{(B+\overline{C})}}}}$$

The answer is (A).

7. The Karnaugh maps are as follows.

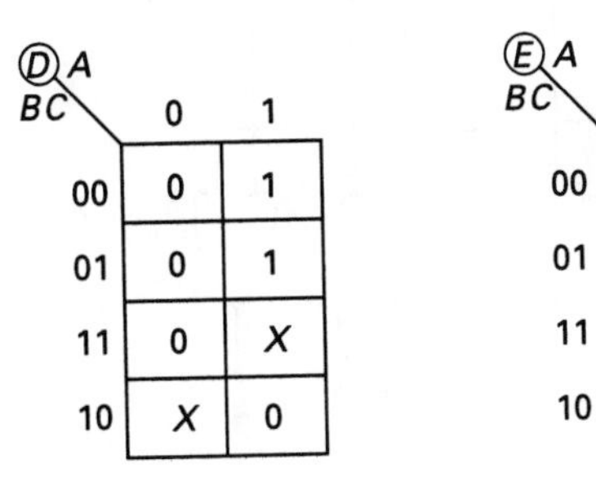

E: BC \ A	0	1
00	1	1
01	X	1
11	0	1
10	1	X

F: BC \ A	0	1
00	1	X
01	1	X
11	0	X
10	0	X

The Veitch diagrams are as follows.

(D)

	A
0	1
0	1
0	X
X	0

(rows 2–3: C; rows 3–4: B)

(E)

	A
1	1
X	1
0	1
1	X

(rows 2–3: C; rows 3–4: B)

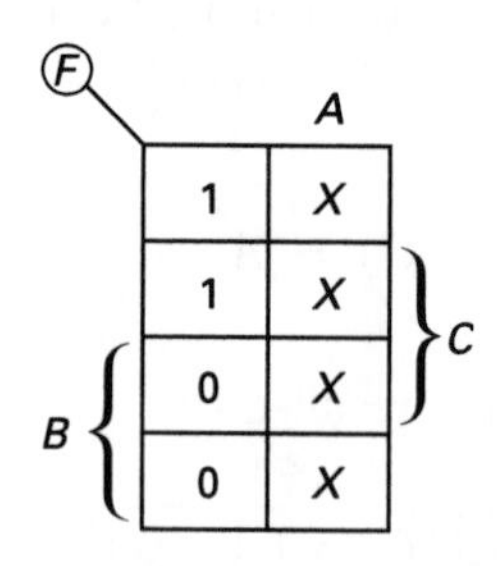

8. After minimization, the output is

$$D = A \cdot \overline{B}$$

$$\overline{E} = \overline{A} \cdot C$$

$$E = A + \overline{C}$$

$$F = \overline{B}$$

The circuit is realized as

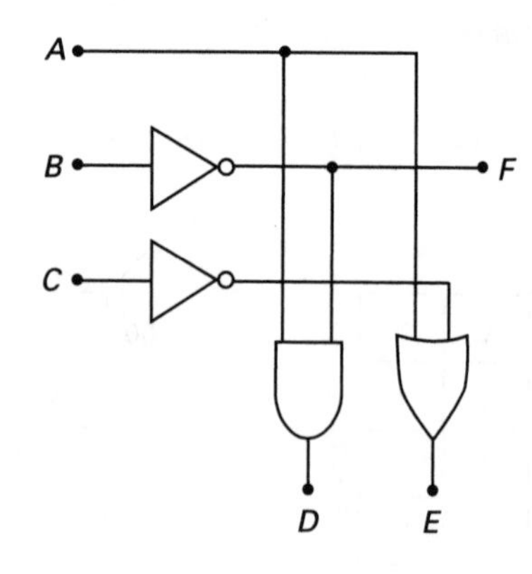

9. The outputs of the decoder are all high, except for the case when $\overline{G}1$ and $\overline{G}2$ are both low. In that case, the output with the binary address corresponding to $A3$, $A2$, $A1$, and $A0$ will go low, thus selecting a particular device.

The objective is to provide logic, using a minimum number of two-input NAND gates, so that the decoder will be active when lines $A4$, $A5$, $A6$, and $A7$ are all high.

This can be done with two NAND gates as shown.

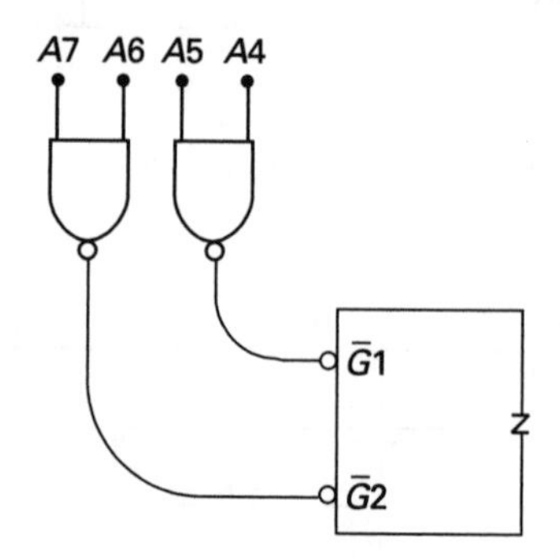

10. The address used is

$$A_7A_6A_5A_4A_3A_2A_1A_0 = 00101101$$

$$A_5A_4A_3A_2 = 1011_2 = 11_{10}$$

The output selected is Y_{11}. Y_{11} is asserted low. The desired logic function is

$$F = \overline{X}\,\overline{Y}_{11}\overline{A}_1A_0$$

The realization with two input gates follows.

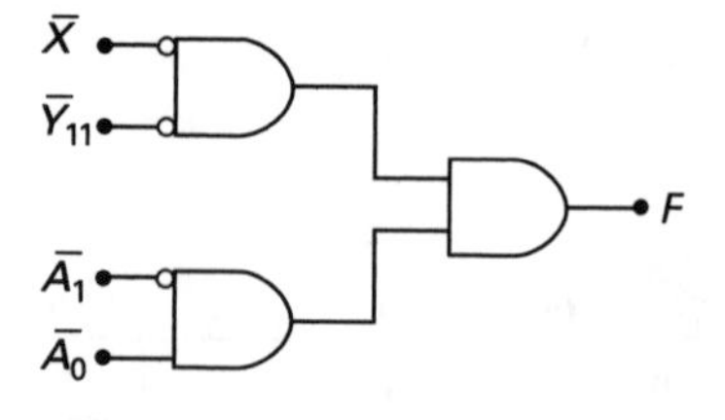

51 Synchronous Sequential Networks

PRACTICE PROBLEMS

1. A register is

(A) a flip-flop
(B) a set of flip-flops
(C) a memory device
(D) (B) and (C)

2. What combination of values is not allowed as inputs to an S-R flip-flop?

(A) $S = 0$, $R = 0$
(B) $S = 0$, $R = 1$
(C) $S = 1$, $R = 0$
(D) $S = 1$, $R = 1$

3. In the S-R flip-flop timing diagram shown, what is the output sequence from point 1 to point 3?

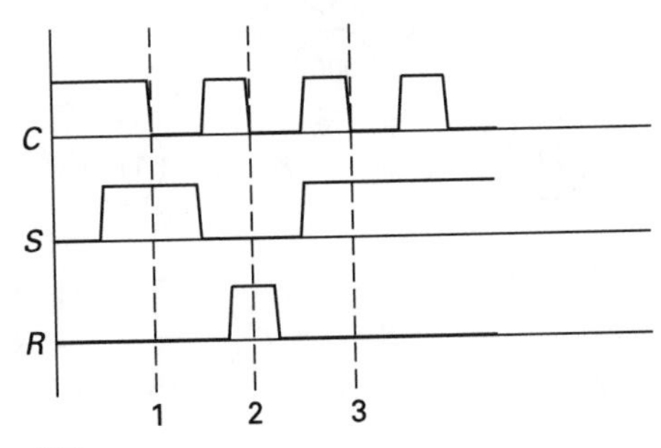

(A) 000
(B) 010
(C) 101
(D) 111

4. What best describes the overall function of the network shown in the figure?

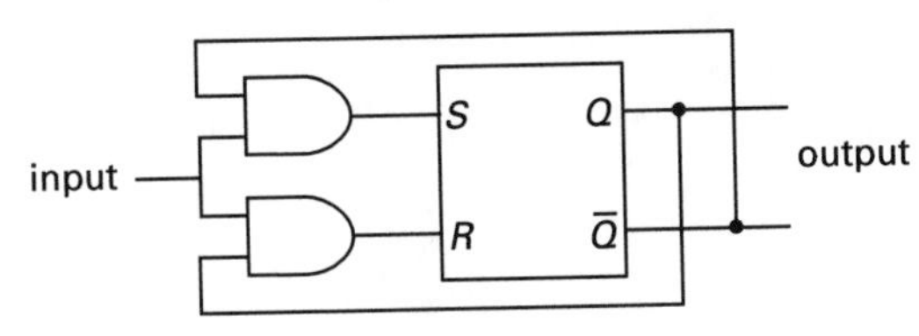

(A) delay
(B) set-reset/single input
(C) toggle
(D) none of the above

5. Of the four principal flip-flops, which can be used as the building block for all the others?

(A) D
(B) J-K
(C) S-R
(D) T

6. Develop the state transition diagram for the synchronous sequential circuit shown. Show all work.

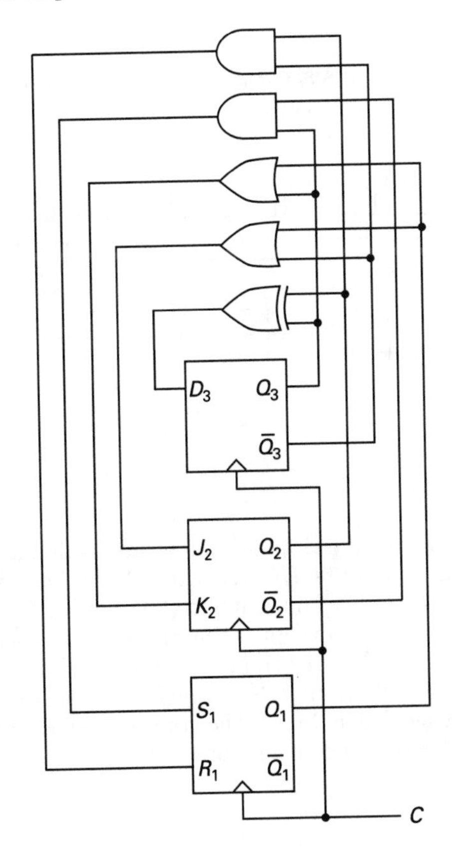

Computers

7. For the counter circuit shown, develop the state table and the state transition diagram.

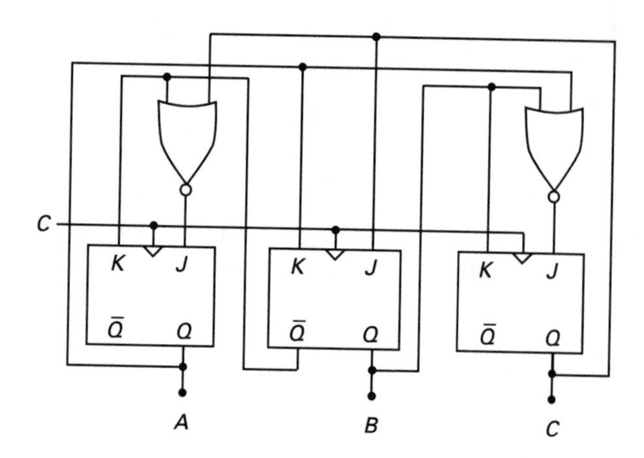

8. Using T flip-flops, design a counter that will produce the following sequence.

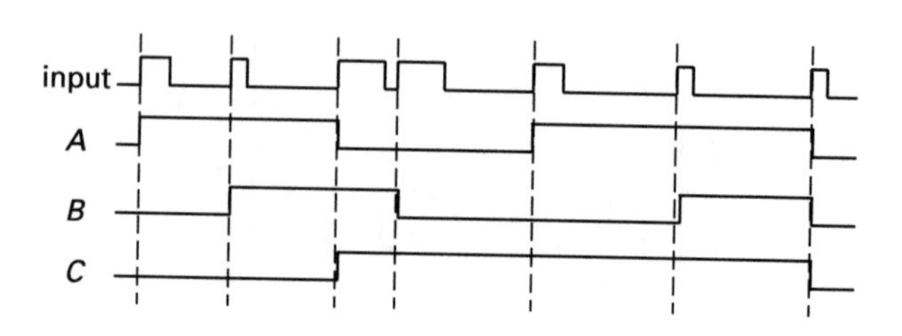

9. Four-bit binary ripple counters are available. Each has parallel outputs $DCBA$, with A the least significant bit (LSB), a count input (CI), and a carry out (CO) that is low except when the counter output is $DCBA$ = HHHH. The ripple counters also have a level-sensitive preset input (PR), which will set $DCBA$ = HHHH when PR is high. The counters respond to the falling edge of the count input.

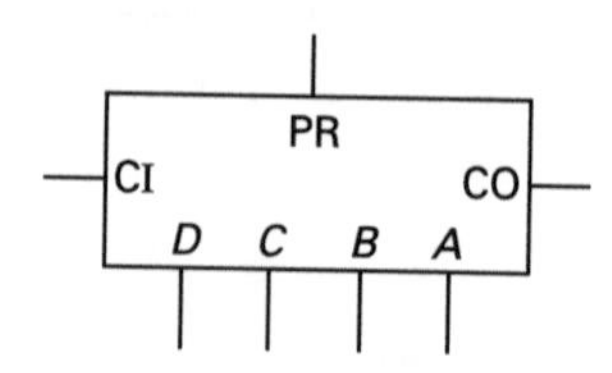

Using two such counters and necessary logic gates, design a circuit that will provide a one (high) after each 100 counts.

10. Given the digital circuit shown, complete the state diagram. Show all transitions and the method of analysis.

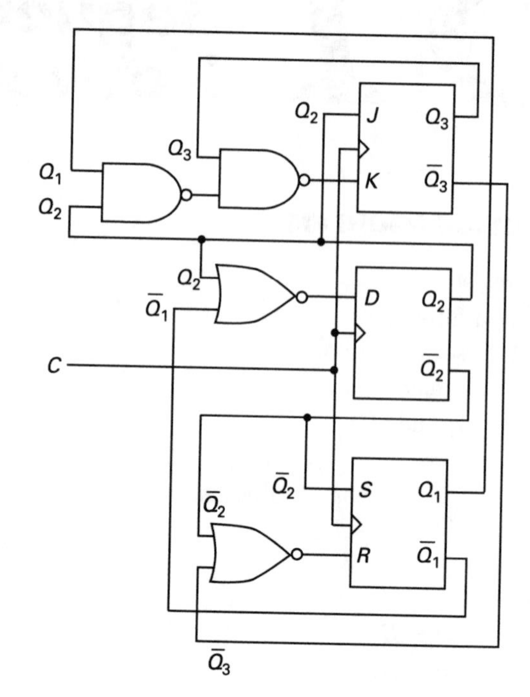

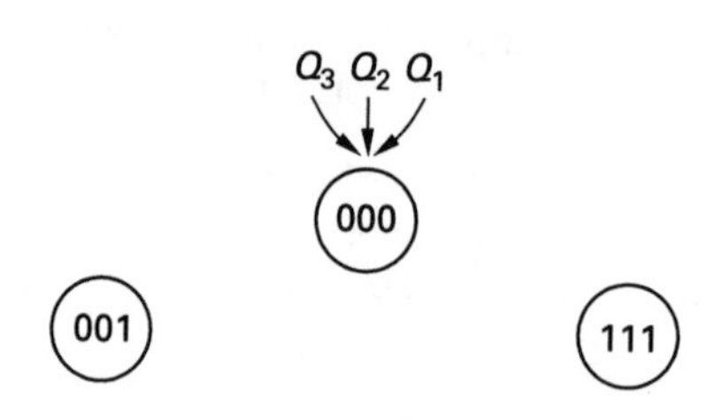

11. A counter is constructed from *J-K* master-slave flip-flops and NAND gates. Show one complete cycle of the counter. Assume $Q_1 = Q_2 = Q_3 = 0$ initially.

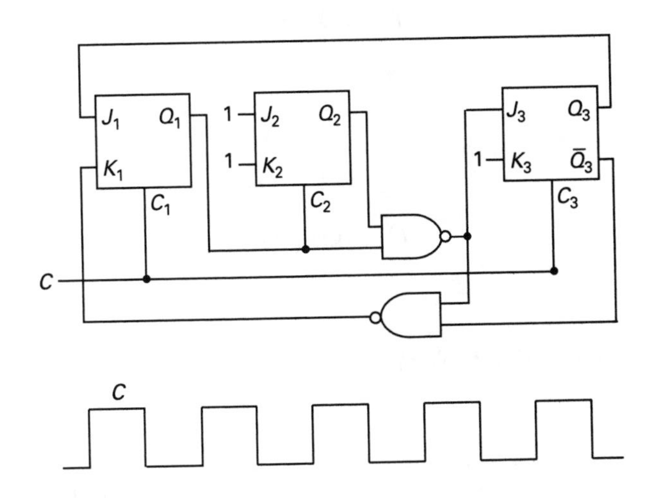

12. Design a digital circuit using only two-input NAND gates and *J-K* master-slave flip-flops. Use as many of each as is necessary, but the design should be minimized with respect to packages. (There are four two-input NAND gates in a package and two *J-K* flip-flops per package.)

The system is to have two inputs: a clock and X. The X is a sequence of binary voltages—high $= 1$ and low $= 0$. Binary data is presented in sequences of one, two, or three ones followed by at least one zero. No other sequence will ever occur (i.e., a sequence of four ones will not occur).

There are to be two outputs, A and B. A is to go high when the first one of a sequence occurs and go low after the second or when the next zero occurs. B is to go high when the second one of a sequence occurs and stay high until the next zero occurs.

13. A shaft is encoded by having a conductor covering from 0° to 180° and an insulator covering from 180° to 360°. A 5 V source is connected to the conductor. One stationary carbon brush sensor contacts the shaft at a reference position, and another contacts the shaft 90° from the first in the positive direction of shaft rotation. When a brush is in contact with the metal sector of the shaft, it delivers a high (approximately 5 V) voltage to one input of the logic circuit, and when not in contact with the metal sector, it provides 0 V to that input.

A sequential logic circuit is to be clocked at a frequency more than ten times the maximum shaft rotational velocity and is to provide input signals to an up-down counter to correspond with the number of rotations the shaft makes. The counter output is then to be the integral number of rotations the shaft makes.

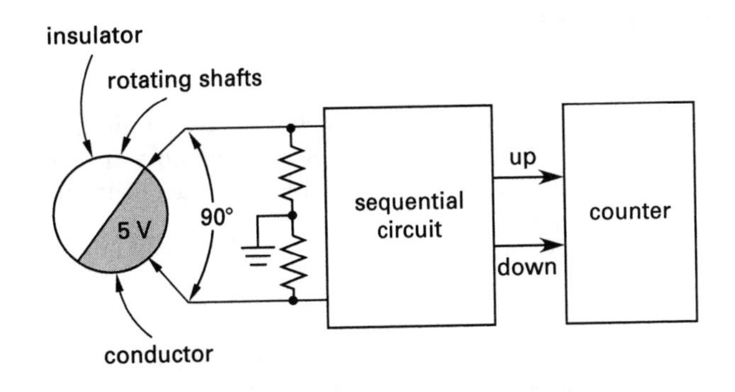

14. The following flowchart deals with three input variables (A, B, and C), and three output variables (X, Y, and Z). Using two-input NAND gates, design combinational logic to realize the flowchart functions.

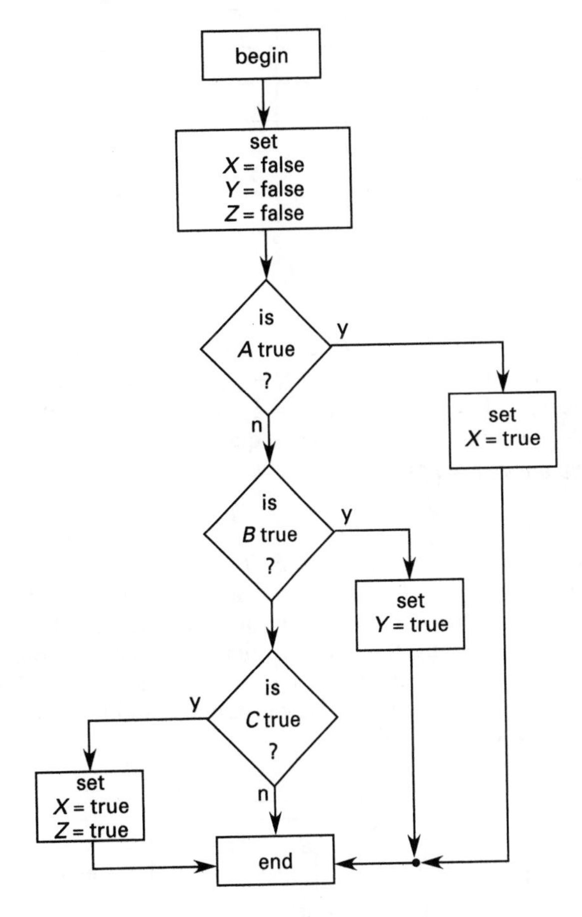

Computers

SOLUTIONS

1. A register is the computer hardware required to store one machine word. It is used as memory in synchronous sequential networks and is constructed from a set of flip-flops.

The answer is (D).

2. The $S = 1$ and $R = 1$ state in an S-R flip-flop is disallowed because the output is indeterminate.

The answer is (D).

3.

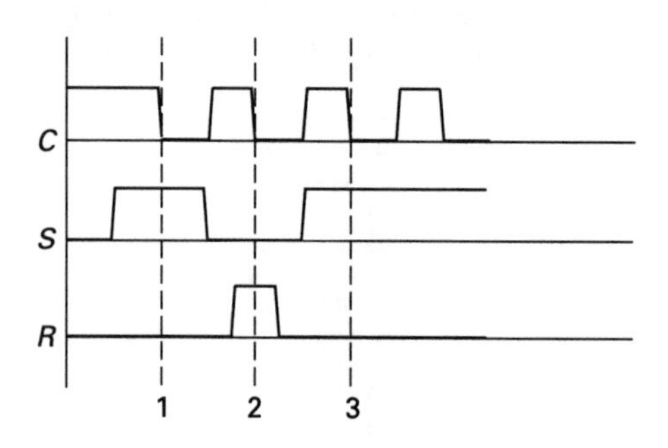

At point 1, $S = 1$, $R = 0$, the flip-flop is set, and $Q = 1$. At point 2, $S = 0$, $R = 1$, the flip-flop is reset, and $Q = 0$. At point 3, $S = 1$, $R = 0$, the flip-flop is set, and $Q = 1$. The output sequence is 101.

The answer is (C).

4. The combination of an S-R flip-flop and logic elements connected as indicated forms a toggle. (This can be discerned by assuming an output, say $Q = 0$, and then varying the input to determine the response. The process is repeated for $Q = 1$. The truth table for the network can then be constructed and the overall response determined.)

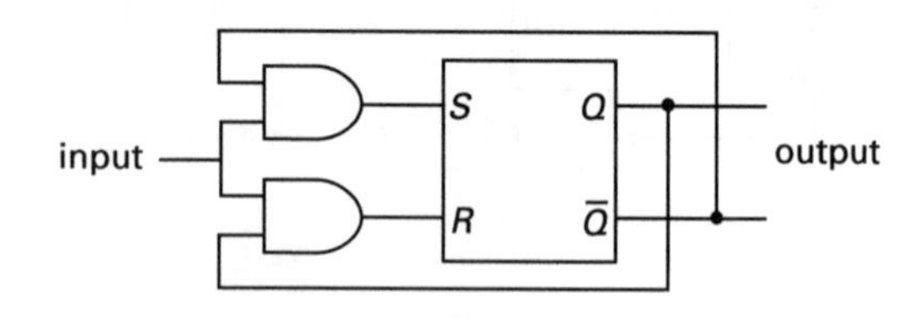

The answer is (C).

5. Of the four principal flip-flops, all can be constructed using an S-R flip-flop as the building block. Thus, an S-R flip-flop requires no additional logic elements to construct.

The answer is (C).

6. Tracing the circuit results in the following equations.

$$R_1 = Q_2 \cdot \overline{Q}_3$$

$$S_1 = \overline{Q}_2 \cdot Q_3$$

$$K_2 = Q_1 + Q_3$$

$$J_2 = Q_1 + \overline{Q}_3$$

$$D_3 = Q_2 \oplus Q_3$$

The exitation table is

Q_1	Q_2	Q_3	R_1	S_1	K_2	J_2	D_3	Q_1^+	Q_2^+	Q_3^+
0	0	0	0	0	0	1	0	0	1	0
0	0	1	0	1	1	0	1	1	0	1
0	1	0	1	0	0	1	1	0	1	1
0	1	1	0	0	1	0	0	0	0	0
1	0	0	0	0	1	1	0	1	1	0
1	0	1	0	1	1	1	1	1	1	1
1	1	0	1	0	1	1	1	0	0	1
1	1	1	0	0	1	1	0	1	0	0

The state transition diagram can be taken directly from the table as

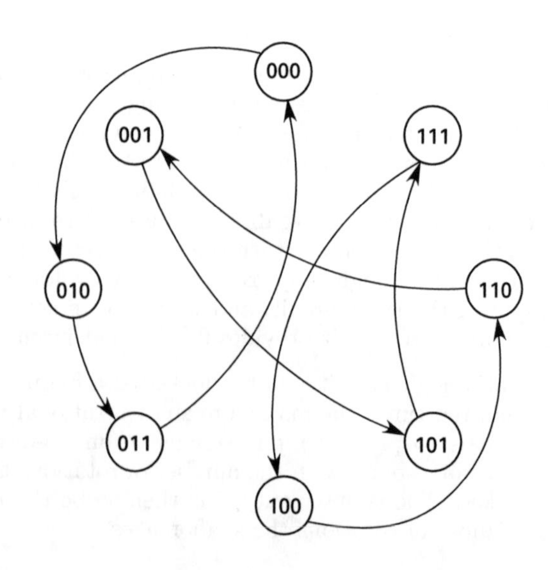

7. Tracing the circuit results in the following equations.

$$K_A = \overline{B}$$
$$K_B = A$$
$$K_C = B$$
$$J_A = \overline{\overline{B} + C} = B \cdot \overline{C}$$
$$J_B = C$$
$$J_C = \overline{A + B} = \overline{A} \cdot \overline{B}$$

The state table for a J-k flip-flop is

J	K	Q	Q^+
0	0	0	0
0	0	1	1
0	1	0	0
0	1	1	0
1	0	0	1
1	0	1	1
1	1	0	1
1	1	1	0

The next state for a J–K is given by

$$Q^+ = Q \cdot \overline{K} + \overline{Q} \cdot J$$

Thus, the outputs A, B, and C are given by

$$A^+ = A \cdot B + \overline{A}B\overline{C} = A \cdot B + B \cdot \overline{C}$$
$$= (110) \text{ or } (111) \text{ or } (010)$$

$$B^+ = \overline{A} \cdot B + \overline{B}C$$
$$= (010) \text{ or } (011) \text{ or } (001) \text{ or } (101)$$

$$C^+ = \overline{B}C + \overline{A} \cdot \overline{B} \cdot \overline{C} = \overline{B}C + \overline{A} \cdot \overline{B}$$
$$= (000) \text{ or } (001) \text{ or } (101)$$

The state table for the counter is

A	B	C	A^+	B^+	C^+
0	0	0	0	0	1
0	0	1	0	1	1
0	1	0	1	1	0
0	1	1	0	1	0
1	0	0	0	0	0
1	0	1	0	1	1
1	1	0	1	0	0
1	1	1	1	0	0

The state transition diagram follows.

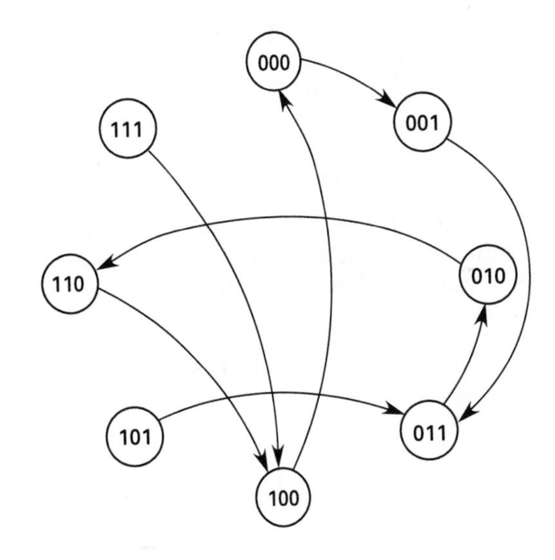

8. From the timing diagram, the sequence triggered by the rising clock edge is

ABC
000 → 100 → 110 → 011 → 001
111 ← 101 ←

The next-state table is

A	B	C	A^+	B^+	C^+	T_A	T_B	T_C
0	0	0	1	0	0	1	0	0
0	0	1	1	0	1	1	0	0
0	1	0	–	–	–	–	–	–
0	1	1	0	0	1	0	1	0
1	0	0	1	1	0	0	1	0
1	0	1	1	1	1	0	1	0
1	1	0	0	1	1	1	0	1
1	1	1	0	0	0	1	1	1

From this,

$$T_A = \overline{A} \cdot \overline{B} + A \cdot B$$

$$T_B = A \cdot \overline{B} + BC$$

$$T_C = A \cdot B$$

This requires six two-input NAND gates.

Something must be done about the 010 state since it results in a hang-up state should it be the state in existence at start-up. This can be done by adding a term $B \cdot \overline{C}$ to T_C.

$$T_C = A \cdot B + B \cdot \overline{C}$$

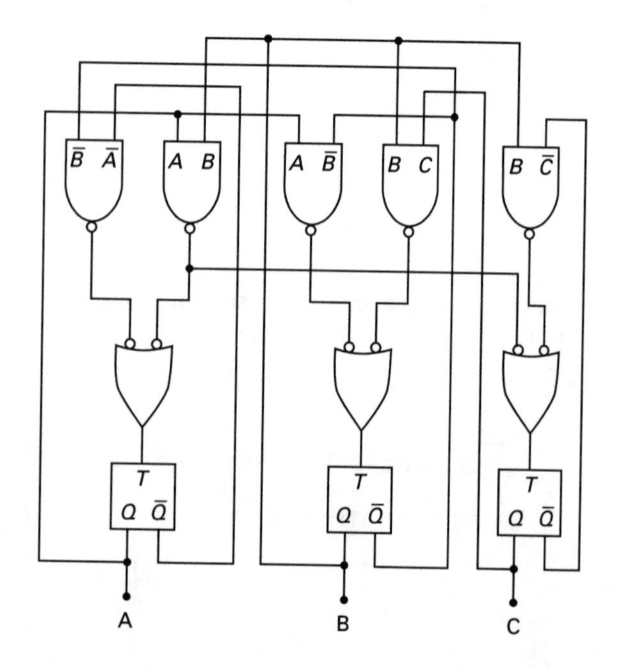

9. Preset must be activated at 99, not 100.

$$99_{10} = 64 + 32 + 2 + 1 = 01100011_2$$

Use an AND gate to obtain the needed preset.

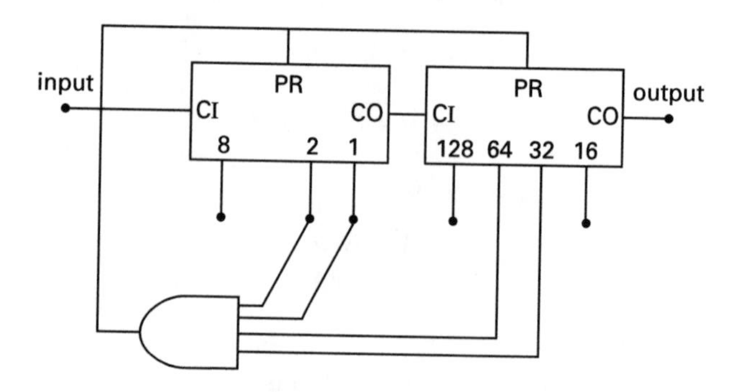

10. The logic for the flip-flop inputs is

$K = Q_1 \cdot Q_2 + \overline{Q}_3$

$D = Q_1 \cdot \overline{Q}_2$

$R = Q_2 \cdot Q_3$

The equations for all the inputs are

$$J = Q_2$$

$$K = Q_1 \cdot Q_2 + \overline{Q}_3$$

$$D = Q_1 \cdot \overline{Q}_2$$

$$S = \overline{Q}_2$$

$$R = Q_2 \cdot Q_3$$

The excitation table is

Q_3	Q_2	Q_1	K	J	D	R	S	Q_3^+	Q_2^+	Q_1^+
0	0	0	1	0	0	0	1	0	0	1
0	0	1	1	0	1	0	1	0	1	1
0	1	0	1	1	0	0	0	1	0	0
0	1	1	1	1	0	0	0	1	0	1
1	0	0	0	0	0	0	1	1	0	1
1	0	1	0	0	1	0	1	1	1	1
1	1	0	0	1	0	1	0	1	0	0
1	1	1	1	1	0	1	0	0	0	0

The state diagram taken directly from the table must be

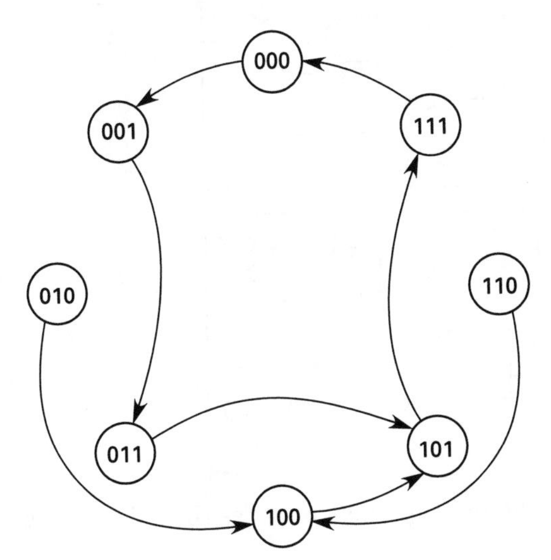

11. Q_2 will toggle when Q_1 falls from high to low (assuming outputs change on the falling edge of the clock).

Q_2
Q_1
$J_3 = \overline{Q_1 \cdot Q_2} = \overline{Q}_1 + \overline{Q}_2$
K_1
$\overline{Q}_3$

The flip-flop input equations are

$$J_1 = Q_3$$

$$K_1 = Q_1 \cdot Q_2 + Q_3$$

$$J_3 = \overline{Q}_1 + \overline{Q}_2$$

$$K_3 = 1$$

The excitation table is

Q_1	Q_2	Q_3	J_1	K_1	$Q_1 \rightarrow Q_1^+$	J_3	Q_1^+	Q_2^+	Q_3^+
0	0	0	0	0	0→0	1	0	0	1
0	0	1	1	1	0→1	1	1	0	0
0	1	0	0	0	0→0	1	0	1	1
0	1	1	1	1	0→1	1	1	1	0
1	0	0	0	0	1→1	1	1	0	1
1	0	1	1	1	1→0	1	0	1	0
1	1	0	0	1	1→0	0	0	0	0
1	1	1	1	1	1→0	0	0	0	0

The state diagram is taken from the table.

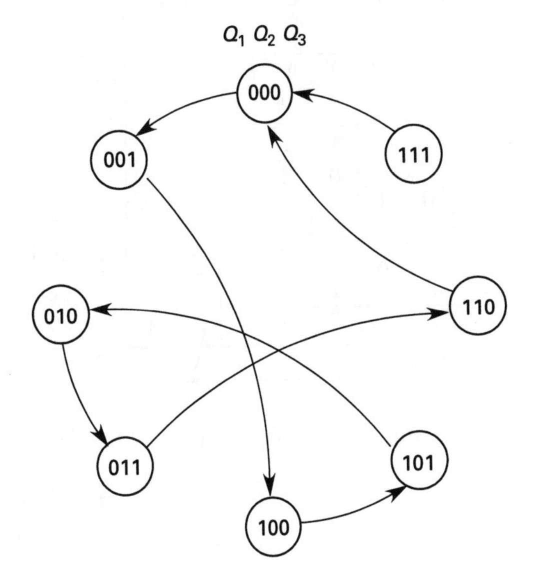

The timing diagram that correlates to the state diagram follows.

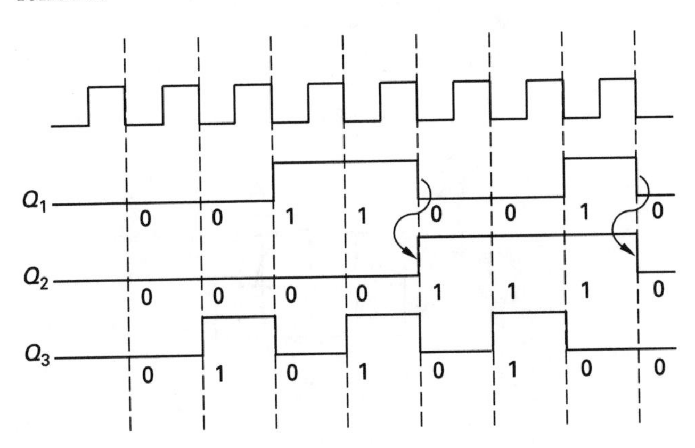

12. The state diagram is

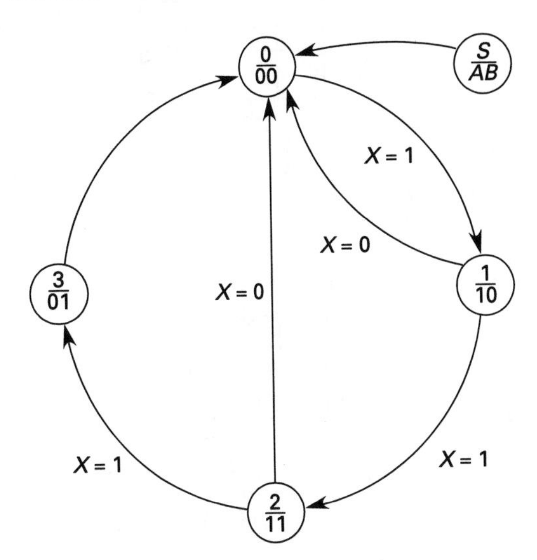

There are four states, so two flip-flops are required. The states are set to minimize output logic.

	A	B	X	A^+	B^+	J_A	K_A	J_B	K_B
S_0	0	0	0	0	0	0	X	0	X
			1	1	0	1	X	0	X
S_3	0	1	0	0	0	0	X	X	1
			1	0	0	0	X	X	1
S_1	1	0	0	0	0	X	1	0	X
			1	1	1	X	0	1	X
S_2	1	1	0	0	0	X	1	X	1
			1	0	1	X	1	X	0

The flip-flop inputs are determined from Veitch diagrams.

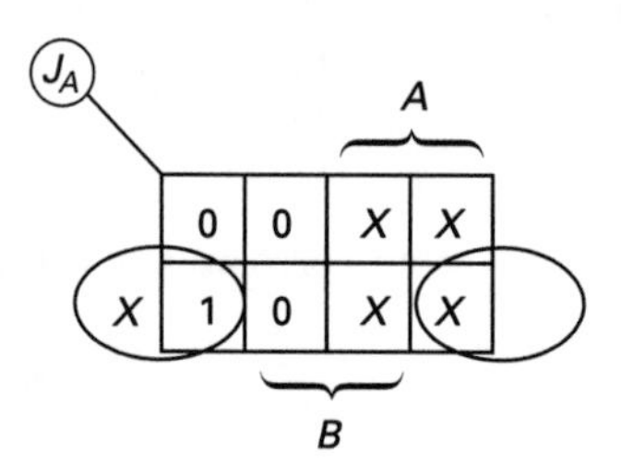

$J_A = \overline{B} \cdot X$

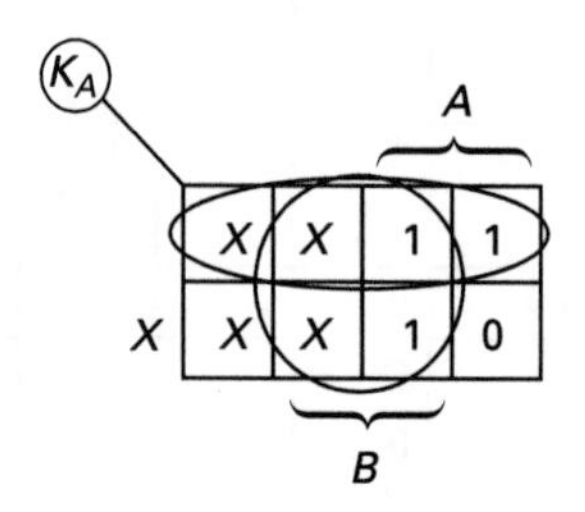

$K_A = B + \overline{X}$

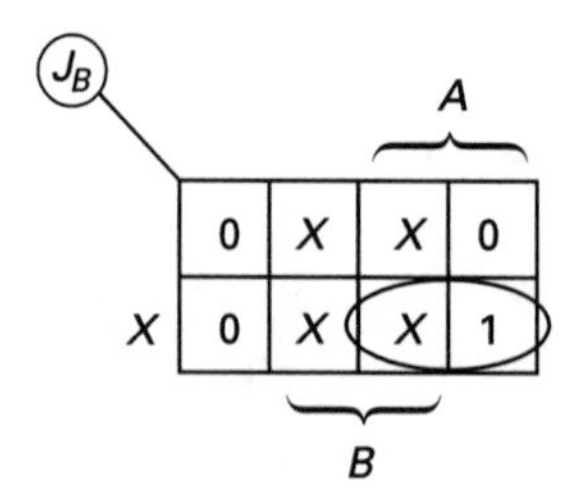

$J_B = A \cdot X$

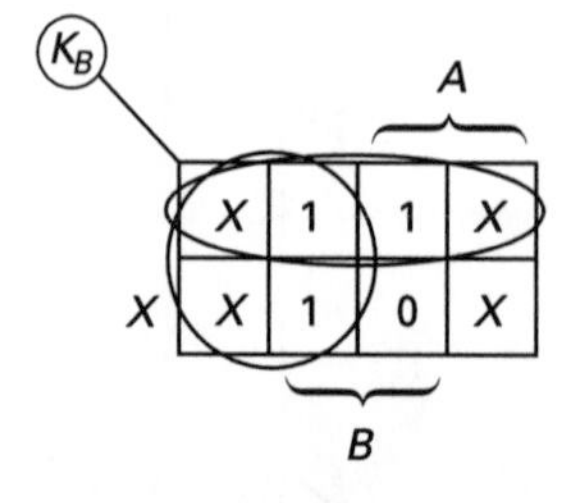

$K_B = \overline{A} + \overline{X}$

The resulting circuit follows.

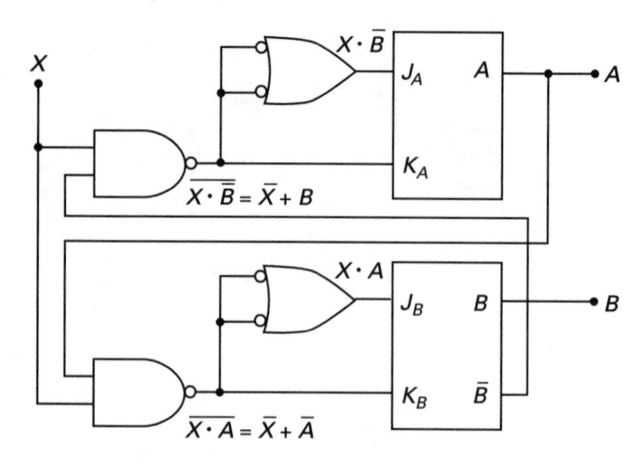

13. The rotating shaft provides the following inputs.

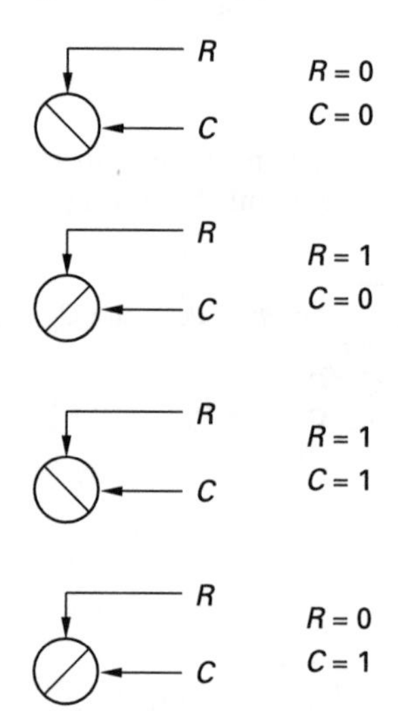

For a positive sequence the resulting input is

R	0 0 0 1 1 1 1 0 0 0 0
C	0 0 0 0 0 1 1 1 1 0 0

For a negative sequence the resulting input is

R	0 0 0 0 1 1 1 1 1 1 0 0 0 0
C	0 1 1 1 1 1 1 0 0 0 0 0 0 0

One solution assigns states to correspond to the four possible combinations of R and C.

S_0 (R, C) has been (00)

S_1 (R, C) has been (01)

S_2 (R, C) has been (10)

S_3 (R, C) has been (11)

The outputs are assigned to the transition between two states, say S_1 and S_3. When the transition is from S_1 to S_3, the output is up. When the transition is from S_3 to S_1, the output is down.

The transition from S_1 to S_3 occurs when

Q_R	Q_C	R	C
0	1	1	1

$$\text{up} = \overline{Q}_R Q_C R C$$

The transition from S_3 to S_1 occurs when

Q_R	Q_C	R	C
1	1	0	1

$$\text{down} = Q_R Q_C \overline{R} C$$

The transition table is

S	Q_R	Q_C	R	C	Q_R^+	Q_C^+	up	down
S_0	0	0	0	0	0	0	0	0
S_0	0	0	0	1	0	1	0	0
S_0	0	0	1	0	1	0	0	0
S_0	0	0	1	1	X	X	0	0
S_1	0	1	0	0	0	0	0	0
S_1	0	1	0	1	0	1	0	0
S_1	0	1	1	0	X	X	0	0
S_1	0	1	1	1	1	1	1	0
S_2	1	0	0	0	0	0	0	0
S_2	1	0	0	1	X	X	0	0
S_2	1	0	1	0	1	0	0	0
S_2	1	0	1	1	1	1	0	0
S_3	1	1	0	0	X	X	0	0
S_3	1	1	0	1	0	1	0	1
S_3	1	1	1	0	1	0	0	0
S_3	1	1	1	1	1	1	0	0

From this, $D_R = R$, and $D_C = C$.

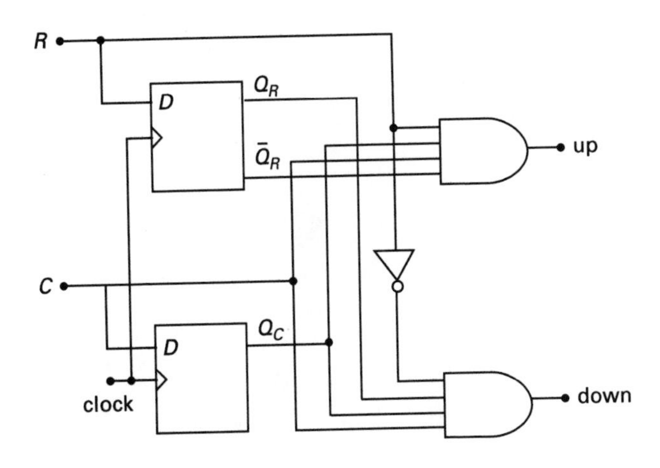

14. From the flow chart, the resulting equations are derived.

$$X = A + \overline{B} \cdot C$$

$$Y = \overline{A} \cdot B$$

$$Z = \overline{A} \cdot \overline{B} \cdot C$$

The logic to realize the equations follows.

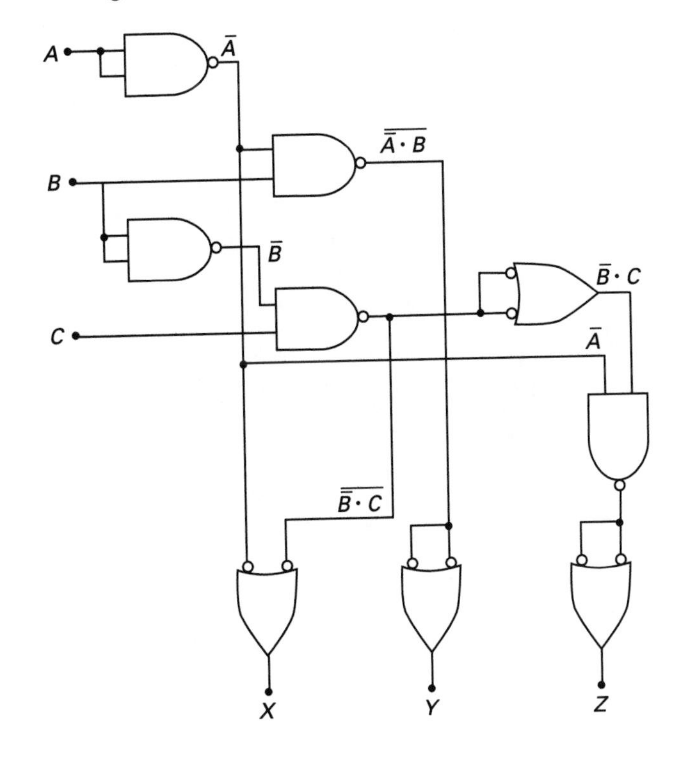

Computers

52 Programming Languages: FORTRAN

PRACTICE PROBLEMS

Arithmetic Operations

1. If J = 3, K = 4, L = 6, and M = 9, what is I = (J*K) + (L − M)?

2. If J = 15, K = 4, L = 9, and M = 19, what is the value of the logical variable X?

(a) X = J.LT.L.OR.K.GE.(M−J)

(b) X = J.GT.L.OR.K.LT.(M−J)

3. What is the value of Q?

```
      R = 18.
      S = 6.
      T = 2.
      Q = R/S**T-T
      IF Q 10, 20, 30
  10  Q = 10
  15  GO TO 40
  20  Q = 100
  25  GO TO 40
  30  Q = 1000
  40  END
```

Program Loops

4. Write a DO loop to multiply ten numbers together until the product exceeds five.

5. Write a DO loop to add the remainders when each of ten numbers is divided by seven.

Input/Output Statements

6. Write the FORTRAN statements that will print the numbers 4104, 6802, and 5798 in scientific notation with two digits to the right of the decimal point, leaving one line between each number.

7. What is the output of the following routine?

```
      F = 50.0*4.6
      WRITE (6,100)F
 100  FORMAT ((' '),T12,E7.3)
```

Statement Functions

8. Write a statement function to find the sine of θ, given that $\sin\theta = \sqrt{\omega t}$.

9. Write a statement function to convert inches to feet.

Subroutines

10. Write a subroutine that can look at a list of numbers and sum only those that are greater than or equal to ten.

11. A main program consists of the following statements.

```
      COMMON ALPHA, BETA
      ALPHA = 4.0
      BETA = 3.0
```

What must be written in the beginning of the subroutine so that A takes the value of ALPHA in the main program and B takes the value of BETA in the main program?

FORTRAN Programs

12. A, B, and C are positive, unequal integers. Write a FORTRAN program to find and print out the smallest integer.

13. A program is to calculate the force required to move springs various distances, using the formula $F = kx$. For each calculation, the user will enter the spring constant and distance. Write a program that performs the calculation, then prints the results to look like the following sample.

```
K = 01.300     X = 14.500     F = 18.900
```

Some lines have been written already.

```
      DATA K,X/1.3,14.5
      READ (5,100)K,X

      F = K*X

 999  END
```

Computers

SOLUTIONS

1. The algebraic translation is

$$\begin{aligned} I &= JK + L - M \\ &= (3)(4) + 6 - 9 = \boxed{9} \end{aligned}$$

2. The algebraic translations are

(a) X is true if $J < L$ or $K \geq (M - J)$. Otherwise, X is false.

Since K is equal to (M-J),

X is true.

(b) X is true if $J > L$ or $K < (M - J)$. Otherwise, X is false.

Since J is greater than L,

X is true.

3. The algebraic translation of the FORTRAN formula for Q is

$$Q = \frac{R}{S^T} - T = \frac{18}{(6)^2} - 2 = -1.5$$

Since Q is negative, go to statement 10.

```
10    Q = 10
```

Go to statement 40.

```
40    END
```

Since the program ends at statement 40 with Q still equal to 10,

Q = 10.

4.

```
    PROD=1.0
    DO 15 I=1, 10
    READ(5,10) B
10  FORMAT(F7.4)
    PROD=PROD*B
    IF (PROD .GT. 5.0) GO TO 20
15  CONTINUE
20  (any subsequent statement)
```

5.

```
    SUM=0.0
    DO 15 I=1,10
    READ(5,10) B
10  FORMAT(F7.4)
    C=AMOD(B/7.0)
    SUM=SUM+C
15  CONTINUE
```

6.

```
    A=4104.
    B=6802.
    C=5798.
    WRITE (6,10)  A,B,C
10  FORMAT ('',3 (E7.2/))
```

7.

2.300 E2 (starting in column 12)

8.

```
    REAL FUNCTION STHETA (OMEGA,T)
    STHETA = SQRT (OMEGA*T)
    RETURN
    END
```

9.

```
    REAL FUNCTION FEET(RINCHS)
    FEET=RINCHS/12.0
    RETURN
    END
```

10.

```
    SUBROUTINE ADD (B,SUM)
    IF (B.LT.10.0) GO TO 5
    SUM=SUM+B
 5  RETURN
    END
```

11.

```
    COMMON A,B
```

12.

```
      INTEGER A,B,C
      READ (5,10) A,B,C
   10 FORMAT (3I4)
      IF(A.LT.B. AND.A.LT.C) I=A
      IF(B.LT.C. AND.B.LT.A) I=B
      IF(C.LT.A. AND.C.LT.B) I=C
      WRITE(6,20) I
   20 FORMAT(' ','THE SMALLEST
          INTEGER IS',I4)
      STOP
      END
```

13.

```
      REAL K
      DATA K,X/1.3,14.5
      F=K*X+0.05
      F=0.1*FLOAT(INT(F*10.0))
      WRITE(6,200) K,X,F
  200 FORMAT(' ','K=',F6.3,' ','X=',F6.3,
          ' ','F=',F6.3)
      STOP
  999 END
```

An equivalent program can be written by substituting the following code for the data statement.

```
      READ(5,100) K,X
  100 FORMAT(2F4.1)
```

Computers

53 Analog Computers: Analog Logic Functions

PRACTICE PROBLEMS

1. Draw the analog computer circuit for the given differential equation.

$$V_0 = -\int (V_1 + 4V_2)\,dt$$

2. Draw the analog simulation for the given equation.

$$7x''' + 14x'' + 42x' + 2x = 49$$

3. What is the analog computer simulation for the mechanical system shown?

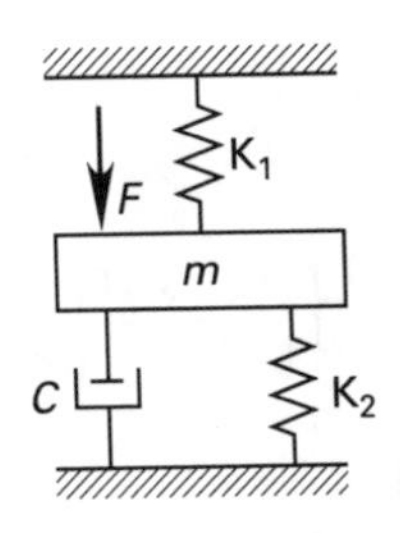

4. Draw an analog circuit to simulate variable x in the mechanical system shown.

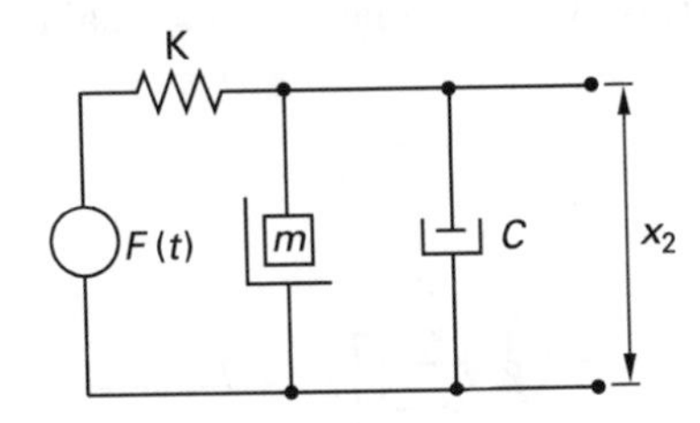

5. Draw an analog circuit to simulate variable x in the mechanical system shown.

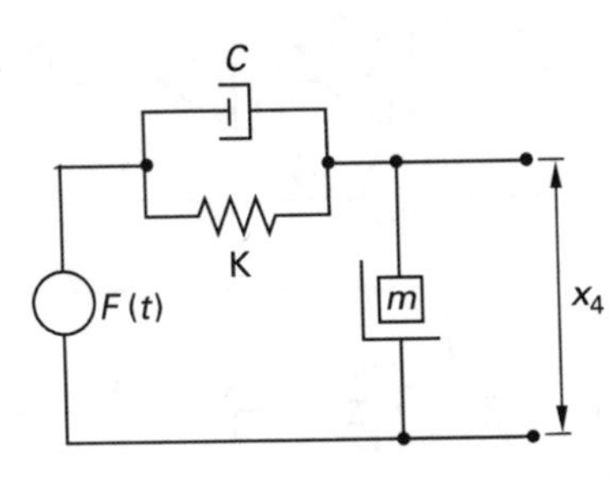

SOLUTIONS

1. Differentiate both sides.

$$V_0' = -V_1 - 4V_2$$

$$-V_0' = V_1 + 4V_2$$

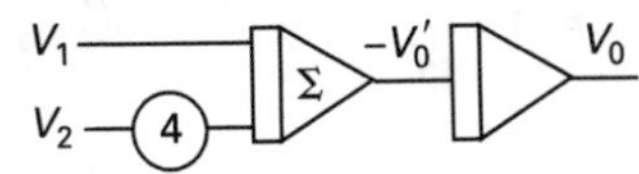

2. *step 1:* $x''' + 2x'' + 6x' + \frac{2}{7}x = 7$

step 2: $-x''' = 2x'' + 6x' + \frac{2}{7}x - 7$

step 3:

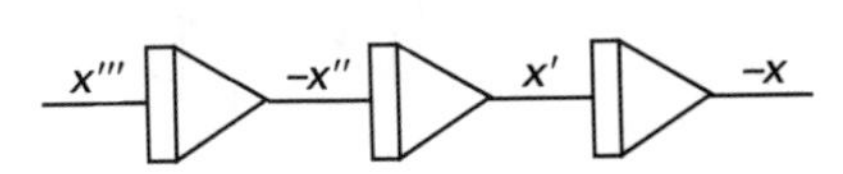

step 4:

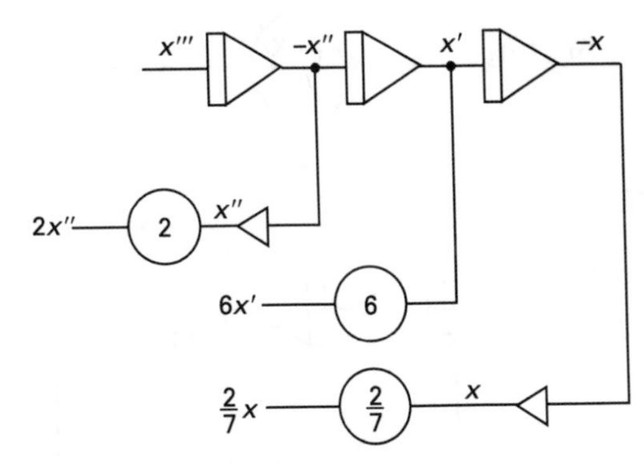

steps 5 and 6:

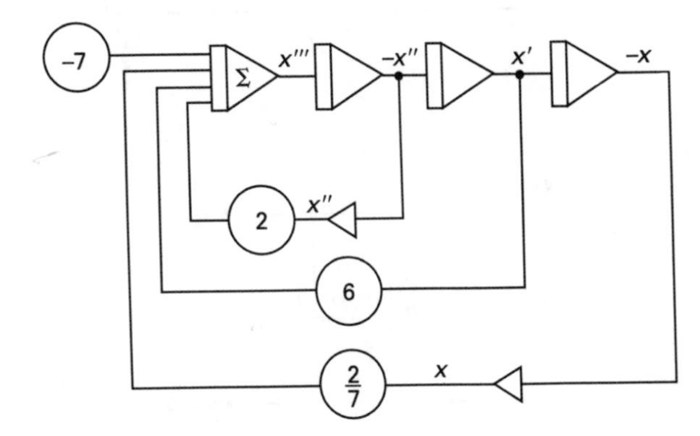

3.

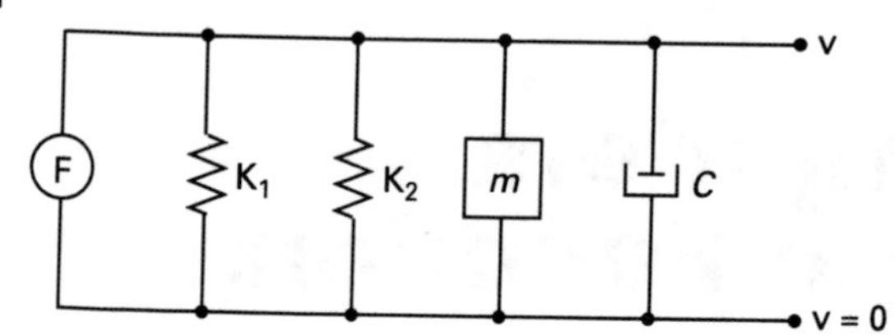

$$F = ma + Cv + (K_1 + K_2)x$$

$$mx'' + Cx' + (K_1 + K_2)x - F = 0$$

step 1: $x'' + \left(\frac{C}{m}\right)x' + \left(\frac{K_1 + K_2}{m}\right)x - \frac{F}{m} = 0$

step 2: $-x'' = \left(\frac{C}{m}\right)x' + \left(\frac{K_1 + K_2}{m}\right)x - \frac{F}{m}$

step 3:

step 4:

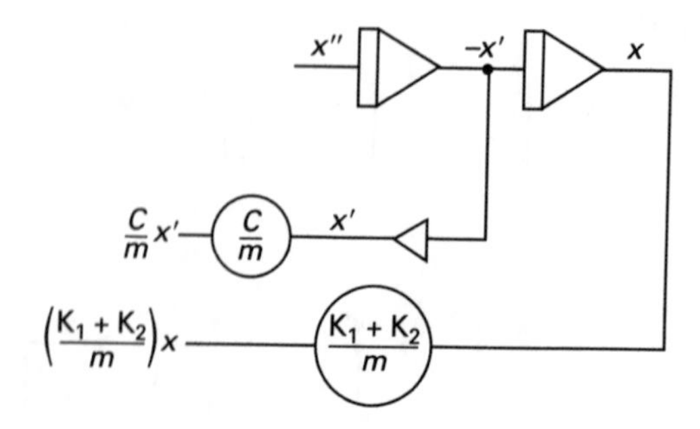

steps 5 and 6:

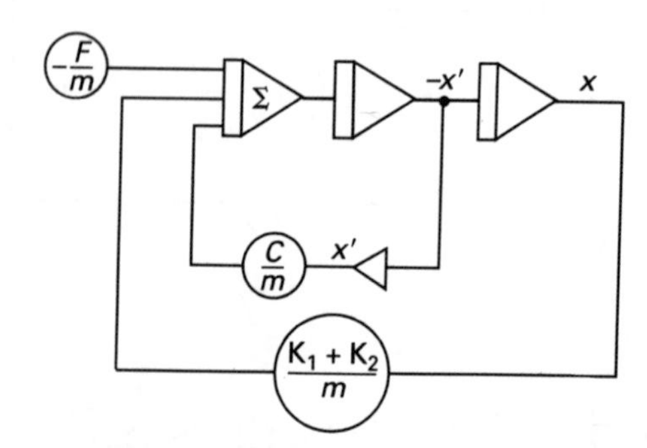

4.

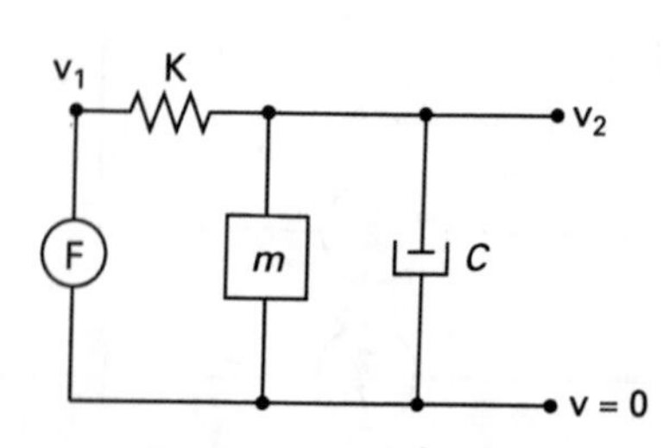

$$F = ma_2 + Cv_2$$

$$mx_2'' + Cx_2' - F = 0$$

step 1: $x_2'' + \left(\frac{C}{m}\right)x_2' - \frac{F}{m} = 0$

step 2: $-x_2'' = \left(\frac{C}{m}\right)x_2' - \frac{F}{m}$

step 3:

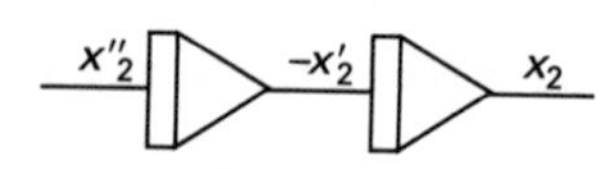

step 4:

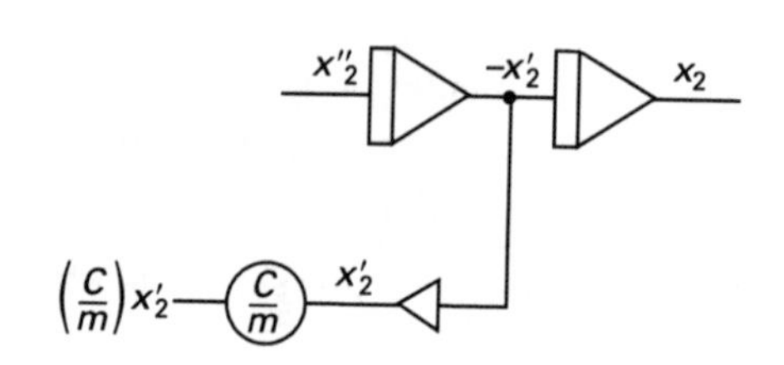

steps 5 and 6:

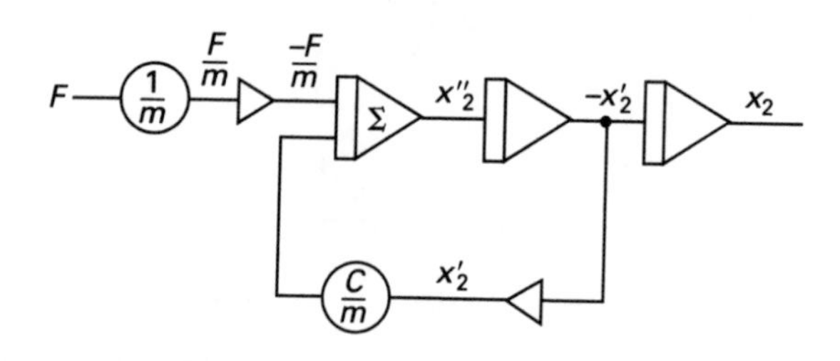

Notice that $F = K(x_1 - x_2)$ is a second equation that applies to this system. Since x_2 is available, x_1 can be found.

$$x_1 = \frac{F}{K} + x_2$$

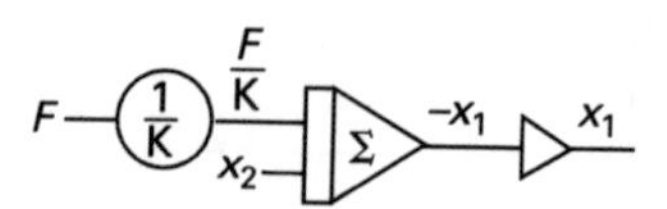

The entire system is

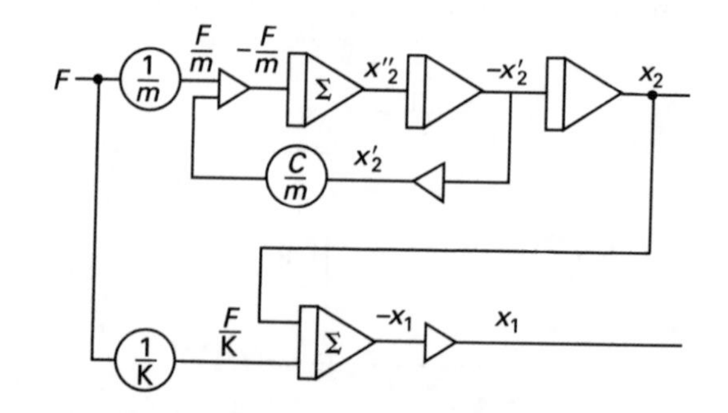

5.

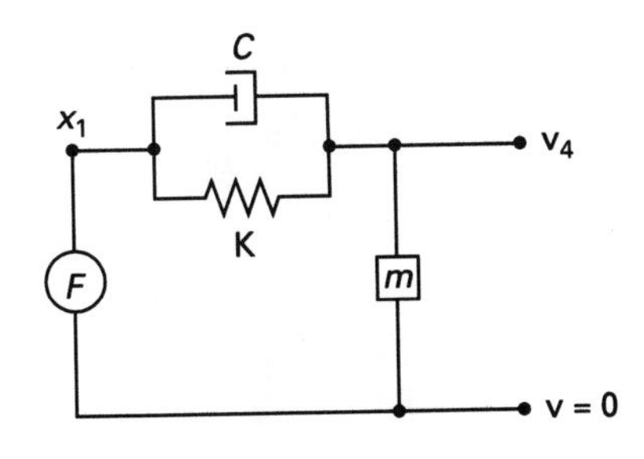

$$F = ma_4$$

$$mx_4'' - F = 0$$

step 1: $x_4'' - \frac{F}{m} = 0$

step 2: $x_4'' = \frac{F}{m}$

step 3:

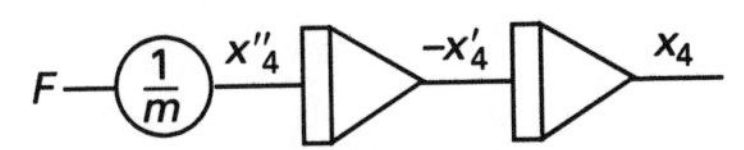

Notice that $F = C(v_1 - v_4) + K(x_1 - x_4)$ is a second equation that applies to the system. Since x_4 and x_4' are available, x_1 can be found.

$$F = Cx_1' - Cx_4' + Kx_1 - Kx_4$$

$$-x_1' = -x_4' + \left(\frac{K}{C}\right)x_1 - \left(\frac{K}{C}\right)x_4 - \frac{F}{C}$$

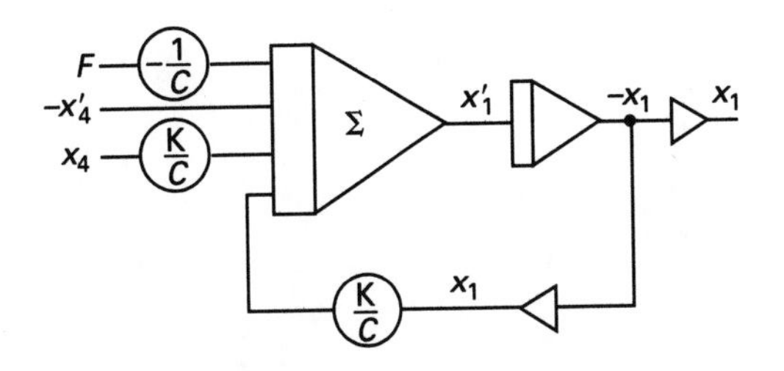

The entire system is

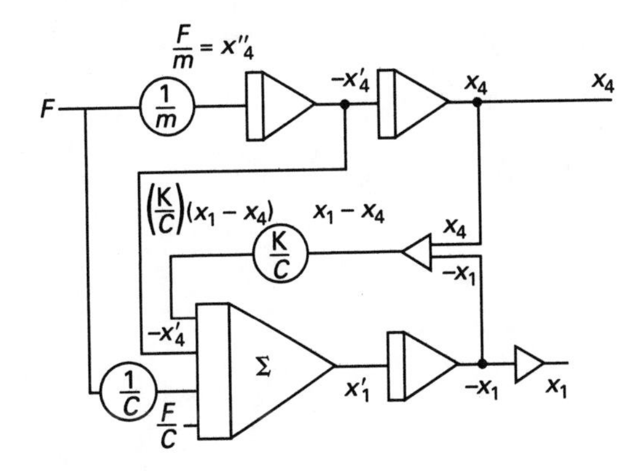

54 Operating Systems, Networking Systems, and Standards

PRACTICE PROBLEMS

There are no practice problems corresponding with Ch. 54 of the *Electrical Engineering Reference Manual*.

55 Digital Systems

PRACTICE PROBLEMS

1. Which of the following components does not exist in a digital system?

(A) central processing unit
(B) control unit
(C) information processing unit
(D) register

2. What term is used for a synchronizing element that temporarily stores information during interactions between electronic devices?

(A) buffer
(B) flag
(C) handshaking unit
(D) interface

3. What is the maximum value that can be represented by a 4-bit binary code?

(A) 8
(B) 15
(C) 16
(D) 32

4. What is the voltage per step of an A/D converter with an input voltage range of −5 to +5 V with a 4-bit output?

(A) 0.32 V/step
(B) 0.34 V/step
(C) 0.63 V/step
(D) 0.67 V/step

5. What is the magnitude of the maximum output voltage of an inverting ladder-type D/A converter operating as a 4-bit system and using a 10 V reference voltage?

(A) 9.4 V
(B) 16.0 V
(C) 18.8 V
(D) 19.4 V

6. A digitally controlled switch is shown here. When the binary input is logic low, what is the state of the PMOS?

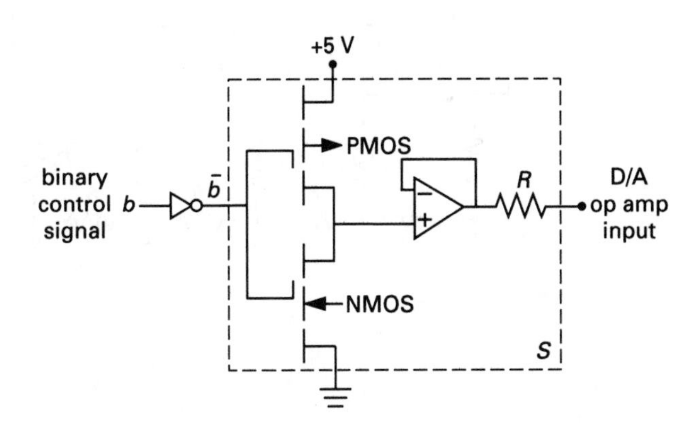

(A) indeterminate
(B) off
(C) on
(D) same as the NMOS

7. The A/D converter shown uses a 3 MHz clock and a 5 V reference voltage. Determine the maximum number of bits for the binary counter if the conversion is to be completed in less than 1 ms. (a) Determine the value of RC to permit full range of the integrator output to −10 V. (b) What is the fundamental accuracy of the conversion of an analog voltage?

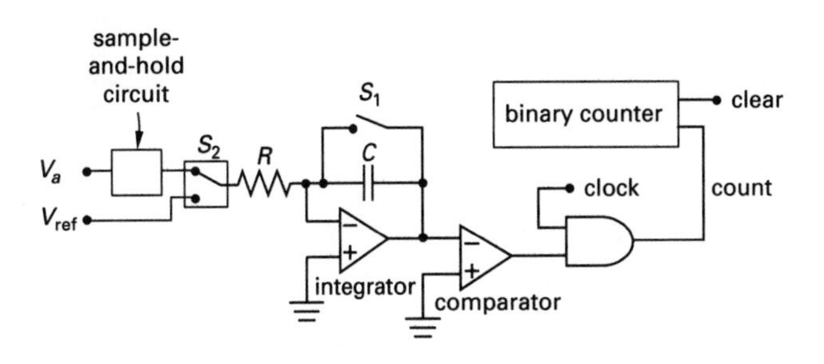

8. The circuit shown is a programmable gain amplifier. It is to be programmable up to a maximum gain of 2.00. It is to be controlled by an 8-bit binary coded data word that controls the electronic switches S_0 through S_7, with S_7 the most significant bit. Specify the values of the resistances to accomplish this.

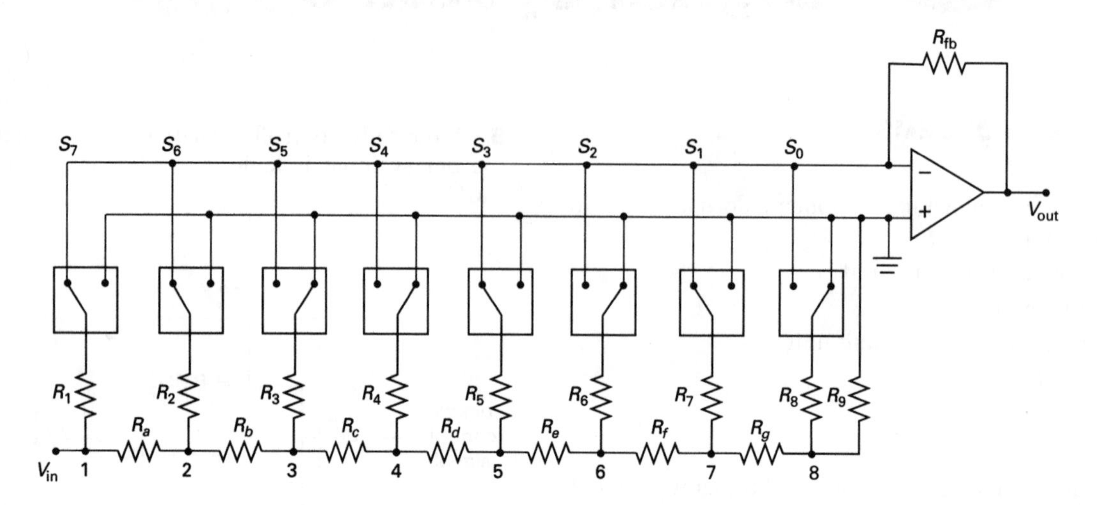

SOLUTIONS

1. The central processing unit, or CPU, is used in computers to allow the system to respond to variations in programming. The digital system has no such unit.

The answer is (A).

2. Temporary storage of information during interactions between electronic devices occurs in memory elements called buffers.

The answer is (A).

3. The maximum value is given by

$$\begin{aligned}\text{maximum value} &= (2)(\text{MSB}) - \text{LSB} \\ &= (2)(2)^3 - (2)^0 \\ &= \boxed{15}\end{aligned}$$

The answer is (B).

4. The voltage range is 10 V. The number of steps for an n-bit number is given by $2^n - 1$. Thus,

$$\begin{aligned}\frac{\text{volts}}{\text{step}} &= \frac{10\text{ V}}{2^n - 1} = \frac{10\text{ V}}{2^4 - 1} = \frac{10\text{ V}}{15} \\ &= \boxed{0.67\text{ V/step}}\end{aligned}$$

The answer is (D).

5. The maximum voltage for an n-bit D/A converter is given by

$$\begin{aligned}|V_{\text{out,max}}| &= V_{\text{ref}}\left(2 - \frac{1}{2^{n-1}}\right) \\ &= (10\text{ V})\left(2 - \frac{1}{(2)^{4-1}}\right) = (10\text{ V})\left(2 - \frac{1}{8}\right) \\ &= \boxed{18.75\text{ V} \quad (18.8\text{ V})}\end{aligned}$$

The answer is (C).

6.

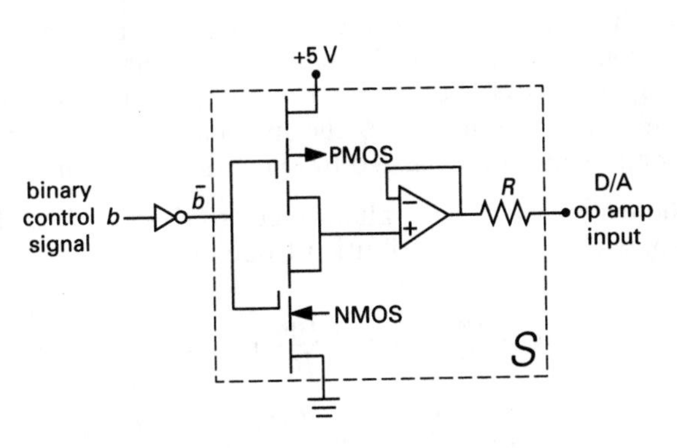

When b is logic low, $\bar{b}$ is logic high. The PMOS is an enhancement type. The $\bar{b}$ signal is a positive voltage on the gate. A negative voltage is required to create the conductivity channel for hole carriers. Therefore, the PMOS is off. (The NMOS is on, placing 0 V on the noninverting terminal and causing the op amp output to be low.)

The answer is (B).

7. (a) In 1 ms, a 3 MHz clock goes through 3000 clock pulses. The closest binary number less than 3000 is 2048, so the full count will be 2048. This is obtained with an 11-bit counter with place values.

$$\begin{aligned}&(2)^{10} + (2)^9 + (2)^8 + (2)^7 + (2)^6 + (2)^5 \\ &\quad + (2)^4 + (2)^3 + (2)^2 + (2)^1 + 1 = (2)^{11} - 1\end{aligned}$$

The duration of a full count is

$$\frac{(2)^{11}\text{ counts}}{3 \times 10^6\ \frac{\text{counts}}{\text{s}}} = 0.6827\text{ ms}$$

$$\frac{v_{\text{full range}}}{RC}\int_0^{682.7\ \mu\text{s}} dt = v_{\text{max}}$$

$$RC = \left(\frac{v_{\text{full range}}}{v_{\text{max}}}\right) \times (682.7 \times 10^{-6}\text{ s})$$

For $|v_{\text{full range}}| = |v_{\text{max}}| = 10$ V,

$$\boxed{RC = 682.7 \times 10^{-6}\text{ s}}$$

(b) The accuracy is ± ½ LSB and the LSB has a value of 10 V/2^{11}, so the accuracy is

$$\begin{aligned}\pm\tfrac{1}{2}(\text{LSB}) &= \left(\frac{1}{2}\right)\left(\frac{10\text{ V}}{(2)^{11}}\right) \\ &= \frac{10\text{ V}}{(2)^{12}} = \boxed{2.44\text{ mV}}\end{aligned}$$

Computers

8. The resistors R_1 through R_8 are either connected to ground directly at the positive terminal of the op amp, or connected to ground effectively via the negative terminal. In either case, the impedance of the input circuit remains unchanged by the switch positions.

The overall gain, with all switches to the left, is to be two, with the R_1 contribution to be one.

$$-v_{\text{out}} = v_{\text{in}}\left(\frac{R_{\text{fb}}}{R_1}\right) + v_2\left(\frac{R_{\text{fb}}}{R_2}\right) + v_3\left(\frac{R_{\text{fb}}}{R_3}\right) + \cdots$$
$$\cdots + v_6\left(\frac{R_{\text{fb}}}{R_6}\right) + v_7\left(\frac{R_{\text{fb}}}{R_7}\right) + v_8\left(\frac{R_{\text{fb}}}{R_8}\right)$$

The most straightforward solution is to make $R_1 = R_2 = \cdots = R_7 = R_{\text{fb}}$, and then design the circuit so that

$$v_2 = \frac{v_{\text{in}}}{2}$$
$$v_3 = \frac{v_{\text{in}}}{4}$$
$$v_4 = \frac{v_{\text{in}}}{8}$$
$$\vdots$$
$$v_8 = \frac{v_{\text{in}}}{128}$$

At the LSB position,

$$v_8 = \frac{v_{\text{in}}}{128}$$
$$i_g = \left(\frac{v_{\text{in}}}{128}\right)\left(\frac{1}{R_8} + \frac{1}{R_9}\right)$$

If $R_8 = R_9 = R$,

$$i_g = \frac{v_{\text{in}}}{64R}$$
$$v_7 = v_8 + R_g i_g = \frac{v_{\text{in}}}{128} + \frac{2R_g v_{\text{in}}}{128R}$$
$$v_7 = \left(\frac{v_{\text{in}}}{128}\right)\left(1 + \frac{2R_g}{R}\right)$$
$$v_7 \rightarrow \frac{v_{\text{in}}}{64}$$
$$R_g = \frac{R}{2}$$

Then,

$$i_f = i_g + \frac{v_7}{R}$$
$$= \frac{v_{\text{in}}}{64R} + \left(\frac{1}{R}\right)\left(\frac{v_{\text{in}}}{64}\right) = \frac{v_{\text{in}}}{32R}$$
$$v_6 = v_7 + R_f i_f = \frac{v_{\text{in}}}{64} + \left(\frac{R_f}{32R}\right)v_{\text{in}}$$
$$= \left(\frac{v_{\text{in}}}{64}\right)\left(1 + \frac{2R_f}{R}\right) \rightarrow \frac{v_{\text{in}}}{32}$$

Then $R_f = R/2 = R_g$.

This process can be repeated to show

$$R_g, R_f, \ldots, R_a = \frac{R}{2}$$

$$R_a = R_b = R_c = R_d = R_e = R_f = \boxed{R_{\text{fb}}/2}$$

$$R_1 = R_2 = R_3 = R_4 = R_5 = R_6 = R_7 = R_8 = \boxed{R_{\text{fb}}}$$

56 High-Frequency Transmission

PRACTICE PROBLEMS

1. A transmission line with a characteristic impedance of 50 Ω and a termination resistance of 70 Ω has $\beta l = \pi/2$. Determine the (a) reflection coefficient, (b) input impedance, and (c) VSWR.

2. A transmission line with $Z_0 = 50\ \Omega$ is terminated with a load impedance of $25 - j25\ \Omega$. (a) Find the electrical angle (βl) where a compensating capacitor can be inserted in the line to cause the impedance seen at that point to be 50 Ω. (b) Determine the reactance of the capacitor.

3. Given a coaxial cable with $Z_0 = 50\ \Omega$, determine the length of cable (in wavelengths) necessary to produce (a) a reactance of −25 Ω, and (b) a reactance of +50 Ω. (The stubs may be either open or shorted.)

4. A transmission line with $Z_0 = 72\ \Omega$ and $\beta = 0.5$ rad/m is 7.5 m long. It is terminated with a pure 50 Ω resistance. Determine the impedance seen from the source.

5. A transmission line with $Z_0 = 100\ \Omega$ is terminated with $Z_L = 25 + j25\ \Omega$. This is to be compensated with a 72 Ω line section. Determine the length of the compensating line section and specify its placement from the load on the 100 Ω line for (a) series compensation and (b) parallel compensation.

SOLUTIONS

1. (a)

$$\Gamma_{\text{load}} = \frac{Z_L - Z_0}{Z_L + Z_0} = \frac{70 - 50}{70 + 50}$$

$$p = \boxed{\frac{1}{6}}$$

(b)

$$Z_{\text{in}} = Z_0\left(\frac{Z_L\cos\beta l + jZ_0\sin\beta l}{Z_0\cos\beta l + jZ_L\sin\beta l}\right)$$

$$= (50\ \Omega)\left(\frac{70\cos\left(\frac{\pi}{2}\right) + j50\sin\left(\frac{\pi}{2}\right)}{50\cos\left(\frac{\pi}{2}\right) + j70\sin\left(\frac{\pi}{2}\right)}\right)$$

$$= \boxed{35.71\ \Omega}$$

(c)

$$\text{VSWR} = \frac{1 + |\Gamma|}{1 - |\Gamma|} = \frac{1 + \frac{1}{6}}{1 - \frac{1}{6}} = \boxed{1.40}$$

2. (a)

$$\frac{Z_L}{Z_0} = \frac{25 - j25\ \Omega}{50\ \Omega} = \frac{1}{2} - j\frac{1}{2}\ \text{pu}$$

The load point is shown on the Smith chart. Impedance locus is shown as a dashed circle.

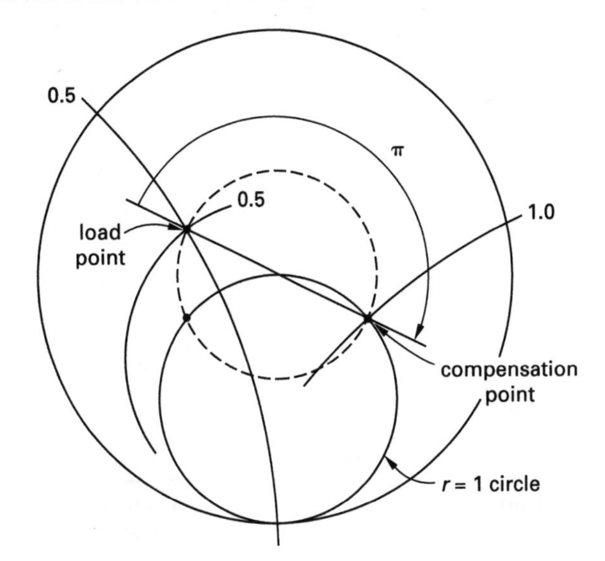

The compensation point for series capacitance is where the impedance locus intersects the $r = 1$ circle on the right-hand side of the chart.

For this particular load, the compensation point is 180° (on the chart) from the load point, at

$$\frac{Z_{\text{in}}}{Z_0} = 1 + j1 \text{ pu}$$

180° on the Smith chart is $\lambda/4$ or $\beta l = \boxed{\pi/2 \text{ rad}}$

(b)
$$\frac{Z_{\text{in}}}{Z_0} = \frac{\frac{Z_L}{Z_0}\cos\beta l + j\sin\beta l}{\cos\beta l + j\left(\frac{Z_L}{Z_0}\right)\sin\beta l}$$

$$= \frac{\frac{1}{2}\cos\beta l + j\left(\sin\beta l - \frac{1}{2}\cos\beta l\right)}{\cos\beta l + \frac{1}{2}\sin\beta l + j\frac{1}{2}\sin\beta l}$$

At $\beta l = \pi/2$,

$$\sin\beta l = 1$$

$$\cos\beta l = 0$$

$$\frac{Z_{\text{in}}}{Z_0} = \frac{j}{\frac{1}{2} + j\frac{1}{2}} = \frac{2}{\frac{1}{j} + 1} = 1 + j1$$

The compensating reactance must be $-j1$ or

$$X_C = -Z_0 = \boxed{-50\ \Omega}$$

3. (a)
$$\frac{Z}{Z_o} = \frac{-j25}{50} = -0.5j \text{ pu}$$

The bottom of the chart is open.

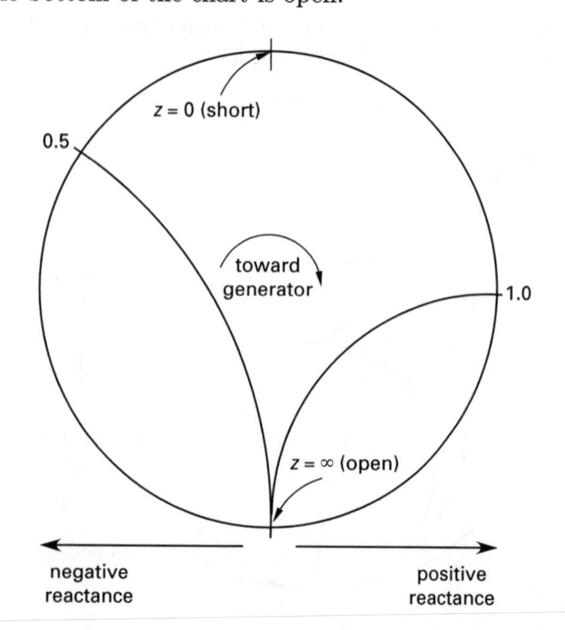

The shortest length to obtain $-j0.5 \times 50$ is to start with an open line, going clockwise 126.5° to the intersection with $x = 0.5$ on the left half of the Smith chart. This corresponds to

$$\beta l = \left(\frac{126.5^\circ}{2}\right)\left(\frac{\pi}{180^\circ}\right) \text{ rad}$$

$$= 1.1 \text{ rad} = (1.1 \text{ rad})\left(\frac{0.5\ \lambda}{\pi \text{ rad}}\right)$$

$$= 0.175\ \lambda$$

(b)
$$\frac{Z}{Z_0} = \frac{j50}{50} = j1 \text{ pu}$$

The shortest line is found for a shorted line (short point at the top of the Smith chart) going clockwise to intersect the $x = 1$ circle (on the right). This is 90° on the chart, corresponding to

$$\beta l = (45^\circ)\left(\frac{\pi}{180^\circ}\right) \text{ rad}$$

$$= (0.785 \text{ rad})\left(\frac{0.5\ \lambda}{\pi \text{ rad}}\right) = \boxed{\lambda/8}$$

4.
$$Z_0 = 72\ \Omega$$

$$Z_L = 50\ \Omega$$

$$\beta = 0.5$$

$$l = 7.5 \text{ m}$$

$$\frac{Z_L}{Z_0} = z_L = \frac{50\ \Omega}{72\ \Omega} = 0.6944 \text{ pu}$$

$$\beta l = 3.75 \text{ rad} = 214.86^\circ$$

$$2\beta l = 429.72^\circ = 360^\circ + 69.72^\circ$$

The load point and source point are shown.

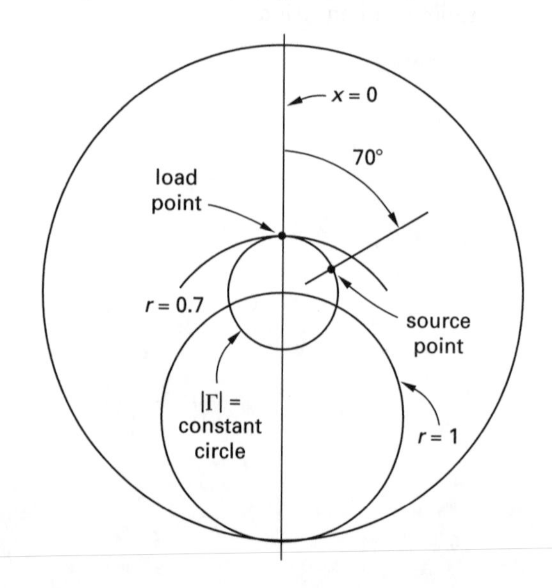

Reading from an enlarged Smith chart, the value is

$$\frac{Z_{\text{in}}}{Z_0} \approx 0.84 + j0.29 \text{ pu}$$

$$Z_{\text{in}} \approx (0.84 + j0.29 \text{ pu})(72\ \Omega) = \boxed{60 + j21}$$

5.

$$Z_0 = 100\ \Omega$$

$$Z_L = 25 + j25\ \Omega$$

$$z_L = 0.25 + j0.25$$

$$y_L = 2 - j2$$

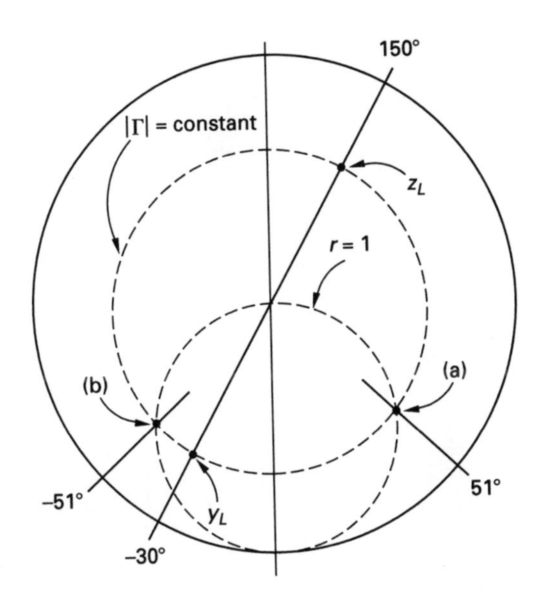

(a) The z_L point, $0.25 + j0.25$, is at 150°. Following a constant $|\Gamma|$ circle toward the generator, the compensation point is at 51°, where $z_{\text{in}} = 1 + j1.6$. A series compensation of $-j1.6$ can be added. The line length from the load is

$$\left(\frac{150° - 51°}{720°}\right)\lambda = \boxed{0.1375\lambda}$$

$$Z_C = (100)(-j1.6) = -j160\ \Omega$$

Using a 72 Ω line for the compensator, its reactance is

$$x_C = -\frac{160}{72} = -2.22 \text{ pu}$$

Starting from the open position (bottom of the Smith chart), the location of $-j2.22$ is about $-48°$ for a compensator length of

$$l_{\text{cp}} = \left(\frac{48°}{720°}\right)\lambda \approx \boxed{\frac{\lambda}{15}}$$

(b) The y_L point is 180° from the z_L point on the same $|\Gamma|$ = constant circle.

The parallel compensation point is at $-51°$, so the distance from the load for compensation is at

$$\left(\frac{51° - 30°}{720°}\right)\lambda = \boxed{0.0292\lambda}$$

At that point,

$$y_L = 1 - j1.6 \text{ pu}$$

$$y = \frac{Z_0}{Z} = Z_0 Y$$

$$Y_{\text{cp}} = \frac{y}{Z_0} = j\left(\frac{1.6}{100}\right) \text{S}$$

For the 72 Ω line,

$$y_{\text{cp}} = j\left(\frac{1.6}{100\ \Omega}\right)(72\ \Omega) = j1.152 \text{ pu}$$

1.152 is on the right side of the chart at 82°. The shortest line is from the top of the chart ($y = 0$, open) at 180°.

$$l = \left(\frac{180° - 82°}{720°}\right)\lambda = \boxed{0.136\lambda}$$

57 Antenna Theory

PRACTICE PROBLEMS

1. What is a group of components arranged to vary the transmission or reception of electromagnetic waves called?

(A) array
(B) Huygens' source
(C) waveguide
(D) dish antenna

2. What is the noise power for an antenna with a bandwidth of 1 GHz at its maximum design temperature of 65°C?

(A) 9×10^{-13} W
(B) 2×10^{-12} W
(C) 5×10^{-12} W
(D) 5×10^{-9} W

3. What is the phase constant for a 500 MHz signal?

(A) 1.05 rad/m
(B) 10.5 rad/m
(C) 32.0 rad/m
(D) 38.0 rad/m

4. The peak electric field of a plane electromagnetic wave 10 km south of a vertical antenna is 0.2×10^{-2} V/cm. What is the magnitude of the magnetic field at that point?

(A) 10 nA/m
(B) 1 μA/m
(C) 0.5 mA/m
(D) 0.05 mA/m

5. A radio station's assigned frequency is 107 MHz. What is the free-space basic transmission loss 50 km from the transmitting antenna?

(A) 3.5 dB
(B) 47 dB
(C) 83 dB
(D) 110 dB

SOLUTIONS

1. An array is used to manipulate the direction of an antenna with respect to incoming or outgoing electromagnetic waves.

The answer is (A).

2. The noise power is given by

$$P_n = \kappa T B$$

The design temperature of 65°C equates to 338K. Substituting gives

$$\begin{aligned} P_n &= \left(1.3805 \times 10^{-23}\ \frac{\text{J}}{\text{K}}\right)(338\text{K})(10^9\ \text{Hz}) \\ &= \boxed{4.67 \times 10^{-12}\ \text{W}} \end{aligned}$$

The answer is (C).

3. The phase constant, β, is given by

$$\beta = \frac{2\pi}{\lambda}$$

The wavelength of a 500 MHz wave is

$$\begin{aligned} \lambda &= \frac{c}{f} = \frac{3 \times 10^8\ \frac{\text{m}}{\text{s}}}{500 \times 10^6\ \frac{1}{\text{s}}} \\ &= 0.6\ \text{m} \end{aligned}$$

Substituting gives

$$\beta = \frac{2\pi}{\lambda} = \frac{2\pi}{0.6\ \text{m}} = \boxed{10.5\ \text{rad/m}}$$

The answer is (B).

Communications

4. From Poynting's vector,

$$\mathbf{S} = \mathbf{E} \times \mathbf{H} = \epsilon c E^2 \mathbf{a}$$

Therefore,

$$|\mathbf{S}| = EH = \epsilon c E^2$$

Air has a relative permittivity, ϵ_r, of approximately one. Thus, $\epsilon = \epsilon_0 \epsilon_r \approx \epsilon_0$. Rearranging and substituting gives

$$\begin{aligned} EH &= \epsilon c E^2 \\ H &= \epsilon c E \approx \epsilon_0 c E \\ &= \left(8.854 \times 10^{-12}\ \frac{\text{F}}{\text{m}}\right)\left(3 \times 10^8\ \frac{\text{m}}{\text{s}}\right) \\ &\quad \times \left(0.2 \times 10^{-2}\ \frac{\text{V}}{\text{cm}}\right)\left(\frac{100\ \text{cm}}{1\ \text{m}}\right) \\ &= 5.31 \times 10^{-4}\ \text{F·V/m·s} \\ &= \boxed{5.31 \times 10^{-4}\ \text{A/m} \quad (0.531\ \text{mA/m})} \end{aligned}$$

The answer is (C).

5. The free-space basic transmission loss is given by

$$L_{b0} = 20 \log \left(\frac{4\pi r}{\lambda}\right)$$

The wavelength is determined from

$$\lambda = \frac{c}{f} = \frac{3 \times 10^8\ \frac{\text{m}}{\text{s}}}{107 \times 10^6\ \frac{1}{\text{s}}} = 2.8\ \text{m}$$

Substituting gives

$$\begin{aligned} L_{b0} &= 20 \log \left(\frac{4\pi(50 \times 10^3\ \text{m})}{2.8\ \text{m}}\right) \\ &= \boxed{107\ \text{dB} \quad (110\ \text{dB})} \end{aligned}$$

The answer is (D).

58 Communication Links

PRACTICE PROBLEMS

1. A radio station transmits 50 kW with a directivity gain of 1.5. What is the power density 10 km from the radiating antenna?

(A) 0.06 mW/m^2
(B) 5.9 W/m^2
(C) 33 W/m^2
(D) 75 W/m^2

2. For the radio station in Prob. 1, what is the electric field strength at 10 km from the radiating antenna?

(A) 0.08 V/m
(B) 0.15 V/m
(C) 50 V/m
(D) 100 V/m

3. An antenna has an effective aperture of 10 m^2. What additional information is required to determine the amplifier power, P_{amp}?

(A) power density
(B) power loss
(C) space loss
(D) total aperture

4. A horn antenna with an efficiency of 90% has a directivity of 25. What is the gain in decibels?

(A) 1 dB
(B) 14 dB
(C) 23 dB
(D) 28 dB

5. Two antennae used for wireless communication at 900 MHz are identical, with physical apertures of 0.1 m^2 and a total efficiency of 50%. The transmitting power is 1 W. The antennae provide adequate reception if the received power density is 10 μW. What is the maximum separation distance?

(A) 15 m
(B) 25 m
(C) 50 m
(D) 150 m

SOLUTIONS

1. The power density is given by

$$p_r = \left(\frac{D_T}{4\pi r^2}\right) P_T$$

Substituting gives

$$\begin{aligned} p_r &= \left(\frac{1.5}{4\pi(10 \times 10^3 \text{ m})^2}\right)(50 \times 10^3 \text{ W}) \\ &= \boxed{5.97 \times 10^{-5} \text{ W/m}^2 \quad (0.06 \text{ mW/m}^2)} \end{aligned}$$

The answer is (A).

2. The power density in terms of the radiating electric field is

$$p_r = \frac{E^2}{Z_0}$$

The characteristic impedance of the atmosphere is approximately that of free space. Thus, rearranging gives

$$\begin{aligned} E &= \sqrt{p_r Z_0} \\ &= \sqrt{(0.06 \times 10^{-3} \ \frac{\text{W}}{\text{m}^2})(120\pi\Omega)} \\ &= \boxed{0.15 \text{ V/m}} \end{aligned}$$

The answer is (B).

3. The amplifier power is given by

$$P_{\text{amp}} = A_{\text{eff}} p_r$$

Thus, the information required is the power density, p_r.

The answer is (A).

4. The gain is given by

$$G = \eta D_g$$

Communications

Substituting the given value gives

$$G = (0.90)(25) = 22.5$$

In decibels, the gain is

$$G_{\text{dB}} = 10\log(22.5) = \boxed{13.5 \text{ dB}}$$

The answer is (B).

5. The transmission equation is

$$P_R = \left(\frac{\lambda}{4\pi r}\right)^2 G_R G_T P_T$$

The gains are unknown. The gains are

$$G = \eta D$$

Substituting gives

$$P_R = \left(\frac{\lambda}{4\pi r}\right)^2 \eta_R D_R \eta_T D_T P_T$$

The two antennae are identical. Thus,

$$P_R = \left(\frac{\lambda}{4\pi r}\right)^2 (\eta D)^2 P_T$$

The directivity is unknown. The directivity is given by

$$D = \left(\frac{4\pi}{\lambda^2}\right) A_{\text{eff}} = \left(\frac{4\pi}{\lambda^2}\right) A_{\text{physical}}$$

The physical aperture can be used since the total efficiency is given, which includes the antenna I^2R efficiencies and is accounted for in ηD. Alternatively, the gain is

$$G = \left(\frac{4\pi}{\lambda^2}\right) \eta_{\text{total}} A_{\text{physical}}$$

Substituting the directivity D into P_R gives

$$\begin{aligned} P_R &= \left(\frac{\lambda}{4\pi r}\right)^2 \left(\eta_{\text{total}} \left(\frac{4\pi}{\lambda^2}\right) A_{\text{physical}}\right)^2 P_T \\ &= \left(\frac{\eta_{\text{total}}^2}{\lambda^2 r^2}\right) A_{\text{physical}}^2 P_T \end{aligned}$$

Now, the only unknown is the radial distance r. Rearranging to solve for r gives

$$r = \sqrt{\frac{\eta_{\text{total}}^2 A_{\text{physical}}^2 P_T}{P_R \lambda^2}}$$

Substituting the definition of the wavelength and the given values results in

$$\begin{aligned} r &= \sqrt{\frac{\eta_{\text{total}}^2 A_{\text{physical}}^2 P_T}{P_R \left(\frac{c}{f}\right)^2}} \\ &= \left(\frac{\eta_{\text{total}} A_{\text{physical}}}{\left(\frac{c}{f}\right)}\right) \sqrt{\frac{P_T}{P_R}} \\ &= \frac{(0.5)(0.1 \text{ m}^2)}{\left(\dfrac{3 \times 10^8 \ \frac{\text{m}}{\text{s}}}{900 \times 10^6 \ \frac{1}{\text{s}}}\right)} \sqrt{\frac{1 \text{ W}}{10 \times 10^{-6} \text{ W}}} \\ &= \boxed{47.4 \text{ m} \quad (50 \text{ m})} \end{aligned}$$

The answer is (C).

59 Signal Formats

PRACTICE PROBLEMS

1. Commercial AM radio generally uses what type of modulation?

(A) DSB-SC
(B) DSB-LC
(C) QAM
(D) VSAM

2. The maximum frequency deviation for U.S. AM radio stations is 5 kHz. The allowable frequency band is 535 to 1705 kHz. How many radio stations can a single geographical area support?

(A) 117
(B) 170
(C) 234
(D) 341

3. A 105 MHz carrier is frequency modulated by a 75 kHz information signal. The information signal has a 1 V amplitude and a frequency modulator constant of 100 Hz/V. What is the bandwidth?

(A) 75 kHz
(B) 100 kHz
(C) 150 kHz
(D) 200 kHz

4. The same carrier in Prob. 3 is now phase modulated, with $k_p = 0.3$ rad/v. What is the bandwidth?

(A) 75 kHz
(B) 100 kHz
(C) 150 kHz
(D) 200 kHz

5. A communication system design calls for multilevel modulation of a digital signal. A mapping space using four-digit binary codes has been selected. How many potential modulations of the carrier are possible?

(A) 2
(B) 4
(C) 8
(D) 16

6. For the NRZ-M baseband data signal shown, what binary signal is transmitted, assuming that the most significant bit is transmitted first?

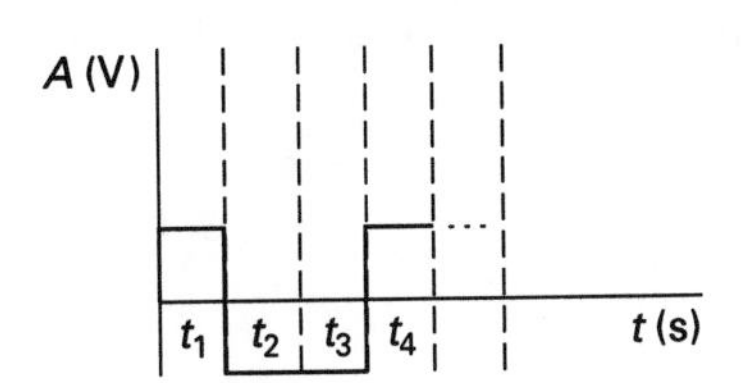

(A) 0010
(B) 0110
(C) 1001
(D) 1101

7. If the signal in Prob. 6 is inverted, what data is sent?

(A) 0010
(B) 0110
(C) 1001
(D) 1101

8. What is the bandwidth efficiency of a one-dimensional 4-ary PAM signal?

(A) 1
(B) 2
(C) 4
(D) 8

9. If the bandwidth in Prob. 8 is 10 kHz, what is the signal rate or bit rate in kilobits per second?

(A) 5 kbps
(B) 10 kbps
(C) 40 kbps
(D) 80 kbps

10. Intersymbol interference is eliminated by applying which of the following?

(A) Nyquist criterion
(B) pulse code modulation
(C) Shannon bandwidth
(D) SSB modulation

11. What is the probability of error for a 4-ary PAM scheme, assuming $E_b/N_0 = 1$?

(A) 0.15
(B) 0.30
(C) 0.60
(D) 0.90

12. What is the bit error rate of the 4-ary PAM scheme in Prob. 11?

(A) 0.15
(B) 0.30
(C) 0.60
(D) 0.90

SOLUTIONS

1. Commercial AM radio uses the double-sideband large-carrier (DSB-LC) signal, because this signal utilizes simple inexpensive receivers.

The answer is (B).

2. The bandwidth is given by

$$\begin{aligned} \text{BW} &= 2\Delta f \\ &= (2)(5 \text{ kHz}) \\ &= 10 \text{ kHz} \end{aligned}$$

The AM bandwith is

$$1705 \text{ kHz} - 535 \text{ kHz} = 1170 \text{ kHz}$$

The number of possible stations is

$$\begin{aligned} \text{stations} &= \frac{\text{AM BW}}{\text{station BW}} = \frac{1170 \text{ kHz}}{\frac{10 \text{ kHz}}{\text{station}}} \\ &= \boxed{117 \text{ stations}} \end{aligned}$$

The answer is (A).

3. The bandwidth of an FM signal is approximated using Carson's rule.

$$\text{BW} \approx 2(\Delta\omega + \omega_{\text{mod}})$$

The maximum frequency deviation for an FM signal is given by

$$\Delta\omega = ak_f$$

Substituting the given values results in

$$\begin{aligned} \text{BW} &\approx 2(\Delta\omega + \omega_{\text{mod}}) = 2(ak_f + \omega_{\text{mod}}) \\ &= (2)\left((1 \text{ V})\left(100 \; \frac{\text{Hz}}{\text{V}}\right) + 75 \times 10^3 \text{ Hz}\right) \\ &= (2)(75{,}100 \text{ Hz}) \\ &= \boxed{150{,}200 \text{ Hz} \quad (150 \text{ kHz})} \end{aligned}$$

The answer is (C).

4. For a phase-modulated signal, Carson's rule still applies, but the maximum frequency deviation is

$$\begin{aligned} \Delta\omega &= ak_p\omega_{\text{mod}} \\ &= (1 \text{ V})\left(0.3 \; \frac{\text{rad}}{\text{V}}\right)(75{,}000 \text{ Hz}) \\ &= 22{,}500 \text{ Hz} \end{aligned}$$

Substituting into Carson's rule gives

$$\begin{aligned}\text{BW} &\approx 2(\Delta\omega + \omega_{\text{mod}}) \\ &= 2(ak_f\omega_{\text{mod}} + \omega_{\text{mod}}) \\ &= (2)(22{,}500\ \text{Hz} + 75{,}000\ \text{Hz}) \\ &= \boxed{195{,}000\ \text{Hz}\quad(200\ \text{kHz})}\end{aligned}$$

The answer is (D).

5. The number of mapping locations, M, for binary digits, n, is given by

$$\begin{aligned}M &= 2^n \\ &= 2^4 = \boxed{16}\end{aligned}$$

The answer is (D).

6.

A (V)

t_1 t_2 t_3 t_4 … t (s)

1 1 0 1

Since the assumption is that the signal starts at zero, the first interval represents a change and thus equals a logic 1.

A change occurs in the consecutive signals from bit interval t_1 to t_2, and thus the second interval equals logic 1.

No change occurs from t_2 to t_3, and thus t_3 is logic 0.

A change occurs from t_3 to t_4, and thus t_4 is logic 1. The net result is a signal of 1101.

The answer is (D).

7. Inverting an NRZ-M signal does not change the transitions from one bit interval to the next. Thus, the logic circuitry is able to detect the same signal, 1101.

The answer is (D).

8. The bandwidth efficiency (noting that $D = 1$) is defined as

$$\begin{aligned}\eta_{\text{BW}} &= \frac{R_s}{\text{BW}} = \frac{\dfrac{\log_2 M}{T}}{\dfrac{D}{2T}} = \frac{2\log_2 M}{D} \\ &= 2\log_2(4) \\ &= \boxed{4}\end{aligned}$$

The answer is (C).

9. The bandwidth efficiency is

$$\eta_{\text{BW}} = \frac{R_s}{\text{BW}}$$

Solving for the signal rate and substituting the given values and the calculated efficiency from Prob. 8 gives

$$\begin{aligned}R_s &= \eta_{\text{BW}}(\text{BW}) \\ &= (4)(10\ \text{kHz}) \\ &= 40\ \text{kHz} \\ &= \boxed{40\ \text{kbps}}\end{aligned}$$

The answer is (C).

10. The Nyquist criterion is specifically designed to achieve zero intersymbol interference.

The answer is (A).

11. The probability of error for a PAM scheme is

$$\mathcal{P}(e) = \left(1 - \frac{1}{M}\right)\text{erfc}\sqrt{\left(\frac{3\log_2 M}{M^2 - 1}\right)\left(\frac{E_b}{N_0}\right)}$$

The number of bits in a symbol for an M-ary modulation scheme is

$$\text{bits} = \log_2 M = \log_2(4) = 2$$

The symbols are 00, 01, 10, and 11 in a 4-ary scheme with 2 bits per symbol.

Substituting $M = 4$ and $E_b/N_0 = 1$ gives

$$\begin{aligned}\mathcal{P}(e) &= \left(1 - \frac{1}{4}\right)\text{erfc}\sqrt{\left(\frac{3\log_2(4)}{(4)^2 - 1}\right)(1)} \\ &= \left(\frac{3}{4}\right)\text{erfc}\sqrt{\frac{6}{15}} \\ &= \left(\frac{3}{4}\right)\text{erfc}\sqrt{\frac{2}{5}} \\ &= (0.75)\,\text{erfc}\,(0.63)\end{aligned}$$

From a table of the error function, also called the probability function or error integral,

$$\begin{aligned}\text{erfc}(0.63) &= 1 - \text{erf}(0.63) \\ &= 1 - 0.6270 = 0.373\end{aligned}$$

Thus,

$$\mathcal{P}(e) = (0.75)(0.373) = \boxed{0.28\quad(0.30)}$$

The answer is (B).

Note: The bit error rate is high because the bit energy, E_b, is low; in this case, it is equal to the noise N_0.

Communications

12. The bit errror rate (BER) is given by

$$\mathcal{P}_b(e) = \frac{\mathcal{P}(e)}{\log_2 M}$$

The probability of error for a 4-ary PAM is calculated in Prob. 11. The number of bits is given by the denominator of the BER equation and is calculated in Prob. 11. Substituting gives

$$\mathcal{P}_b(e) = \frac{\mathcal{P}(e)}{\log_2 M} = \frac{0.28}{2}$$

$$= \boxed{0.14 \quad (0.15)}$$

The answer is (A).

60 Signal Multiplexing

PRACTICE PROBLEMS

There are no practice problems corresponding with Ch. 60 of the *Electrical Engineering Reference Manual.*

61 Communication Systems and Channels

PRACTICE PROBLEMS

1. A microwave cellular telephone system is built with antenna towers 90 m high. What is most nearly the maximum distance between towers?

(A) 20 km
(B) 30 km
(C) 40 km
(D) 50 km

2. A certain communication channel has a signal-to-noise ratio of 7 and a bandwidth of 10 kHz. What is the channel capacity?

(A) 30 bps
(B) 80 bps
(C) 3000 bps
(D) 30,000 bps

3. A copper cable used in a POTS line has a loss of 1.5 dB/km. What is the approximate loss experienced if the user is located 15 km from the central office?

(A) 1.5 dB
(B) 15 dB
(C) 20 dB
(D) 220 dB

4. A superheterodyne AM receiver with an intermediate frequency (IF) of 455 kHz is to be tuned to listen to a station at 540 kHz. What tuning frequency is used in the local oscillator?

(A) 95 kHz
(B) 540 kHz
(C) 550 kHz
(D) 995 kHz

5. When the receiver in Prob. 4 is tuned to the 540 kHz station, what other AM station offers potential interference?

(A) 85 kHz
(B) 995 kHz
(C) 1080 kHz
(D) 1450 kHz

SOLUTIONS

1. Microwave signals are line-of-sight signals. The radio horizon for an antenna is given by

$$d = 4.1\sqrt{h}$$

The distance, d, is derived as follows.

$$d = (4.1)\left(\sqrt{90\ \text{m}}\right) = \boxed{38.9\ \text{km}}$$

The situation is illustrated as shown.

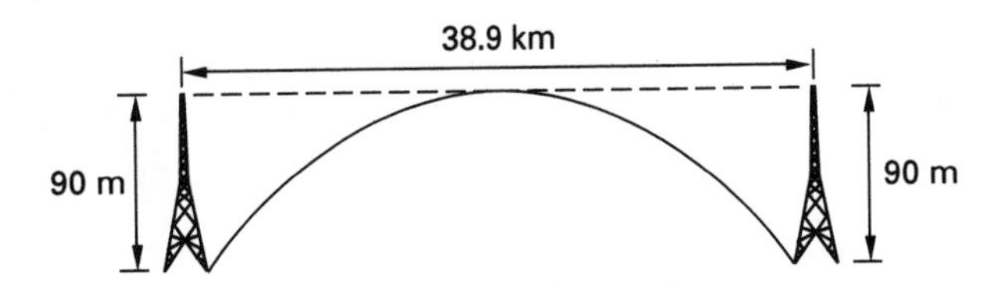

The answer is (C).

2. The channel capacity is given by

$$C = (\text{BW})\ \log_2\left(1 + \frac{S}{N}\right)$$

Substitute the given information.

$$C = (10 \times 10^3\ \text{Hz})\ \log_2(1 + 7) = \boxed{30{,}000\ \text{bps}}$$

The answer is (D).

3. The given loss is 1.5 dB/km. The total loss is

$$L_{\text{dB}} = \left(1.5\ \frac{\text{dB}}{\text{km}}\right)(15\ \text{km})$$
$$= \boxed{22.5\ \text{dB}\quad(20\ \text{dB})}$$

The answer is (C).

4. The process of heterodyning mixes the desired signal with the intermediate frequency (IF), generating the sum and difference. Thus, the heterodyned frequencies are

$$f_H = \text{signal} \pm \text{IF}$$
$$= 540\ \text{kHz} \pm 455\ \text{kHz}$$
$$= \boxed{995\ \text{kHz}}\ \text{and } 85\ \text{kHz}$$

The highest of the two is always used since this leads to a proportionally smaller band over which the oscillator must tune.

The answer is (D).

5. The oscillator frequency in Prob. 4 is 995 kHz, which is also the difference frequency between a 1450 kHz station and a 455 kHz IF. This station, called the *image station*, is located at 2(IF) from the desired station. IF filters remove the image station signal of 1450 kHz.

The answer is (D).

62 Biomedical Electrical Engineering

PRACTICE PROBLEMS

The following information is applicaable to Probs. 1 and 2.

A medical incubator is designed to maintain the temperature, humidity, and CO_2 at the proper levels for the sampled items. The humidifier uses the low-water-level alarm circuitry shown.

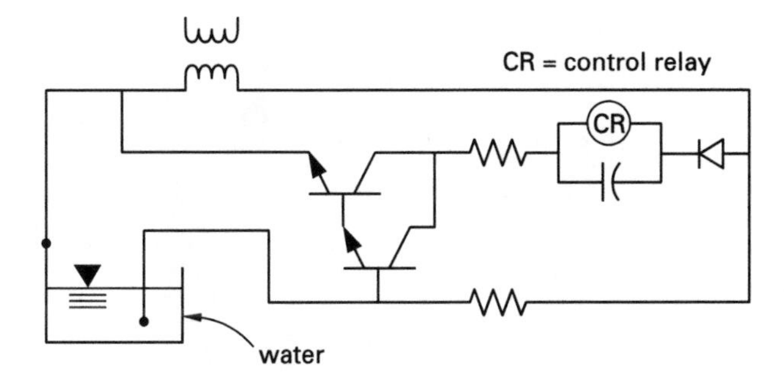

1. What is the name for the configuration of the two transistors?

(A) common collector
(B) common base
(C) common source
(D) Darlington pair

2. Which of the following transistor parameters represents the gain used to drive the control relay?

(A) h_{fe}
(B) h_{ie}
(C) h_{oe}
(D) h_{re}

3. A Class II biological safety cabinet (hood) uses a 115 V single-phase AC motor to maintain the necessary vacuum within the hood to prevent the spread of biological contaminants. With clean filters, the motor draws 10 A at its rated power factor of 0.85 lagging. What is the power consumption of the motor?

(A) 980 W
(B) 1150 W
(C) 1700 W
(D) 1990 W

4. Leakage current standards for biomedical devices are generally in the microamp (μA) range. Which of the following items is (are) used to reduce leakage currents for 60 Hz systems?

(A) DC-DC inverters
(B) rechargeable batteries
(C) motor-generator
(D) all of the above

SOLUTIONS

1. The two transistors are in a common emitter configuration (analogous to a common source configuration for field-effect transistors (FETs)). This specific configuration is more commonly referred to as a Darlington pair.

The answer is (D).

2. The Darlington pair has a high input resistance and a high current gain (h_{fe}). It is this forward current gain that enables operation of the control relay.

The answer is (A).

3. The power in a single-phase motor is given by

$$\begin{aligned} P &= IV(\text{pf}) \\ &= (10\ \text{A})(115\ \text{V})(0.85) \\ &= \boxed{977.5\ \text{W} \quad (980\ \text{W})} \end{aligned}$$

The answer is (A).

4. Leakage currents are minimized by using nonoscillatory power sources (such as DC-DC inverters and rechargeable batteries) or electrical items with nonelectromagnetic coupling (such as motor-generators).

The answer is (D).

63 Analysis of Engineering Systems: Control Systems

PRACTICE PROBLEMS

Graphical Solution

1. Consider a homogeneous second-order linear differential equation with constant coefficients. The damping ratio is 0.3, and the natural frequency is 10 rad/s. Estimate the amplitude at $t = 0.5$ s graphically.

Transfer Functions

2. Determine the transfer functions for the following systems.

(a)

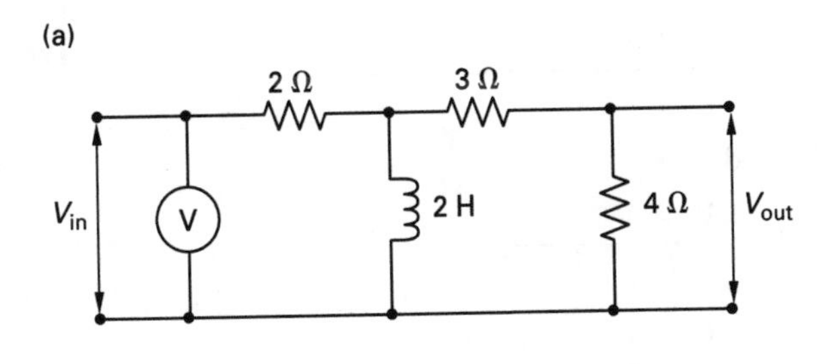

(b)

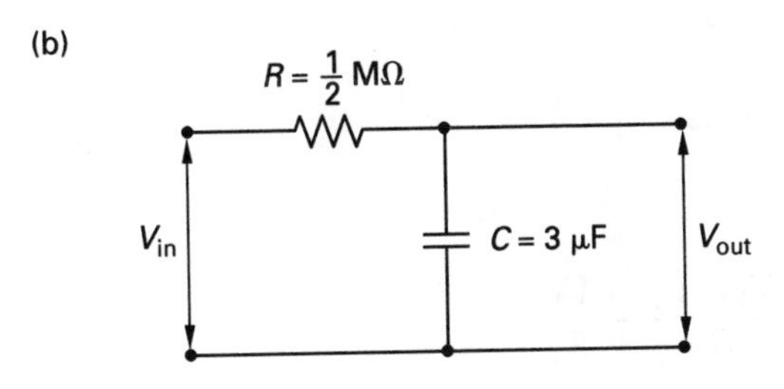

Feedback Theory

3. An amplifier consists of several stages with gains as shown. A voltage divider provides feedback. What is the overall gain with and without feedback?

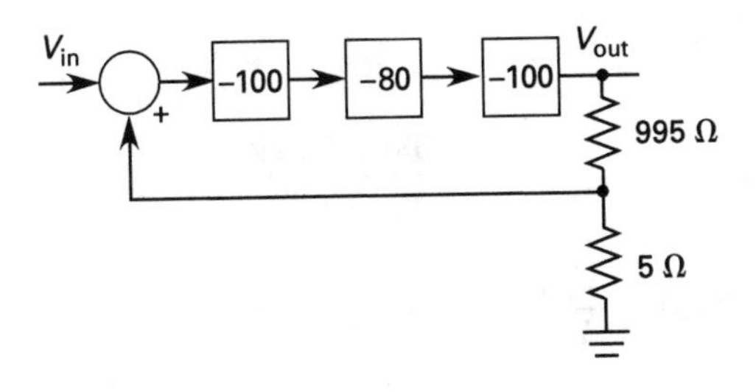

Sensitivity

4. A system is composed of a forward path and a unity feedback loop. The gain in the forward loop is $1000 \pm 10\%$. What is the uncertainty of the output signal?

Block Diagram Algebra

5. (a) Simplify the following block diagrams and (b) determine the overall system gain.

(a)

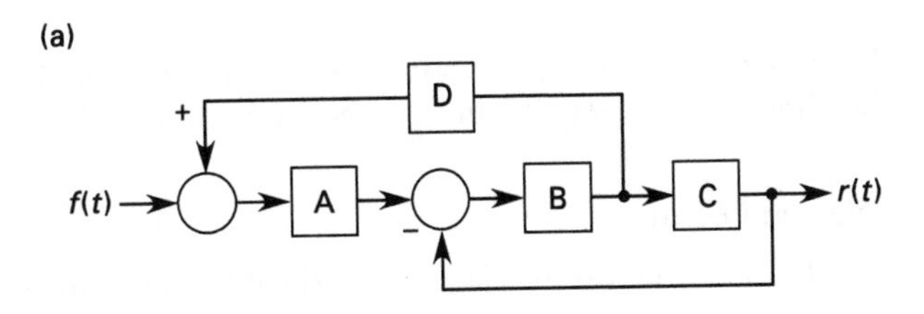

(b)

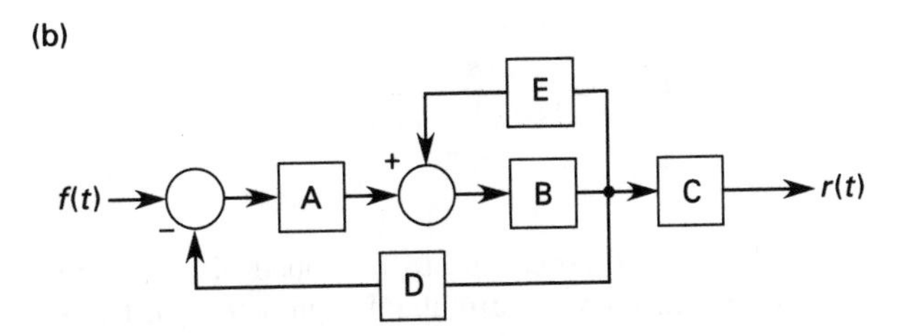

6. (a) Simplify the following block diagram and (b) determine the overall system gain. (c) What is the system sensitivity if $G_1 = -5$, $G_2 = 2$, $G_3 = 4$, and $G_4 = 3$?

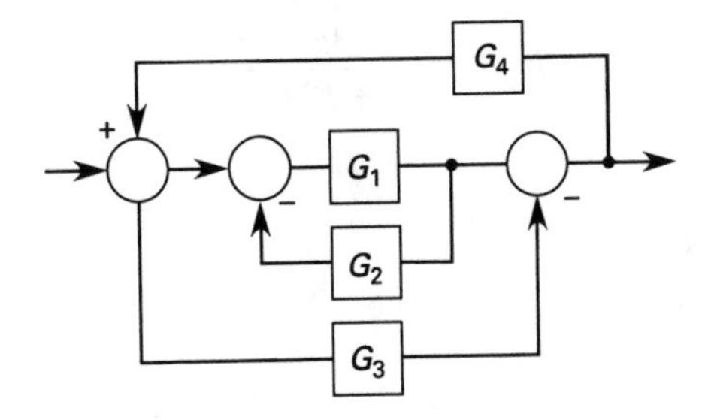

Frequency Response

7. What is the steady-state response to an impulse $\delta(t)$ of the system represented by the following differential equation? Assume all initial conditions are zero.

$$r''(t) + 3r'(t) + r(t) = \delta'(t) + \delta(t)$$

8. Find the steady-state response to a step input for a system in which the transfer function is

$$T(s) = \frac{b_0 s^p + b_1 s^{p-1} + \cdots + b_p}{s^n + a_1 s^{n-1} + \cdots + a_n} \quad [a_n \neq 0]$$

9. Consider the following transfer function.

$$T(s) = \frac{1 + s}{1 + 2s + 2s^2}$$

What is the (a) amplitude and (b) phase of the steady-state response to a sinusoidal input of $\sin \omega t$?

10. Consider a system with the following transfer function.

$$T(s) = \frac{1}{(s + a)(s + b)}$$

What is the steady-state response to a unit step (a) in the time domain and (b) in the frequency domain?

Poles and Zeros

11. Draw the pole-zero diagram for the following transfer function.

$$T(s) = \frac{(s^2 + 4)(s - 2)}{s^2(s^2 + 4s + 5)(s + 1)}$$

12. What is the steady-state response of a system to a sinusoidal input with angular frequency ω_0 if the system's transfer function has a zero at $s = j\omega_0$?

Frequency Response to Sinusoidal Forcing Functions

13. Find the (a) bandwidth, (b) peak frequency, (c) half-power points, and (d) quality factor for

$$T(s) = \frac{3s + 18}{s^2 + 12s + 3200}$$

Stability

14. Is a system with the following transfer function stable?

$$T(s) = \frac{(s - 2)(s + 3)}{(s^2 + s - 2)(s + 1)}$$

SOLUTIONS

1. At $t = 0.5$ s,

$$\omega t = \left(10\ \frac{\text{rad}}{\text{s}}\right)(0.5\ \text{s}) = 5\ \text{rad}$$

For a normalized curve of natural response corresponding to $\zeta = 0.3$,

$$\frac{x(t)}{\omega} \approx -0.22\ \text{s/rad} \quad [\text{at } t = 0.5]$$

$$x(t = 0.5\ \text{s}) = \left(-0.22\ \frac{\text{s}}{\text{rad}}\right)\left(10\ \frac{\text{rad}}{\text{s}}\right) = \boxed{-2.2}$$

2. (a)

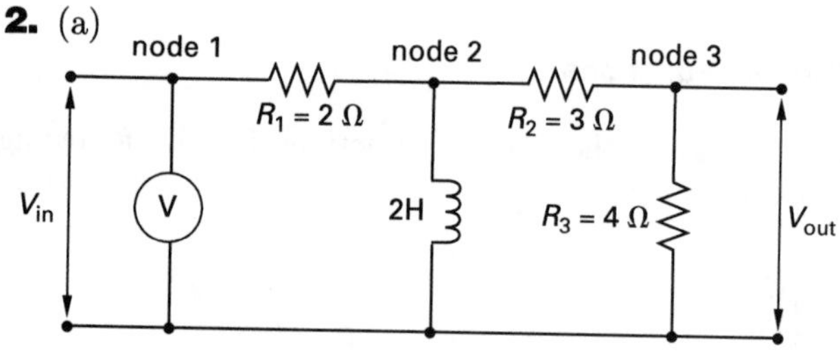

node 1 node 2 node 3

$R_1 = 2\ \Omega$, $R_2 = 3\ \Omega$, V_{in}, V, 2H, $R_3 = 4\ \Omega$, V_{out}

At node 2,

$$0 = \frac{V_2 - V_1}{R_1} + \frac{V_2 - V_3}{R_2} + \frac{1}{L}\int V_2 dt$$

$$= \frac{V_2 - V_1}{2} + \frac{V_2 - V_3}{3} + \frac{1}{2}\int V_2 dt$$

$$= \frac{V_2 - V_1}{2} + \frac{V_2 - V_3}{3} + \frac{V_2}{2s} \quad \text{[Eq. 1]}$$

At node 3,

$$0 = \frac{V_3 - V_2}{R_2} + \frac{V_3}{R_3}$$

$$= \frac{V_3 - V_2}{3} + \frac{V_3}{4} \quad \text{[Eq. 2]}$$

From Eq. 1,

$$V_2 = \frac{3sV_1 + 2sV_3}{5s + 3}$$

From Eq. 2,

$$V_2 = \frac{7V_3}{4}$$

$$\frac{3sV_1 + 2sV_3}{5s + 3} = \frac{7V_3}{4}$$

$$T(s) = \frac{V_{out}}{V_{in}} = \frac{V_3}{V_1} = \frac{12s}{27s + 21} = \boxed{\frac{4s}{9s + 7}}$$

The black box representation of this system is

$$V_{in} \rightarrow \boxed{\frac{4s}{9s+7}} \rightarrow V_{out}$$

(b)

$$0 = \frac{V_{out} - V_{in}}{R} + C\left(\frac{dV_{out}}{dt}\right)$$

$$0 = \frac{V_{out} - V_{in}}{R} + CsV_{out}$$

$$0 = V_{out} - V_{in} + CRsV_{out}$$

$$T(s) = \frac{V_{out}}{V_{in}} = \frac{1}{1 + CRs}$$

$$= \frac{1}{1 + (3 \times 10^{-6}\ \text{F})\left(\frac{1}{2} \times 10^{6}\ \Omega\right)s}$$

$$= \boxed{\frac{2}{3s+2}}$$

$$V_{in} \rightarrow \boxed{\frac{2}{3s+2}} \rightarrow V_{out}$$

3.

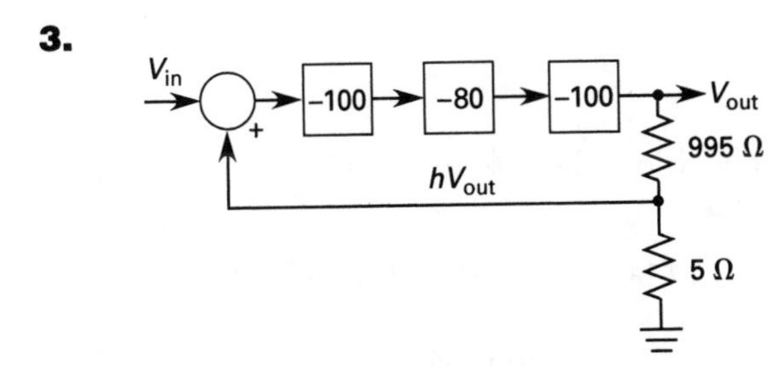

The overall gain without feedback is

$$K = (-100)(-80)(-100) = \boxed{-800{,}000}$$

$$\frac{V_{out} - hV_{out}}{995\ \Omega} = \frac{hV_{out}}{5\ \Omega}$$

$$h = \frac{5\ \Omega}{995\ \Omega + 5\ \Omega} = 0.005$$

$K < 0$, but the feedback adds, so the feedback is negative.

$$K_{loop} = \frac{K}{1 - Kh} = \frac{-800{,}000}{1 - (-800{,}000)(0.005)} = \boxed{-200}$$

4.

$$v_{out} = G_{loop}v_i$$

$$\frac{\Delta v_{out}}{v_{out}} = \frac{\Delta G_{loop}}{G_{loop}}$$

(Assume the input signal has no uncertainty.)

$$\left(\frac{\Delta G_{loop}}{G_{loop}}\right)\left(\frac{G}{\Delta G}\right) = \frac{1}{1+G}$$

$$\frac{\Delta v_{out}}{v_{out}} = \frac{\Delta G_{loop}}{G_{loop}} = \left(\frac{\Delta G}{G}\right)\left(\frac{1}{1+G}\right)$$

$$\frac{\Delta G}{G} = 0.1$$

$$\frac{\Delta v_{out}}{v_{out}} = (0.1)\left(\frac{1}{1 + 1000}\right) = 9.99 \times 10^{-5}$$

$$\approx \boxed{0.01\%}$$

5. (a)

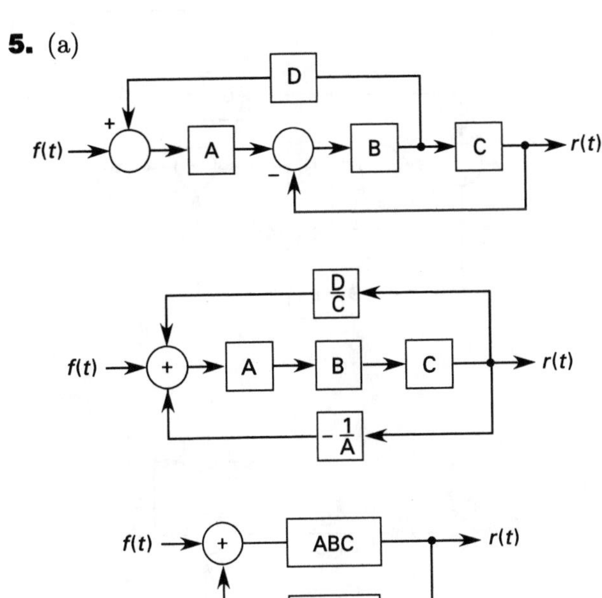

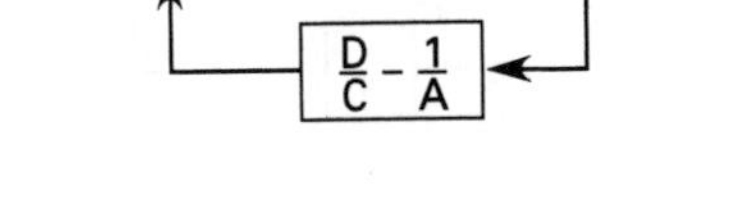

$$\text{overall system gain} = \frac{r(t)}{f(t)} = \boxed{\frac{\text{ABC}}{1 - \text{ABD} + \text{BC}}}$$

(b)

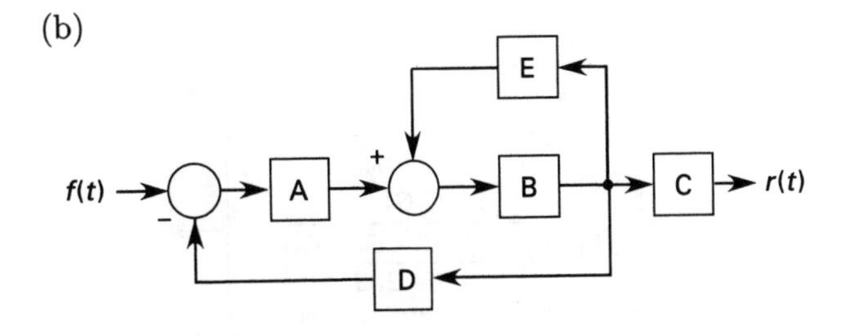

Control Systems

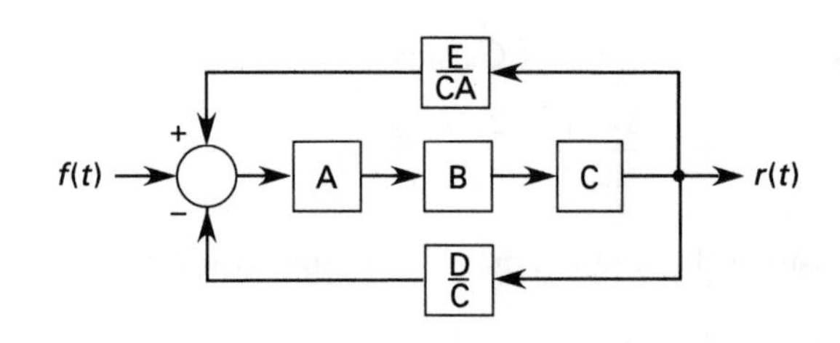

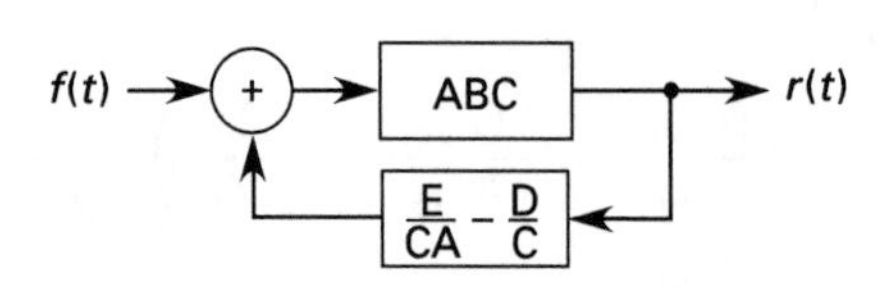

$$f(t) \rightarrow \boxed{\frac{\text{ABC}}{1-\text{BE}+\text{ABD}}} \rightarrow r(t)$$

$$\text{overall system gain} = \frac{r(t)}{f(t)} = \boxed{\frac{\text{ABC}}{1-\text{BE}+\text{ABD}}}$$

6. (a)

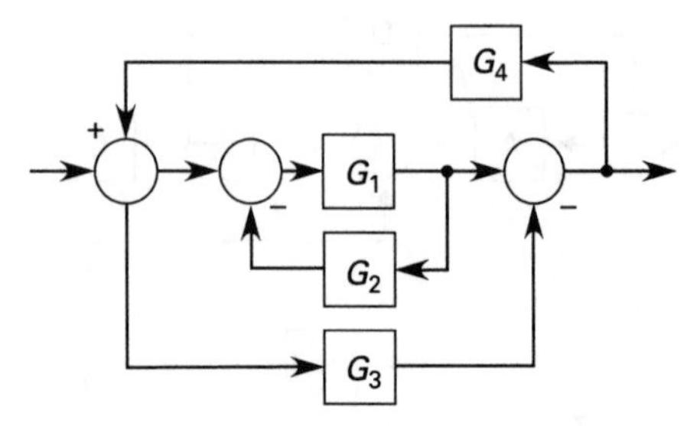

step 1:

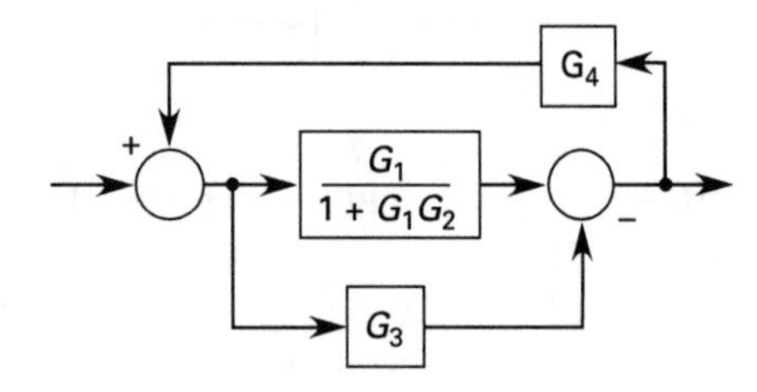

step 2:

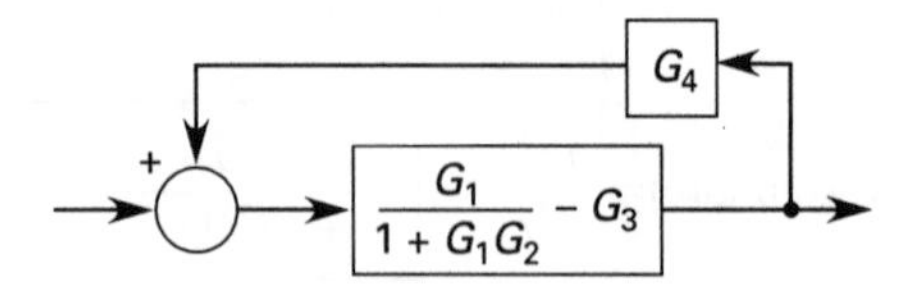

step 3:

$$\rightarrow \boxed{\frac{\dfrac{G_1 - G_3 - G_1G_2G_3}{1+G_1G_2}}{1 - G_4\left(\dfrac{G_1 - G_3 - G_1G_2G_3}{1+G_1G_2}\right)}} \rightarrow$$

$$\rightarrow \boxed{\frac{G_1 - G_3 - G_1G_2G_3}{1 + G_1G_2 - G_1G_4 + G_3G_4 + G_1G_2G_3G_4}} \rightarrow$$

(b) The overall system gain is

$$\boxed{\frac{G_1 - G_3 - G_1G_2G_3}{1 + G_1G_2 - G_1G_4 + G_3G_4 + G_1G_2G_3G_4}}$$

From step 2,

$$\begin{aligned} G(s) &= \frac{G_1}{1+G_1G_2} - G_3 \\ &= \frac{-5}{1+(-5)(2)} - 4 = -\frac{31}{9} \\ H(s) &= G_4 = 3 \end{aligned}$$

$$G(s)H(s) = \left(-\frac{31}{9}\right)(3) = -\frac{31}{3} < 0$$

(c) The system has negative feedback; therefore,

$$\begin{aligned} \text{sensitivity } S &= \frac{1}{1+G(s)H(s)} \\ &= \frac{1}{1+\left(-\dfrac{31}{3}\right)} = \boxed{-\frac{3}{28}} \end{aligned}$$

7. By definition, the system transfer function is

$$\begin{aligned} T(s) &= \mathcal{L}\left(\frac{r_{\text{sys}}(t)}{f_{\text{sys}}(t)}\right) = \frac{R_{\text{sys}}(s)}{F_{\text{sys}}(s)} \\ &= \frac{\text{output signal}}{\text{input signal}} \end{aligned}$$

The output signal is

$$\begin{aligned} r_{\text{sys}}(t) &= \delta'(t) + \delta(t) \\ R_{\text{sys}}(s) &= s + 1 \end{aligned}$$

The input signal is

$$\begin{aligned} f_{\text{sys}}(t) &= r''(t) + 3r'(t) + r(t) \\ F_{\text{sys}}(s) &= s^2 + 3s + 1 \\ T(s) &= \frac{R_{\text{sys}}(s)}{F_{\text{sys}}(s)} = \frac{s+1}{s^2+3s+1} \end{aligned}$$

The forcing function, $f(t)$, acting on this system is

$$f(t) = \delta(t)$$

$$F(s) = \mathcal{L}(f(t)) = \mathcal{L}(\delta(t)) = 1$$

$$R(s) = T(s)F(s) = (T(s))(1)$$

$$= T(s) = \boxed{\frac{s+1}{s^2+3s+1}}$$

8. Using the final value theorem, obtain the steady-state step response by substituting zero for s in the transfer function.

If $b_p \neq 0$,

$$\boxed{R(s) = T(0) = \frac{b_p}{a_n}}$$

If $b_p = 0$, the numerator is zero.

$$\boxed{R(s) = 0}$$

9. (a) Substitute $j\omega$ for s in the transfer function to obtain the steady-state response for a sinusoidal input.

$$R(s) = T(j\omega) = \frac{1+j\omega}{1+2j\omega-2\omega^2} = \frac{1+j\omega}{(1-2\omega^2)+2j\omega}$$

The amplitude is the absolute value of $R(s)$.

$$|R(s)| = \sqrt{T^2(j\omega)} = \frac{\sqrt{1+\omega^2}}{\sqrt{(1-2\omega^2)^2+4\omega^2}}$$

$$= \boxed{\sqrt{\frac{1+\omega^2}{1+4\omega^4}}}$$

(b) The phase angle of the steady-state response is $\text{Arg}(R(s))$. Find this by substituting $T(j\omega)$ into the form $(a+jb)/c$, whose argument is

$$\arctan\left(\frac{\frac{b}{c}}{\frac{a}{c}}\right) = \arctan\left(\frac{b}{a}\right)$$

First eliminate j from the denominator by multiplying by its complex conjugate $(1+2\omega^2) - 2j\omega$.

$$R(s) = T(j\omega) = \frac{(1+j\omega)(1-2\omega^2-2j\omega)}{1+4\omega^4}$$

$$= \frac{1-2\omega^2+2\omega^2+j(-\omega-2\omega^3)}{1+4\omega^4}$$

$$= \frac{1+j(-\omega-2\omega^3)}{1+4\omega^4}$$

$$\text{Arg}[R(s)] = \text{Arg}[T(j\omega)]$$

$$= \boxed{\arctan\left(\frac{-(\omega+2\omega^3)}{1}\right)}$$

10. (a) $R(s) = T(s)F(s)$

$$f(t) = \mu_0 \quad \text{[unit step]}$$

$$F(s) = \mathcal{L}(f(t)) = \frac{1}{s}$$

$$R(s) = T(s)\left(\frac{1}{s}\right) = \frac{1}{s(s+a)(s+b)}$$

From the transform table,

$$r(t) = \mathcal{L}^{-1}[R(s)] = \frac{1}{ab} + \frac{be^{-at}-ae^{-bt}}{ab(a-b)}$$

$$= r_1 + r_2(t)$$

The steady-state response is $\lim_{t\to\infty} r(t)$. Since $\lim_{t\to\infty} r_2(t) = 0$, r_1 is the steady-state response.

$$r_1 = \boxed{\frac{1}{ab}}$$

(b) The steady-state response in the frequency domain for a step input of magnitude h is $R(s) = hT(0)$. Substituting $h = 1$ and $s = 0$ in $T(s)$,

$$R(s) = (h)(T(0)) = \frac{1}{(0+a)(0+b)} = \boxed{\frac{1}{ab}}$$

11.

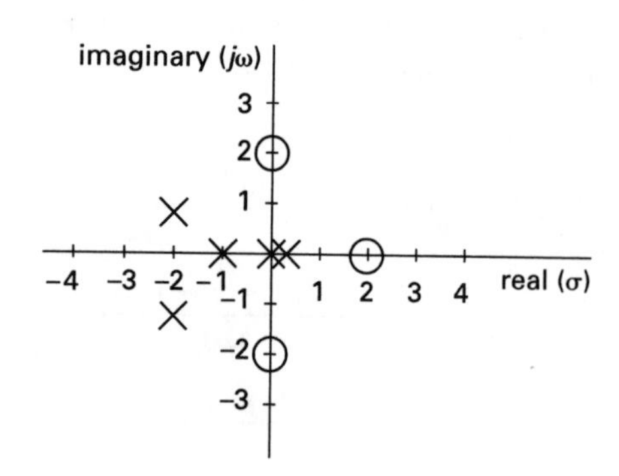

$$\frac{(s^2+4)(s-2)}{s^2(s^2+4s+5)(s+1)} = \frac{(s-2j)(s+2j)(s-2)}{s^2(s+1)(s+2-j)(s+2+j)}$$

For zeros, set the numerator equal to zero. For poles, set the denominator equal to zero.

12. The amplitude and phase of the steady-state response are given by $T(j\omega_o)$, where $T(s)$ is the transfer function. Since $j\omega_o$ is a zero of $T(s)$,

$$R(s) = T(j\omega_o) = 0$$

Therefore, the steady-state response is zero.

The system entirely blocks the angular frequency, ω_o.

13. (a)

$$T(s) = \frac{3s+18}{s^2+12s+3200} = \frac{as+b}{s^2+(\text{BW})s+\omega_n{}^2}$$

$$\text{bandwidth} = \text{BW} = \boxed{12 \text{ rad/s}}$$

(b)

$$\text{peak frequency} = \omega_n = \sqrt{3200} = \boxed{56.57 \text{ rad/s}}$$

(c)

$$\text{half-power points} = \omega_n \pm \frac{\text{BW}}{2} = 56.57 \ \frac{\text{rad}}{\text{s}} \pm \frac{12 \ \frac{\text{rad}}{\text{s}}}{2} = \boxed{\begin{array}{l}62.57 \text{ rad/s and}\\ 50.57 \text{ rad/s}\end{array}}$$

(d)

$$\text{quality factor} = Q = \frac{\omega_n}{\text{BW}} = \frac{56.57 \ \frac{\text{rad}}{\text{s}}}{12 \ \frac{\text{rad}}{\text{s}}} = \boxed{4.71}$$

14. The system has a pole in the right half of the s-plane ($s = 1$).

$$s^2 + s - 2 = (s-1)(s+2)$$

The system is unstable.

64 Electrical Materials

PRACTICE PROBLEMS

There are no practice problems corresponding with Ch. 64 of the *Electrical Engineering Reference Manual*.

65 National Electrical Code

PRACTICE PROBLEMS

1. An unknown wire gage must be determined during an electrical inspection. The wire diameter is measured as 0.2591 cm. What is the AWG standard designation?

(A) AWG 4/0
(B) AWG 4
(C) AWG 8
(D) AWG 10

2. A given circuit is meant to carry a continuous lighting load of 16 A. In addition, four loads designed for permanent display stands are fastened in place and require 2 A each when operating. What is the rating of the overcurrent protective device (OCPD) on the branch circuit?

(A) 20 A
(B) 23 A
(C) 28 A
(D) 30 A

3. If standard three-conductor cable is used for the branch circuit in Prob. 2, what is the minimum conductor size?

(A) AWG 8
(B) AWG 10
(C) AWG 12
(D) AWG 14

4. A dwelling two-wire resistive heating circuit is designed for 240 V and 30 A. AWG 8 copper wire is used with a resistance of 0.809 Ω/1000 ft. What is the voltage drop for a one-way circuit length of 75 ft?

(A) 1.0%
(B) 1.5%
(C) 3.6%
(D) 5.0%

5. A three-wire feeder, AWG 1/0, carries 100 A of continuous load from a 480 V, 3ϕ (three-phase), three-wire source. The THHN copper conductors operate at 75°C and are enclosed within aluminum conduit. The expected power factor is 0.8. The total circuit length is 150 m. What is the voltage drop?

(A) 4.0 V
(B) 7.0 V
(C) 11.0 V
(D) 14.0 V

6. A 120 V dwelling branch circuit supplies four outlets, one of which has four receptacles. What is the total volt-ampere load?

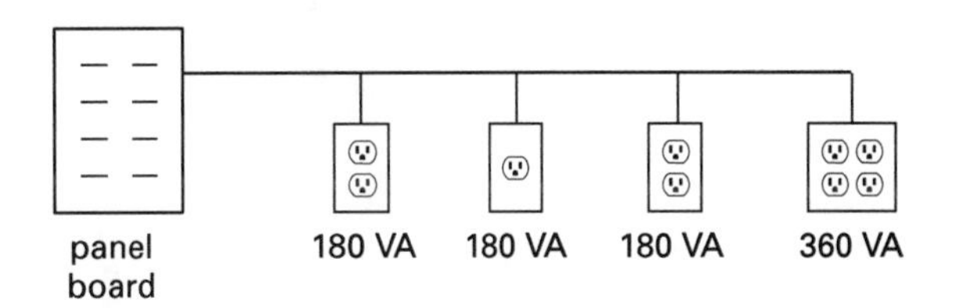

(A) 180 VA
(B) 360 VA
(C) 720 VA
(D) 900 VA

7. A three-phase, four-wire feeder with a full-sized neutral carries 14 A continuous and 40 A noncontinuous loads. The feeder uses an overcurrent device with a terminal rating of 75°C. What is the minimum copper conductor size?

(A) AWG 3
(B) AWG 4
(C) AWG 6
(D) AWG 8

8. All of the following are properties of fuses rather than circuit breakers with one exception. Which item is the property of circuit breakers?

(A) adjustable time-current characteristics
(B) low initial costs
(C) longer downtime after operation
(D) simplicity

9. Installed equipment within a dwelling is attached to an overcurrent device (breaker) rated at 15 A. What is the required size of the equipment's grounding copper conductor?

(A) AWG 10
(B) AWG 12
(C) AWG 14
(D) AWG 18

10. A three-phase, four-wire feeder carrying less than 100 A uses THWN AWG 8 copper wire in an ambient operating environment of 50°C. The derating factor for four conductors in a single raceway is 0.8 from NEC Table 310-15(b)(2)(a). What is the allowable ampacity?

(A) 19 A
(B) 23 A
(C) 32 A
(D) 40 A

SOLUTIONS

1. The wire diameter is

$$d = (0.2591 \text{ cm})\left(\frac{1 \text{ in}}{2.54 \text{ cm}}\right) = 0.1020 \text{ in}$$

Using this to enter a table of wire gages, such as the one in App. 65A, indicates a 10 gage wire. That is, AWG 10.

The answer is (D).

2. From Article 210-20(a) of the National Electric Code® (NEC®), the overcurrent protective device (OCPD) must be rated at 100% of the noncontinuous load plus 125% of the continuous load. Thus,

$$\begin{aligned} \text{OCPD} &= (1.00)(2 \text{ A} + 2 \text{ A} + 2 \text{ A} + 2 \text{ A}) + (1.25)(16 \text{ A}) \\ &= 28 \text{ A minimum} \end{aligned}$$

A standard fixed-trip circuit breaker or a fuse rated at 30 A can be used (see Sec. 240-6).

The answer is (D).

3. Conductor size is based on 100% of the noncontinuous load plus 125% of the continuous load [Sec. 210-19(a)]. Thus, the total load for Prob. 2 is 28 A. The ampacity cannot be less than the rating of the overcurrent device, 30 A in this case [Sec. 310-15(b)(2)]. The ampacity of the conductor for a circuit of less than 100 A should be based on the 60°C column of NEC Table 310-16 (App. 65.B) [Sec. 110-14(c)(1)].

Using App. 65.B, an AWG 10 conductor is chosen. (Checking Article 240-3 of the Code, as is called for by the asterisk in App. 65.B, indicates that 10 gage wire cannot have overcurrent protection greater than 30 A. Thus, all requirements are met.) Note that any derating of the conductor ampacity requires the use of small gage (larger diameter) conductor. See the conductor factors at the bottom of App. 65.B.

The answer is (B).

4. A two-wire circuit is single phase or DC. The heating is resistive, so pf = 1 if the circuit is AC, which is the most likely scenario. Regardless, the voltage drop is given by Ohm's law as

$$\begin{aligned} V &= IR \\ &= IR_l l \end{aligned}$$

The voltage drop occurs on both wires, so a factor of two must be used.

$$V = IR_l 2l$$

Substituting gives

$$\begin{aligned} V &= 2IR_l l \\ &= (2)(30\ \text{A})\left(\frac{0.809\ \Omega}{1000\ \text{ft}}\right)(75\ \text{ft}) \\ &= 3.64\ \text{V} \end{aligned}$$

As a percentage of the 240 V nominal voltage, the voltage drop is

$$\left(\frac{3.64\ \text{V}}{240\ \text{V}}\right)(100\%) = \boxed{1.516\%}$$

The answer is (B).

Note that the conductor ampacities can be found in App. 65.B (NEC Table 310-16). Conductor properties, such as Ω per 1000 ft, are found in App. 65.C (NEC Ch. 9, Table 8).

5. Note that although the continuous load is 100 A and the voltage drop is calculated using that value, the conductors and the overcurrent protector must be able to carry 125 A, that is 125% of the continuous load [Secs. 215-2(a) and 215-3].

The AC voltage drop is given by

$$V_{\text{drop}} = IZ$$

The effective impedance, corrected for the power factor, is calculated using App. 65.D (NEC Ch. 9, Table 9). For a 1/0 gage feeder, the resistance to neutral per 1000 ft in aluminum conduit is 0.13 Ω.

The inductive reactance, X_L, to neutral per 1000 ft is 0.044 Ω. The columns titled "alternating-current resistance for uncoated copper wires" and "X_L (reactance) for all wires" are used. Note that the uncoated copper wire column is used, although the THHN conductors are insulated. Also, the capacitive reactance is ignored. Thus, the voltage drop calculated will be an approximate value.

Since the power factor is not 0.85, the effective impedance cannot be taken directly from the table. The corrected impedance, Z_c, is

$$\begin{aligned} Z_c &= R_l(\text{pf}) + X_{L,l}\sin(\arccos \text{pf}) \\ &= \left(0.13\ \frac{\Omega}{1000\ \text{ft}}\right)(0.8) \\ &\quad + \left(0.044\ \frac{\Omega}{1000\ \text{ft}}\right)\sin(36.86) \\ &= \left(0.1040\ \frac{\Omega}{1000\ \text{ft}} + 0.0264\ \frac{\Omega}{1000\ \text{ft}}\right) \\ &= 0.1304\ \Omega/1000\ \text{ft} \end{aligned}$$

Since this is an "ohms-to-neutral" value, the line-to-neutral, or phase, voltage drop is

$$\begin{aligned} V_{\text{drop (line-to-neutral)}} &= V_\phi = IZ_c \\ &= (100\ \text{A})\left(0.1304\ \frac{\Omega}{1000\ \text{ft}}\right) \\ &= 13.04\ \text{V}/1000\ \text{ft} \end{aligned}$$

Adjusting for the total circuit length gives

$$\begin{aligned} V_\phi &= \left(13.04\ \frac{\text{V}}{1000\ \text{ft}}\right)(150\ \text{m})\left(\frac{1\ \text{ft}}{0.3048\ \text{m}}\right) \\ &= 6.417\ \text{V} \end{aligned}$$

Voltage drops are given in terms of line quantities for a 3ϕ, three-wire system. Thus,

$$\begin{aligned} V_{\text{line}} &= V_\phi\sqrt{3} \\ &= (6.417\ \text{V})(\sqrt{3}) \\ &= \boxed{11.1\ \text{V}} \end{aligned}$$

This is approximately 25% of the 480 V source and is thus within NEC recommendations.

The answer is (C).

6. The situation is as shown.

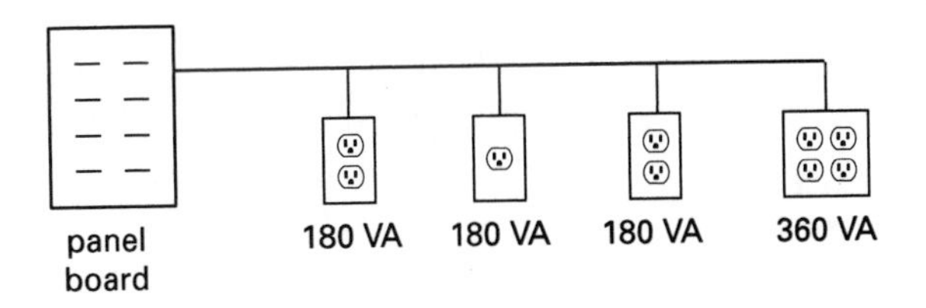

On any outlet with less than four receptacles, the VA load is 180 VA [Sec. 220-3(b)(9)]. For four or more receptacles, the load is

$$\begin{aligned} \text{VA} &= (\text{numbers of receptacles})\left(90\ \frac{\text{VA}}{\text{receptacles}}\right) \\ &= (4)(90\ \text{VA}) \\ &= 360\ \text{VA} \end{aligned}$$

The total load is $\boxed{900\ \text{VA}}$

The answer is (D).

7. Feeder conductor size, before derating, is based on 100% of the noncontinuous load and 125% of the continuous load [Art. 215-2(a)]. Thus,

$$\begin{aligned} A_{\text{load}} &= (1.00)(40\ \text{A}) + (1.25)(14\ \text{A}) \\ &= 40\ \text{A} + 17.5\ \text{A} \\ &= 57.5\ \text{A} \approx 58\ \text{A} \end{aligned}$$

The total of 58 A is used since an ampacity of 0.5 A or greater is rounded up [Sec. 220-2(b)]. Using App. 65.B (NEC Table 310-16), AWG 6 from the 75°C column is picked. But, for all circuits less than 100 A, the 60°C column is used [Sec. 110-14(c)(1)]. (Because of this article, the column primarily used in dwelling calculations is the one titled "60°C (140°F)".) Therefore, the AWG 6 conductor cannot be used as it is rated for 55 A at 60°C. Instead, AWG 4 with a 70 A rating is used.

The answer is (B).

8. The electrical characteristics of fuses are set by the material used, and no external adjustments can occur. Thus, the time-current characteristics cannot change.

The answer is (A).

9. Equipment grounding conductor sizes are given in NEC Table 250-122. For a 15 A overcurrent device, the grounding conductor must be AWG 14 copper.

The answer is (C).

10. Because the circuit carries less than 100 A, the 60°C column of App. 65.B must be used [see Sec. 110-14(c)(1)]. Thus, 40 A is the allowable ampacity. The operating environment is above 30°C for which the values are calculated. Thus, using the correction factor table in App. 65.B, 0.58 is found in the 46-50°C row. The allowable ampacity in a given environment is determined by multiplying the rated ampacity in the table by all the correction factors—in this case, one for temperature and one for more than three conductors in a raceway.

$$\begin{aligned} A_{\text{total}} &= A_{\text{rated}} F_{\text{temp}} F_{\text{cond}} \\ &= (40\ \text{A})(0.58)(0.8) \\ &= \boxed{18.56\ \text{A} \quad (19\ \text{A})} \end{aligned}$$

The answer is (A).

66 Engineering Economic Analysis

PRACTICE PROBLEMS

1. At 6% effective annual interest, how much will be accumulated if $1000 is invested for ten years?

2. At 6% effective annual interest, what is the present worth of $2000 that becomes available in four years?

3. At 6% effective annual interest, how much should be invested to accumulate $2000 in 20 years?

4. At 6% effective annual interest, what year-end annual amount deposited over seven years is equivalent to $500 invested now?

5. At 6% effective annual interest, what will be the accumulated amount at the end of ten years if $50 is invested at the end of each year for ten years?

6. At 6% effective annual interest, how much should be deposited at the start of each year for ten years (a total of ten deposits) in order to empty the fund by drawing out $200 at the end of each year for ten years (a total of ten withdrawals)?

7. At 6% effective annual interest, how much should be deposited at the start of each year for five years to accumulate $2000 on the date of the last deposit?

8. At 6% effective annual interest, how much will be accumulated in ten years if three payments of $100 are deposited every other year for four years, with the first payment occurring at $t = 0$?

9. $500 is compounded monthly at a 6% effective annual interest rate. How much will have accumulated in five years?

10. What is the effective annual rate of return on an $80 investment that pays back $120 in seven years?

11. A new machine will cost $17,000 and will have a resale value of $14,000 after five years. Special tooling will cost $5000. The tooling will have a resale value of $2500 after five years. Maintenance will be $2000 per year. The effective annual interest rate is 6%. What will be the average annual cost of ownership during the next five years?

12. An old covered wooden bridge can be strengthened at a cost of $9000, or it can be replaced for $40,000. The present salvage value of the old bridge is $13,000. It is estimated that the reinforced bridge will last for 20 years, will have an annual cost of $500, and will have a salvage value of $10,000 at the end of 20 years. The estimated salvage value of the new bridge after 25 years is $15,000. Maintenance for the new bridge would cost $100 annually. The effective annual interest rate is 8%. Which is the best alternative?

13. A firm expects to receive $32,000 each year for 15 years from sales of a product. An initial investment of $150,000 will be required to manufacture the product. Expenses will run $7530 per year. Salvage value is zero, and straight line depreciation is used. The income tax rate is 48%. What is the after-tax rate of return?

14. A public works project has initial costs of $1,000,000, benefits of $1,500,000, and disbenefits of $300,000. (a) What is the benefit/cost ratio? (b) What is the excess of benefits over costs?

15. A speculator in land pays $14,000 for property that he expects to hold for ten years. $1000 is spent in renovation, and a monthly rent of $75 is collected from the tenants. (Use the year-end convention.) Taxes are $150 per year, and maintenance costs are $250 per year. What must be the sale price in ten years to realize a 10% rate of return?

16. What is the effective annual interest rate for a payment plan of 30 equal payments of $89.30 per month when a lump sum payment of $2000 would have been an outright purchase?

17. A depreciable item is purchased for $500,000. The salvage value at the end of 25 years is estimated at $100,000. What is the depreciation in each of the first three years using the (a) straight line, (b) sum-of-the-years' digits, and (c) double declining balance methods?

18. Equipment that is purchased for $12,000 now is expected to be sold after ten years for $2000. The estimated maintenance is $1000 for the first year, but it is expected to increase $200 each year thereafter. The effective annual interest rate is 10%. What are the (a) present worth and (b) annual cost?

19. A new grain combine with a 20-year life can remove seven pounds of rocks from its harvest per hour. Any rocks left in its output hopper will cause $25,000 damage in subsequent processes. Several investments are available to increase the rock-removal capacity, as listed in the table. The effective annual interest rate is 10%. What should be done?

rock removal rate	probability of exceeding rock removal rate	required investment to achieve removal rate
7	0.15	0
8	0.10	$15,000
9	0.07	$20,000
10	0.03	$30,000

20. A mechanism that costs $10,000 has operating costs and salvage values as given. An effective annual interest rate of 20% is to be used.

year	operating cost	salvage value
1	$2000	$8000
2	$3000	$7000
3	$4000	$6000
4	$5000	$5000
5	$6000	$4000

(a) What is the economic life of the mechanism? (b) Assuming that the mechanism has been owned and operated for four years already, what is the cost of owning and operating the mechanism for one more year?

21. A salesperson intends to purchase a car for $50,000 for personal use, driving 15,000 miles per year. Insurance for personal use costs $2000 per year, and maintenance costs $1500 per year. The car gets 15 miles per gallon, and gasoline costs $1.50 per gallon. The resale value after five years will be $10,000. The salesperson's employer has asked that the car be used for business driving of 50,000 miles per year and has offered a reimbursement of $0.30 per mile. Using the car for business would increase the insurance cost to $3000 per year and maintenance to $2000 per year. The salvage value after five years would be reduced to $5000. If the employer purchased a car for the salesperson to use, the initial cost would be the same, but insurance, maintenance, and salvage would be $2500, $2000, and $8000, respectively. The salesperson's effective annual interest rate is 10%. (a) Is the reimbursement offer adequate? (b) With a reimbursement of $0.30 per mile, how many miles must the car be driven per year to justify the employer buying the car for the salesperson to use?

22. Alternatives A and B are being evaluated. The effective annual interest rate is 10%. Which alternative is economically superior?

	alternative A	alternative B
first cost	$80,000	$35,000
life	20 years	10 years
salvage value	$7000	0
annual costs		
years 1–5	$1000	$3000
years 6–10	$1500	$4000
years 11–20	$2000	0
additional cost		
year 10	$5000	0

23. A car is needed for three years. Plans A and B for acquiring the car are being evaluated. An effective annual interest rate of 10% is to be used. Which plan is economically superior?

plan A: Lease the car for $0.25/mile (all inclusive).

plan B: Purchase the car for $30,000, keep the car for three years, and sell the car after three years for $7200. Pay $0.14 per mile for oil and gas. Pay other costs of $500 per year.

24. Two methods are being considered to meet strict air pollution control requirements over the next ten years. Method A uses equipment with a life of ten years. Method B uses equipment with a life of five years that will be replaced with new equipment with an additional life of five years. Capacities of the two methods are different, but operating costs do not depend on the throughput. Operation is 24 hours per day, 365 days per year. The effective annual interest rate for this evaluation is 7%.

	method A	method B	
	years 1–10	years 1–5	years 6–10
installation cost	$13,000	$6000	$7000
equipment cost	$10,000	$2000	$2200
operating cost per hour	$10.50	$8.00	$8.00
salvage value	$5000	$2000	$2000
capacity (tons/yr)	50	20	20
life	10 years	5 years	5 years

(a) What is the uniform annual cost per ton for each method? (b) Over what range of throughput (in tons/yr) does each method have the minimum cost?

25. A transit district has asked for assistance in determining the proper fare for its bus system. An effective annual interest rate of 7% is to be used. The following additional information was compiled for the study.

cost per bus	$60,000
bus life	20 years
salvage value	$10,000
miles driven per year	37,440
number of passengers per year	80,000
operating cost	$1.00 per mile in the first year, increasing $0.10 per mile each year thereafter

(a) If the fare is to remain constant for the next 20 years, what is the break-even fare per passenger? (b) If the transit district decides to set the per-passenger fare at $0.35 for the first year, by what amount should the per-passenger fare go up each year thereafter such that the district can break even in 20 years? (c) If the transit district decides to set the per-passenger fare at $0.35 for the first year and the per-passenger fare goes up $0.05 each year thereafter, what additional governmental subsidy (per passenger) is needed for the district to break even in 20 years?

26. Make a recommendation to a client to accept one of the following alternatives. Use the present worth comparison method. (Initial costs are the same.)

alternative A: a 25-year annuity paying $4800 at the end of each year, where the interest rate is a nominal 12% per annum

alternative B: a 25-year annuity paying $1200 every quarter at 12% nominal annual interest

27. A firm has two alternatives for improvement of its existing production line. The data are as follows.

	alternative A	alternative B
initial installation cost	$1500	$2500
annual operating cost	$800	$650
service life	5 years	8 years
salvage value	0	0

Determine the best alternative using an interest rate of 15%.

28. Two mutually exclusive alternatives requiring different investments are being considered. The life of both alternatives is estimated at 20 years with no salvage values. The minimum rate of return that is considered acceptable is 4%. Which alternative is best?

	alternative A	alternative B
investment required	$70,000	$40,000
net income per year	$5620	$4075
rate of return on total investment	5%	8%

29. Compare the costs of two plant renovation schemes, alternatives A and B. Assume equal lives of 25 years, no salvage values, and interest at 25%. Make the comparison on the basis of (a) present worth, (b) capitalized cost, and (c) annual cost.

	alternative A	alternative B
first cost	$20,000	$25,000
annual expenditure	$3000	$2500

30. With interest at 8%, obtain the solutions to the following to the nearest dollar. (a) A machine costs $18,000 and has a salvage value of $2000. It has a useful life of eight years. What is its book value at the end of five years using straight line depreciation? (b) Using data from part (a), find the depreciation in the first three years using the sinking fund method. (c) Repeat part (a) using double declining balance depreciation to find the first five years' depreciation.

31. A chemical pump motor unit is purchased for $14,000. The estimated life is eight years, after which it will be sold for $1800. Find the depreciation in the first two years by the sum-of-the-years' digits method. Calculate the after-tax depreciation recovery using 15% interest with 52% income tax.

32. A soda ash plant has the water effluent from processing equipment treated in a large settling basin. The settling basin eventually discharges into a river that runs alongside the basin. Recently enacted environmental regulations require all rainfall on the plant to be diverted and treated in the settling basin. A heavy rainfall will cause the entire basin to overflow. An uncontrolled overflow will cause environmental damage and heavy fines. The construction of additional height on the existing basic walls is under consideration.

Data on the costs of construction and expected costs for environmental cleanup and fines are shown. Data on 50 typical winters have been collected. The soda ash plant management considers 12% to be their minimum rate of return, and it is felt that after 15 years the plant will be closed. The company wants to select the alternative that minimizes its total expected costs.

additional basin height (ft)	number of winters with basin overflow	expense for environmental cleanup per year	construction cost
0	24	$550,000	0
5	14	$600,000	$600,000
10	8	$650,000	$710,000
15	3	$700,000	$900,000
20	1	$800,000	$1,000,000
	50		

33. A wood processing plant installed a waste gas scrubber at a cost of $30,000 to remove pollutants from the exhaust discharged into the atmosphere. The scrubber has no salvage value and will cost $18,700 to operate next year, with operating costs expected to increase at the rate of $1200 per year thereafter. When should the company consider replacing the scrubber? Money can be borrowed at 12%.

34. Two alternative piping schemes are being considered by a water treatment facility. On the basis of a ten-year life and an interest rate of 12%, determine the number of hours of operation for which the two installations will be equivalent.

	alternative A	alternative B
pipe diameter	4 in	6 in
head loss for required flow	48 ft	26 ft
size motor required	20 hp	7 hp
energy cost per hour of operation	$0.30	$0.10
cost of motor installed	$3600	$2800
cost of pipes and fittings	$3050	$5010
salvage value at end of ten years	$200	$280

35. An 88% learning curve is used with an item whose first production time was six weeks. (a) How long will it take to produce the fourth item? (b) How long will it take to produce the sixth through fourteenth items?

36. A company is considering two alternatives, only one of which can be selected.

alternative	initial investment	salvage value	annual net profit	life
A	$120,000	$15,000	$57,000	5 yr
B	$170,000	$20,000	$67,000	5 yr

The net profit is after operating and maintenance costs but before taxes. The company pays 45% of its year-end profit as income taxes. Use straight line depreciation. Do not use investment tax credit. Find the best alternative if the company's minimum attractive rate of return is 15%.

37. A company is considering the purchase of equipment to expand its capacity. The equipment cost is $300,000. The equipment is needed for five years, after which it will be sold for $50,000. The company's before-tax cash flow will be improved $90,000 annually by the purchase of the asset.

The corporate tax rate if 48%, and straight line depreciation will be used. The company will take an investment tax credit of 6.67%. What is the after-tax rate of return associated with this equipment purchase?

38. A 120-room hotel is purchased for $2,500,000. A 25-year loan is available for 12%. A study was conducted to determine the various occupancy rates.

occupancy	probability
65% full	0.40
70%	0.30
75%	0.20
80%	0.10

The operating costs of the hotel are as follows.

taxes and insurance	$20,000 annually
maintenance	$50,000 annually
operating	$200,000 annually

The life of the hotel is figured to be 25 years when operating 365 days per year. The salvage value after 25 years is $500,000.

Neglect tax credit and income taxes. Determine the average rate that should be charged per room per night to return 15% of the initial cost each year.

39. A company is insured for $3,500,000 against fire. The insurance rate is $0.69/1000. The insurance company will decrease the rate to $0.47/1000 if fire sprinklers are installed. The initial cost of the sprinklers is $7500. Annual costs are $200; additional taxes are $100 annually. The system life is 25 years. What is the rate of return on this investment?

40. Heat losses through the walls in an existing building cost a company $1,300,000 per year. This amount is considered excessive, and two other options are being evaluated. Neither of the options will increase the life of the existing building beyond the current expected life of six years, and neither of the alternatives will produce a salvage value.

alternative A: Do nothing. Continue with current losses.

alternative B: Spend \$2,000,000 immediately to upgrade the building and reduce the loss by 80%. This alternative will require annual maintenance of \$150,000.

alternative C: Spend \$1,200,000 immediately. Repeat the \$1,200,000 expenditure three years from now. Heat loss the first year will be reduced 80%. Due to deterioration, the reduction will be 55% and 20% in the second and third years. (The pattern is repeated starting after the second expenditure.) There are no maintenance costs.

All energy and maintenance costs are considered expenses for tax purposes. The company's tax rate is 48%, and straight line depreciation is used. 15% is regarded as the effective annual interest rate. Evaluate each alternative on an after-tax basis, and recommend the best alternative.

41. Determine if a seven-year-old machine should be replaced. Give a full explanation. Use a before-tax interest rate of 15%.

The existing machine is presumed to have a ten-year life. It has been depreciated on a straight line basis from its original value of \$1,250,000 to a current book value of \$620,000. Its ultimate salvage value was assumed to be \$350,000 for purposes of depreciation. Its present salvage value is estimated at \$400,000, and this is not expected to change over the next three years. The current operating costs are not expected to change from \$200,000 per year.

A new machine costs \$800,000, with operating costs of \$40,000 the first year, and increasing by \$30,000 each year thereafter. The new machine has an expected life of ten years. The salvage value depends on the year the new machine is retired.

year retired	salvage
1	\$600,000
2	\$500,000
3	\$450,000
4	\$400,000
5	\$350,000
6	\$300,000
7	\$250,000
8	\$200,000
9	\$150,000
10	\$100,000

SOLUTIONS

1.

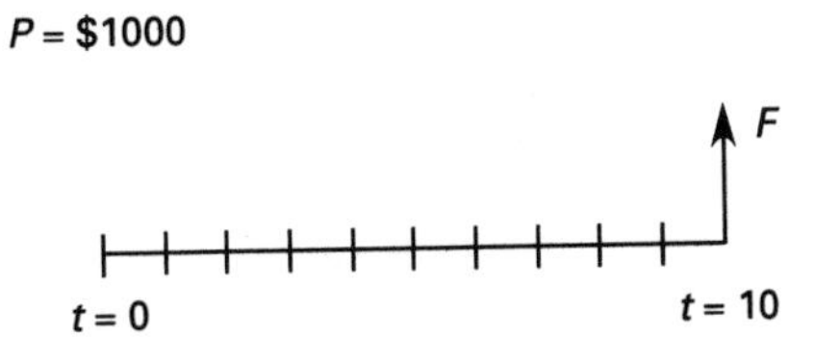

i = 6% a year

By the formula from Table 66.1,

$$F = P(1+i)^n = (\$1000)(1+0.06)^{10}$$

$$= \boxed{\$1790.85}$$

By the factor converting P to F, $(F/P,i,n) = 1.7908$ for $i = 6\%$ per year and $n = 10$ years.

$$F = P(F/P,6\%,10) = (\$1000)(1.7908)$$

$$= \boxed{\$1790.80}$$

2.

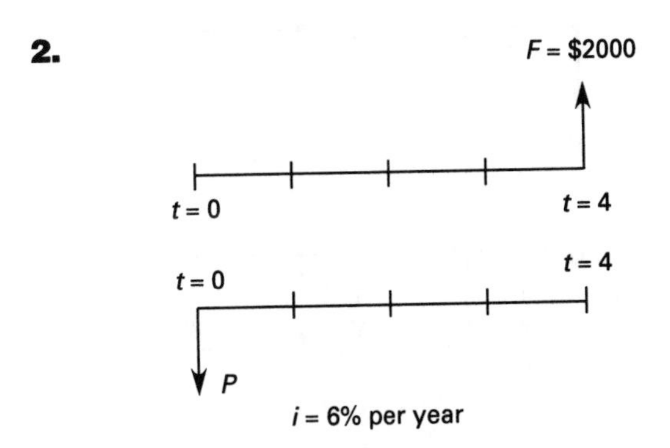

By the formula from Table 66.1,

$$P = \frac{F}{(1+i)^n} = \frac{\$2000}{(1+0.06)^4}$$

$$= \boxed{\$1584.19}$$

From the factor converting F to P, $(P/F,i,n) = 0.7921$ for $i = 6\%$ per year and $n = 4$ years.

$$\text{P} = F(P/F,6\%,4) = (\$2000)(0.7921)$$

$$= \boxed{\$1584.20}$$

3.

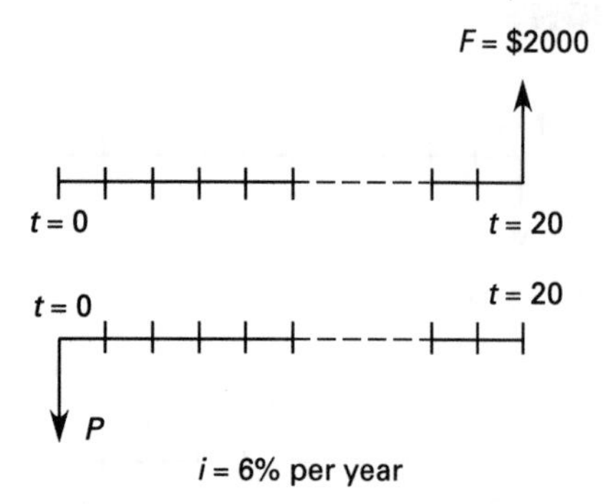

By the formula from Table 66.1,

$$P = \frac{F}{(1+i)^n} = \frac{\$2000}{(1+0.06)^{20}}$$

$$= \boxed{\$623.61}$$

From the factor converting F to P, $(P/F, i, n) = 0.3118$ for $i = 6\%$ per year and $n = 20$ years.

$$P = F(P/F, 6\%, 20) = (\$2000)(0.3118)$$

$$= \boxed{\$623.60}$$

4.

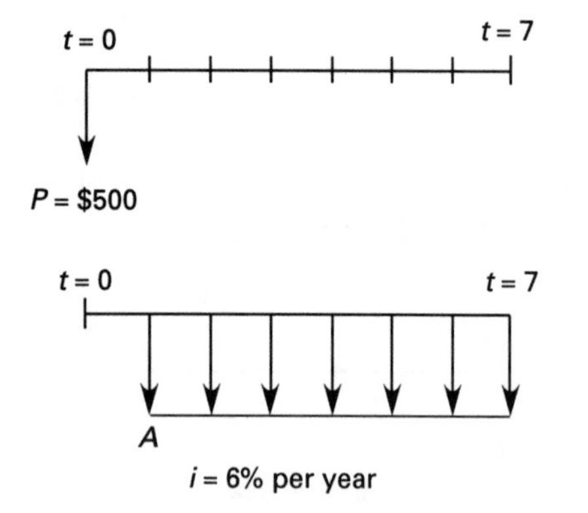

By the formula from Table 66.1,

$$A = P\left(\frac{i(1+i)^n}{(1+i)^n - 1}\right) = (\$500)\left(\frac{(0.06)(1+0.06)^7}{(1+0.06)^7 - 1}\right)$$

$$= \boxed{\$89.57}$$

By the factor converting P to A, $(A/P, i, n) = 0.17914$ for $i = 6\%$ per year and $n = 7$ years.

$$A = P(A/P, 6\%, 7) = (\$500)(0.17914)$$

$$= \boxed{\$89.57}$$

5.

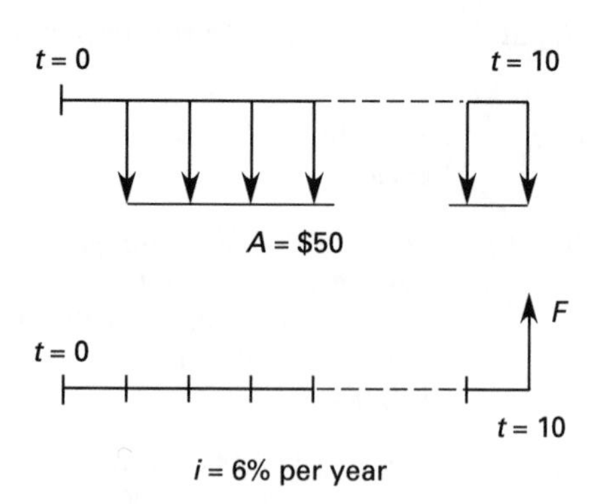

By the formula from Table 66.1,

$$F = A\left(\frac{(1+i)^n - 1}{i}\right) = (\$50)\left(\frac{(1+0.06)^{10} - 1}{0.06}\right)$$

$$= \boxed{\$659.04}$$

By the factor converting A to F, $(F/A, i, n) = 13.181$ for $i = 6\%$ per year and $n = 10$ years.

$$F = A(F/A, 6\%, 10) = (\$50)(13.181)$$

$$= \boxed{\$659.05}$$

6.

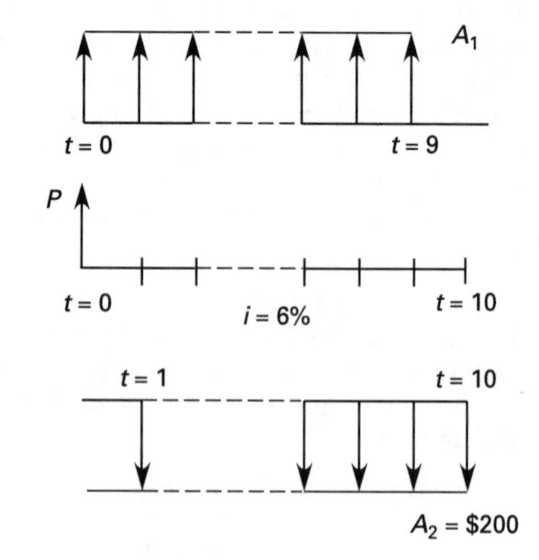

By the formula from Table 66.1, for each cash flow diagram,

$$P = A_1 + A_1\left(\frac{(1+0.06)^9 - 1}{(0.06)(1+0.06)^9}\right)$$

$$= A_2\left(\frac{(1+0.06)^{10} - 1}{(0.06)(1+0.06)^{10}}\right)$$

Professional

Therefore for $A_2 = \$200$,

$$A_1 + A_1\left(\frac{(1+0.06)^9 - 1}{(0.06)(1+0.06)^9}\right) = (\$200)\left(\frac{(1+0.06)^{10} - 1}{(0.06)(1+0.06)^{10}}\right)$$

$$7.80A_1 = \$1472.02$$

$$A_1 = \boxed{\$188.72}$$

By the factor converting A to P,

$$(P/A, 6\%, 9) = 6.8017$$

$$(P/A, 6\%, 10) = 7.3601$$

$$A_1 + A_1(6.8017) = (\$200)(7.3601)$$

$$7.8017A_1 = \$1472.02$$

$$A_1 = \frac{\$1472.02}{7.8017} = \boxed{\$188.68}$$

7.

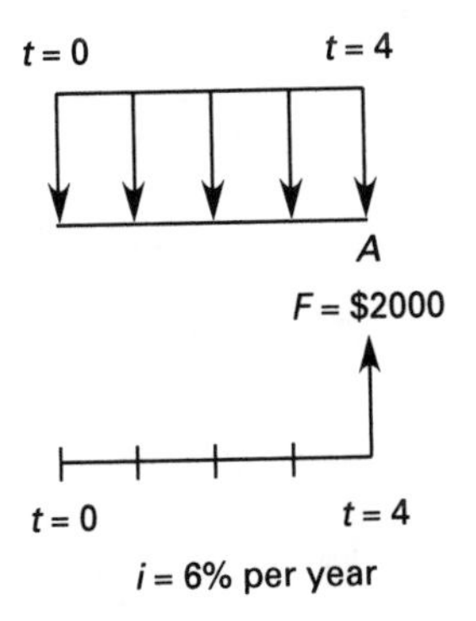

By the formula from Table 66.1,

$$F = A\left(\frac{(1+i)^n - 1}{i}\right)$$

Since the deposits start at the start of each year, $n = 4 + 1$, for a total of five deposits.

$$F = A\left(\frac{(1+i)^{n+1} - 1}{i}\right) = \$2000$$

$$= A\left(\frac{(1+0.06)^5 - 1}{0.06}\right)$$

$$\$2000 = 5.6371A$$

$$A = \frac{\$2000}{5.6371} = \boxed{\$354.79}$$

By the factor converting P and A to F,

$$F = A\big((F/P, 6\%, 4) + (F/A, 6\%, 4)\big)$$

$$\$2000 = A(1.2625 + 4.3746)$$

$$A = \boxed{\$354.79}$$

8.

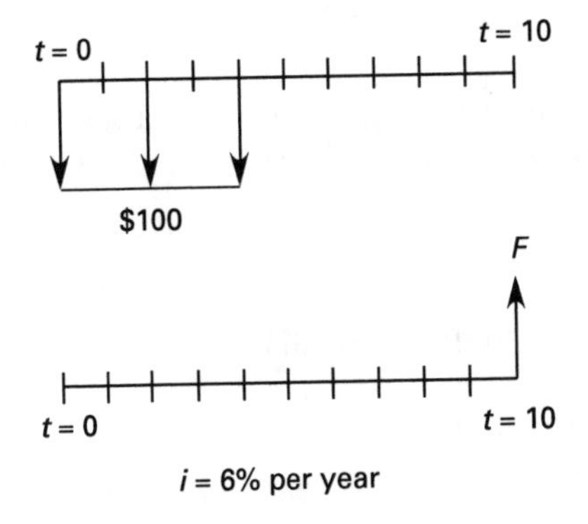

By the formula from Table 66.1, $F = P(1+i)^n$. If each deposit is considered as P, each will accumulate interest for periods of ten, eight, and six years.

Therefore,

$$F = (\$100)(1+0.06)^{10} + (\$100)(1+0.06)^8 + (\$100)(1+0.06)^6$$

$$= (\$100)(1.7908 + 1.5938 + 1.4185)$$

$$= (\$100)(4.8031)$$

$$= \boxed{\$480.31}$$

By the factor converting P to F,

$$(F/P, i, n) = 1.7908 \text{ for } i = 6\% \text{ and } n = 10$$

$$= 1.5938 \text{ for } i = 6\% \text{ and } n = 8$$

$$= 1.4185 \text{ for } i = 6\% \text{ and } n = 6$$

By summation,

$$F = (\$100)(1.7908 + 1.5938 + 1.4185)$$

$$= (\$100)(4.8031)$$

$$= \boxed{\$480.31}$$

9.

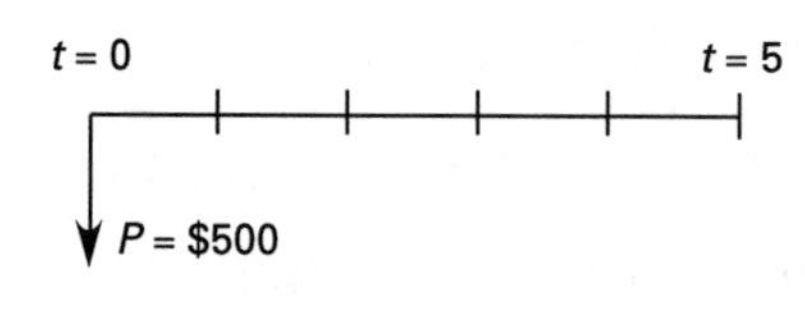

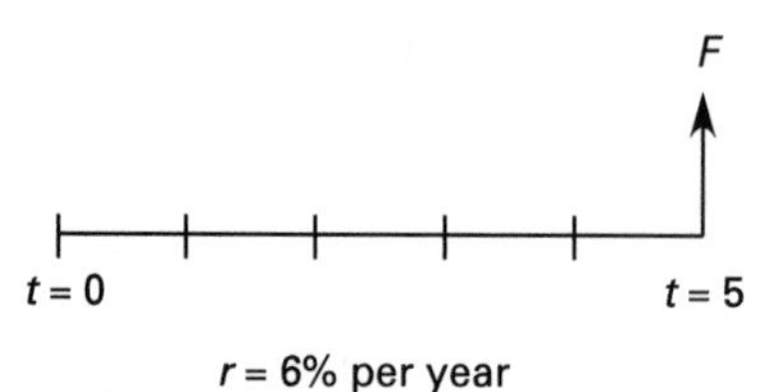

Since the deposit is compounded monthly, the effective interest rate should be calculated as shown by Eq. 66.54.

$$i = \left(1+\frac{r}{k}\right)^k - 1 = \left(1+\frac{0.06}{12}\right)^{12} - 1$$

$$= 0.061677 \quad (6.1677\%)$$

By the formula from Table 66.1,

$$F = P(1+i)^n = (\$500)(1+0.061677)^5$$

$$= \boxed{\$674.42}$$

To use a table of factors, interpolation is required.

$i\%$	factor F/P
6	1.3382
6.1677	desired
7	1.4026

$$i = \left(\frac{6.1677-6}{7-6}\right)(1.4026-1.3382)$$

$$= 0.0108$$

Therefore,

$$F/P = 1.3382 + 0.0108 = 1.3490$$

$$F = P(F/P, 6.1677\%, 5) = (\$500)(1.3490)$$

$$= \boxed{\$674.50}$$

10.

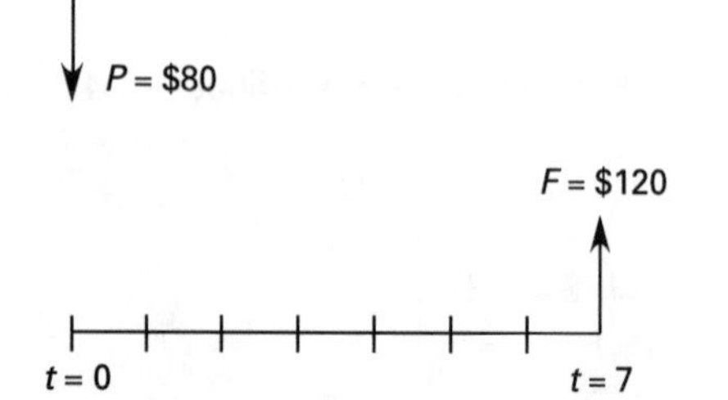

By the formula from Table 66.1,

$$F = P(1+i)^n$$

Therefore,

$$(1+i)^n = F/P$$

$$i = (F/P)^{\frac{1}{n}} - 1 = \left(\frac{\$120}{\$80}\right)^{\frac{1}{7}} - 1$$

$$= 0.059 \approx \boxed{6\%}$$

By the factor coverting P to F,

$$F = P(F/P, i\%, 7)$$

$$(F/P, i\%, 7) = F/P = \frac{\$120}{\$80} = 1.5$$

By checking the interest tables,

$$(F/P, i\%, 7) = 1.4071 \text{ for } i = 5\%$$

$$= 1.5036 \text{ for } i = 6\%$$

$$= 1.6058 \text{ for } i = 7\%$$

Therefore, $i \approx \boxed{6\%.}$

11.

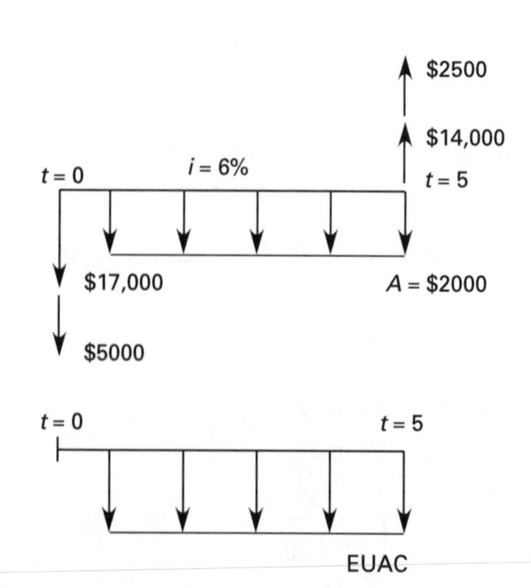

Annual cost of ownership, EUAC, can be obtained by the factors converting P to A and F to A.

$$P = \$17{,}000 + \$5000$$
$$= \$22{,}000$$

$$F = \$14{,}000 + \$2500$$
$$= \$16{,}500$$

$$\text{EUAC} = A + P(A/P, 6\%, 5) - F(A/F, 6\%, 5)$$

$$(A/P, 6\%, 5) = 0.23740$$

$$(A/F, 6\%, 5) = 0.17740$$

$$\begin{aligned}\text{EUAC} &= \$2000 + (\$22{,}000)(0.23740) \\ &\quad - (\$16{,}500)(0.17740) \\ &= \boxed{\$4295.70}\end{aligned}$$

12.

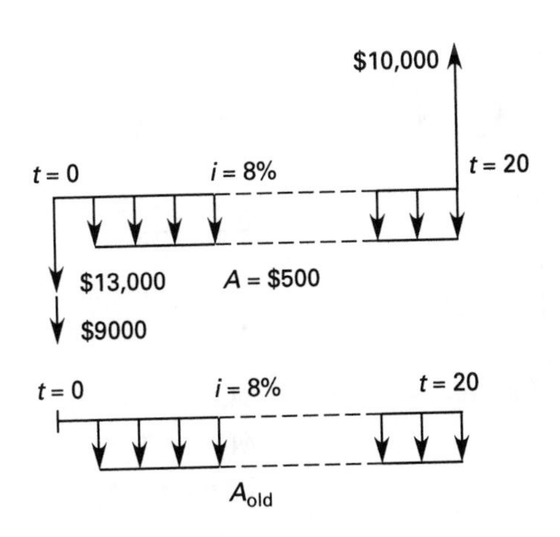

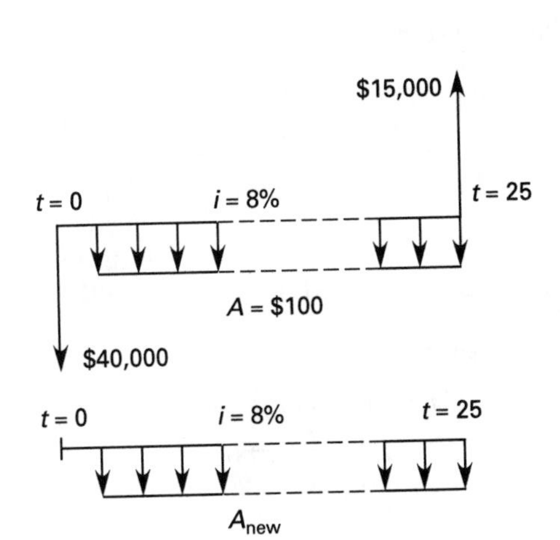

Consider the salvage value as a benefit lost (cost).

$$\begin{aligned}\text{EUAC}_{\text{old}} &= \$500 + (\$22{,}000)(A/P, 8\%, 20) \\ &\quad - (\$10{,}000)(A/F, 8\%, 20)\end{aligned}$$

$$(A/P, 8\%, 20) = 0.1019$$

$$(A/F, 8\%, 20) = 0.0219$$

$$\begin{aligned}\text{EUAC}_{\text{old}} &= \$500 + (\$22{,}000)(0.1019) \\ &\quad - (\$10{,}000)(0.0219) \\ &= \$2522.80\end{aligned}$$

Similarly,

$$\begin{aligned}\text{EUAC}_{\text{new}} &= \$100 + (\$40{,}000)(A/P, 8\%, 25) \\ &\quad - (\$15{,}000)(A/F, 8\%, 25)\end{aligned}$$

$$(A/P, 8\%, 25) = 0.0937$$

$$(A/F, 8\%, 25) = 0.0137$$

$$\begin{aligned}\text{EUAC}_{\text{new}} &= \$100 + (\$40{,}000)(0.0937) \\ &\quad - (\$15{,}000)(0.0137) \\ &= \$3642.50\end{aligned}$$

Therefore, the new bridge is going to be more costly.

The best alternative is to strengthen the old bridge.

13.

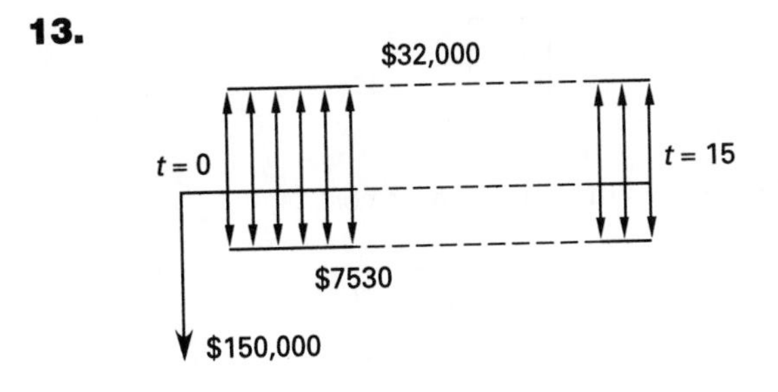

The annual depreciation is

$$\begin{aligned}D &= \frac{C - S_n}{n} = \frac{\$150{,}000}{15} \\ &= \$10{,}000/\text{year}\end{aligned}$$

The taxable income is

$$\$32{,}000 - \$7530 - \$10{,}000 = \$14{,}470/\text{year}$$

Taxes paid are

$$(\$14{,}470)(0.48) = \$6945.60/\text{year}$$

The after-tax cash flow is

$$\$24{,}470 - \$6945.60 = \$17{,}524.40$$

The present worth of the alternate is zero when evaluated at its ROR.

$$0 = -\$150{,}000 + (\$17{,}524.40)(P/A, i\%, 15)$$

Therefore,

$$(P/A, i\%, 15) = \frac{\$150{,}000}{\$17{,}524.40} = 8.55949$$

By checking the tables, this factor matches $i = 8\%$.

$$\boxed{\text{ROR} = 8\%}$$

14. The conventional benefit/cost ratio is

$$B/C = \frac{B - D}{C}$$

(a) The benefit/cost ratio is

$$B/C = \frac{\$1{,}500{,}000 - \$300{,}000}{\$1{,}000{,}000} = \boxed{1.2}$$

(b) The excess of benefits over cost is

$$\boxed{\$200{,}000.}$$

15.

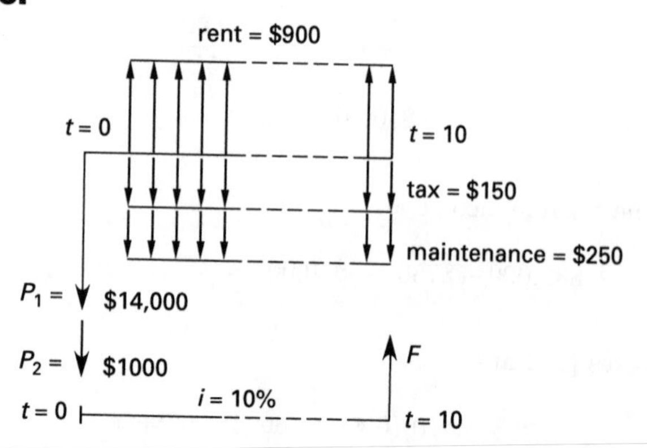

The annual rent is

$$(\$75)\left(12\ \frac{\text{months}}{\text{year}}\right) = \$900$$

$$P = P_1 + P_2 = \$15{,}000$$

$$A_1 = -\$900$$

$$A_2 = \$250 + \$150 = \$400$$

By the factors converting P to F and A to F,

$$\begin{aligned} F &= (\$15{,}000)(F/P, 10\%, 10) \\ &\quad + (\$400)(F/A, 10\%, 10) \\ &\quad - (\$900)(F/A, 10\%, 10) \end{aligned}$$

$$(F/P, 10\%, 10) = 2.5937$$

$$(F/A, 10\%, 10) = 15.937$$

$$\begin{aligned} F &= (\$15{,}000)(2.5937) \\ &\quad + (\$400)(15.937) \\ &\quad - (\$900)(15.937) \\ &= \boxed{\$30{,}937} \end{aligned}$$

16.

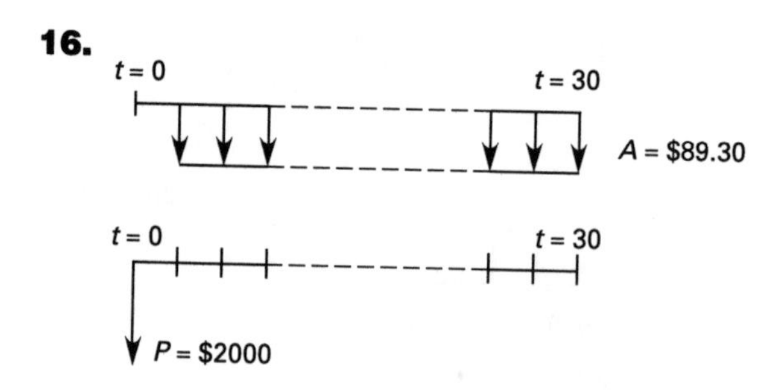

By the formula relating P to A,

$$P = A\left(\frac{(1+i)^n - 1}{i(1+i)^n}\right)$$

$$\frac{(1+i)^{30} - 1}{i(1+i)^{30}} = \frac{\$2000}{\$89.30} = 22.40$$

By trial and error,

$i(\%)$	$(1+i)^{30}$	$\frac{(1+i)^{30}-1}{i(1+i)^{30}}$
10	17.45	9.42
6	5.74	13.76
4	3.24	17.28
2	1.81	22.37

2% per month is close.

$$\begin{aligned} i &= (1 + 0.02)^{12} - 1 \\ &= \boxed{0.2682\quad(26.82\%)} \end{aligned}$$

Professional

17. (a) Use the straight line method, Eq. 66.25.

$$D = \frac{C - S_n}{n}$$

Each year depreciation will remain the same.

$$D = \frac{\$500{,}000 - \$100{,}000}{25}$$
$$= \boxed{\$16{,}000}$$

(b) Sum-of-the-years' digits (SOYD) can be calculated as shown by Eq. 66.28.

$$D_j = \frac{(C - S_n)(n - j + 1)}{T}$$

Use Eq. 66.27.

$$T = \tfrac{1}{2}n(n+1) = \left(\tfrac{1}{2}\right)(25)(25+1) = 325$$

$$D_1 = \frac{(\$500{,}000 - \$100{,}000)(25 - 1 + 1)}{325} = \boxed{\$30{,}769}$$

$$D_2 = \frac{(\$500{,}000 - \$100{,}000)(25 - 2 + 1)}{325} = \boxed{\$29{,}538}$$

$$D_3 = \frac{(\$500{,}000 - \$100{,}000)(25 - 3 + 1)}{325} = \boxed{\$28{,}308}$$

(c) The double declining balance (DDB) method can be used. By Eq. 66.32,

$$D_j = dC(1 - d)^{j-1}$$

Use Eq. 66.31.

$$d = \frac{2}{n} = \frac{2}{25}$$

$$D_1 = \left(\frac{2}{25}\right)(\$500{,}000)\left(1 - \frac{2}{25}\right)^0 = \boxed{\$40{,}000}$$

$$D_2 = \left(\frac{2}{25}\right)(\$500{,}000)\left(1 - \frac{2}{25}\right)^1 = \boxed{\$36{,}800}$$

$$D_3 = \left(\frac{2}{25}\right)(\$500{,}000)\left(1 - \frac{2}{25}\right)^2 = \boxed{\$33{,}856}$$

18.

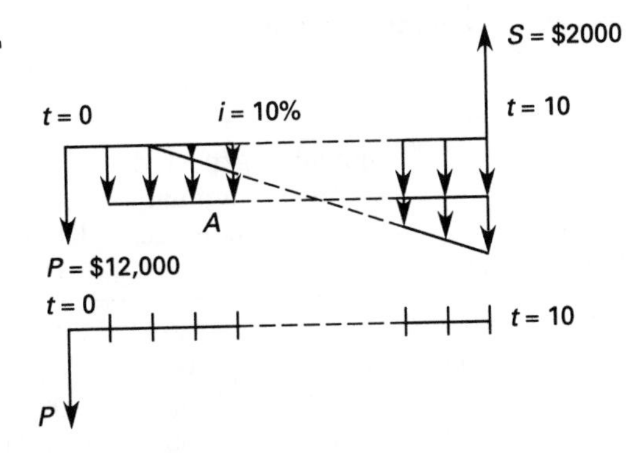

(a) $A = \$1000$ and $G = \$200$ for $t = n - 1 = 9$ years.

$$F = S = \$2000$$

$$P = \$12{,}000 + A(P/A, 10\%, 10) + G(P/G, 10\%, 10) - F(P/F, 10\%, 10)$$
$$= \$12{,}000 + (\$1000)(6.1446) + (\$200)(22.8913) - (\$2000)(0.3855)$$
$$= \boxed{\$21{,}952}$$

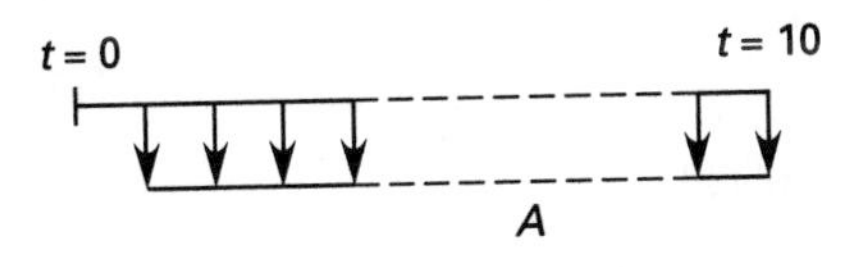

(b)
$$A = (\$12{,}000)(A/P, 10\%, 10) + \$1000 + (\$200)(A/G, 10\%, 10) - (\$2000)(A/F, 10\%, 10)$$
$$= (\$12{,}000)(0.1627) + \$1000 + (\$200)(3.7255) - (\$2000)(0.0627)$$
$$= \boxed{\$3572.70}$$

19. An increase in rock removal capacity can be achieved by a 20-year loan (investment). Different cases available can be compared by equivalent uniform annual cost (EUAC).

$$\text{EUAC} = \text{annual loan cost} + \text{expected annual damage}$$
$$= \text{cost } (A/P, 10\%, 20) + (\$25{,}000)(\text{probability})$$

$$(A/P, 10\%, 20) = 0.1175$$

A table can be prepared for different cases.

rock removal rate	cost ($)	annual loan cost ($)	expected annual damage ($)	EUAC ($)
7	0	0	3750	3750.00
8	15,000	1761.90	2500	4261.90
9	20,000	2349.20	1750	4099.20
10	30,000	3523.80	750	4273.80

It is cheapest to do nothing.

20. Calculate the cost of owning and operating for each year.

$$A_1 = (\$10{,}000)(A/P, 20\%, 1) + \$2000 - (\$8000)(A/F, 20\%, 1)$$

$$(A/P, 20\%, 1) = 1.2$$

$$(A/F, 20\%, 1) = 1.0$$

$$\begin{aligned} A_1 &= (\$10{,}000)(1.2) \\ &\quad + \$2000 \\ &\quad - (\$8000)(1.0) \\ &= \$6000 \end{aligned}$$

$$\begin{aligned} A_2 &= (\$10{,}000)(A/P, 20\%, 2) + \$2000 \\ &\quad + (\$1000)(A/G, 20\%, 2) \\ &\quad - (\$7000)(A/F, 20\%, 2) \end{aligned}$$

$$(A/P, 20\%, 2) = 0.6545$$

$$(A/G, 20\%, 2) = 0.4545$$

$$(A/F, 20\%, 2) = 0.4545$$

$$\begin{aligned} A_2 &= (\$10{,}000)(0.6545) + \$2000 \\ &\quad + (\$1000)(0.4545) \\ &\quad - (\$7000)(0.4545) \\ &= \$5818 \end{aligned}$$

$$\begin{aligned} A_3 &= (\$10{,}000)(A/P, 20\%, 3) + \$2000 \\ &\quad + (\$1000)(A/G, 20\%, 3) \\ &\quad - (\$6000)(A/F, 20\%, 3) \end{aligned}$$

$$(A/P, 20\%, 3) = 0.4747$$

$$(A/G, 20\%, 3) = 0.8791$$

$$(A/F, 20\%, 3) = 0.2747$$

$$\begin{aligned} A_3 &= (\$10{,}000)(0.4747) + \$2000 \\ &\quad + (\$1000)(0.8791) \\ &\quad - (\$6000)(0.2747) \\ &= \$5977.90 \end{aligned}$$

$$\begin{aligned} A_4 &= (\$10{,}000)(A/P, 20\%, 4) \\ &\quad + \$2000 + (\$1000)(A/G, 20\%, 4) \\ &\quad - (\$5000)(A/F, 20\%, 4) \end{aligned}$$

$$(A/P, 20\%, 4) = 0.3863$$

$$(A/G, 20\%, 4) = 1.2762$$

$$(A/F, 20\%, 4) = 0.1863$$

$$\begin{aligned} A_4 &= (\$10{,}000)(0.3863) + \$2000 \\ &\quad + (\$1000)(1.2762) \\ &\quad - (\$5000)(0.1863) \\ &= \$6207.70 \end{aligned}$$

$$\begin{aligned} A_5 &= (\$10{,}000)(A/P, 20\%, 5) + \$2000 \\ &\quad + (\$1000)(A/G, 20\%, 5) \\ &\quad - (\$4000)(A/F, 20\%, 5) \end{aligned}$$

$$(A/P, 20\%, 5) = 0.3344$$

$$(A/G, 20\%, 5) = 1.6405$$

$$(A/F, 20\%, 5) = 0.1344$$

$$\begin{aligned} A_5 &= (\$10{,}000)(0.3344) + \$2000 \\ &\quad + (\$1000)(1.6405) \\ &\quad - (\$4000)(0.1344) \\ &= \$6446.90 \end{aligned}$$

(a) Since the annual owning and operating cost is smallest after two years of operation, it is advantageous to sell the mechanism after the second year.

The economic life is two years.

Professional

(b) After four years of operation, the owning and operating cost of the mechanism for one more year is

$$A = \$6000 + (\$5000)(1+i) - \$4000$$
$$i = 0.2 \quad (20\%)$$
$$A = \$6000 + (\$5000)(1.2) - \$4000$$
$$= \boxed{\$8000}$$

21. To find out if the reimbursement is adequate, calculate the business-related expense.

Charge the company for business travel.

$$\text{insurance: } \$3000 - \$2000 = \$1000$$
$$\text{maintenance: } \$2000 - \$1500 = \$500$$
$$\text{drop in salvage value: } \$10{,}000 - \$5000 = \$5000$$

The annual portion of the drop in salvage value is

$$A = (\$5000)(A/F, 10\%, 5)$$
$$(A/F, 10\%, 5) = 0.1638$$
$$A = (\$5000)(0.1638) = \$819/\text{year}$$

(a) The annual cost of gas is

$$\left(\frac{50{,}000 \text{ mi}}{15 \ \frac{\text{mi}}{\text{gal}}}\right)\left(\frac{\$1.50}{\text{gal}}\right) = \$5000$$

$$\text{EUAC per mile} = \frac{\$1000 + \$500 + \$819 + \$5000}{50{,}000 \text{ mi}}$$
$$= \boxed{\$0.14638/\text{mi}}$$

Since the reimbursement per mile was \$0.30 and since \$0.30 > \$0.14638, the reimbursement is adequate.

(b) Next, determine (with reimbursement) how many miles the car must be driven to break even.

If the car is driven M miles per year,

$$\left(\frac{\$0.30}{1 \text{ mi}}\right) M = (\$50{,}000)(A/P, 10\%, 5) + \$2500 + \$2000 - (\$8000)(A/F, 10\%, 5) + \left(\frac{M}{15 \ \frac{\text{mi}}{\text{gal}}}\right)(\$1.50)$$

$$(A/P, 10\%, 5) = 0.2638$$
$$(A/F, 10\%, 5) = 0.1638$$
$$0.3M = (\$50{,}000)(0.2638) + \$2500 + \$2000 - (\$8000)(0.1638) + 0.1M$$
$$0.2M = \$16{,}379.60$$
$$M = \frac{\$16{,}379.60}{0.2 \ \frac{\$}{\text{mi}}}$$
$$= \boxed{81{,}898 \text{ mi}}$$

22.

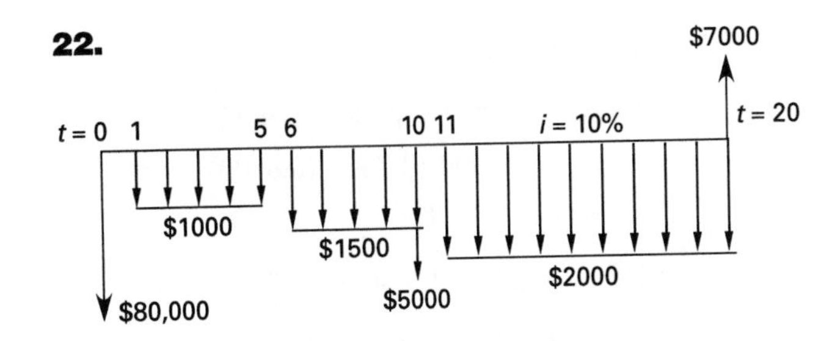

$$P_A = \$80{,}000 + (\$1000)(P/A, 10\%, 5) + (\$1500)(P/A, 10\%, 5) \times (P/F, 10\%, 5) + (\$2000)(P/A, 10\%, 10) \times (P/F, 10\%, 10) + (\$5000)(P/F, 10\%, 10) - (\$7000)(P/F, 10\%, 20)$$

$$(P/A, 10\%, 5) = 3.7908$$
$$(P/F, 10\%, 5) = 0.6209$$
$$(P/A, 10\%, 10) = 6.1446$$
$$(P/F, 10\%, 10) = 0.3855$$
$$(P/F, 10\%, 20) = 0.1486$$

$$P_A = \$80{,}000 + (\$1000)(3.7908) + (\$1500)(3.7908)(0.6209) + (\$2000)(6.1446)(0.3855) + (\$5000)(0.3855) - (\$7000)(0.1486)$$
$$= \$92{,}946.15$$

Since the lives are different, compare by EUAC.

$$\text{EUAC(A)} = (\$92{,}946.14)(A/P, 10\%, 20)$$
$$= (\$92{,}946.14)(0.1175) = \$10{,}921$$

Similarly, evaluate alternative B.

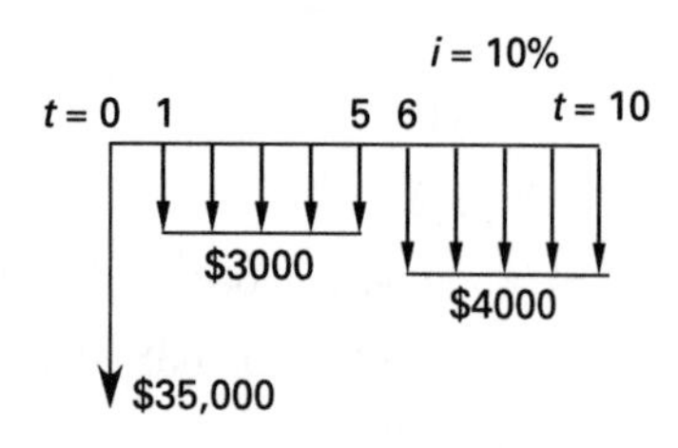

$$P_B = \$35{,}000 + (\$3000)(P/A, 10\%, 5) + (\$4000)(P/A, 10\%, 5)(P/F, 10\%, 5)$$

$$(P/A, 10\%, 5) = 3.7908$$

$$(P/F, 10\%, 5) = 0.6209$$

$$P_B = \$35{,}000 + (\$3000)(3.7908) + (\$4000)(3.7908)(0.6209) = \$55{,}787.23$$

$$\text{EUAC(B)} = (\$55{,}787.23)(A/P, 10\%, 10)$$
$$= (\$55{,}787.23)(0.1627) = \$9077$$

Since EUAC(B) < EUAC(A),

Alternative B is economically superior.

23. For both cases, if the annual cost is compared with a total annual mileage of M,

$$A_A = \$0.25M$$

$$A_B = (\$30{,}000)(A/P, 10\%, 3) + \$0.14M + \$500 - (\$7200)(A/F, 10\%, 3)$$

$$(A/P, 10\%, 3) = 0.4021$$

$$(A/F, 10\%, 3) = 0.3021$$

$$A_B = (\$30{,}000)(0.4021) + \$0.14M + \$500 - (\$7200)(0.3021)$$
$$= \$12{,}063 + \$0.14M + \$500 - \$2175.12$$
$$= \$10{,}387.88 + \$0.14M$$

For an equal annual cost $A_A = A_B$,

$$\$0.25M = \$10{,}387.88 + \$0.14M$$

An annual mileage would be $M = 94{,}435$ mi.

For an annual mileage less than that, $A_A < A_B$.

Plan A is economically superior until that mileage is exceeded.

24. (a) For method A,

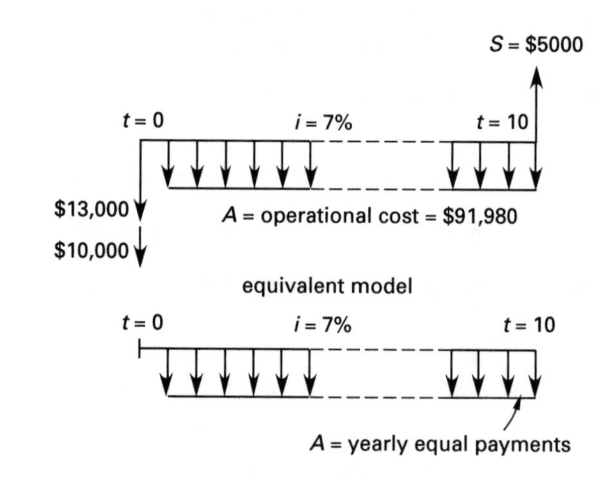

24 hr/day
365 days/yr
total of (24)(365) = 8760 hr/yr
$10.50 operational cost/hr
total of (8760)($10.50) = $91,980 operational cost/yr

$$A = \$91{,}980 + (\$23{,}000)(A/P, 7\%, 10) - (\$5000)(A/F, 7\%, 10)$$

$$(A/P, 7\%, 10) = 0.1424$$

$$(A/F, 7\%, 10) = 0.0724$$

$$A = \$91{,}980 + (\$23{,}000)(0.1424) - (\$5000)(0.0724) = \$94{,}893.30/\text{yr}$$

Therefore, the uniform annual cost per ton each year will be

$$\frac{\$94{,}893.30}{50 \text{ ton}} = \boxed{\$1897.87}$$

For method B,

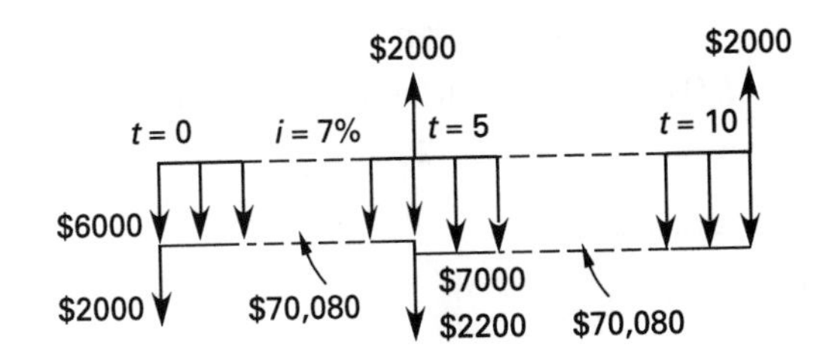

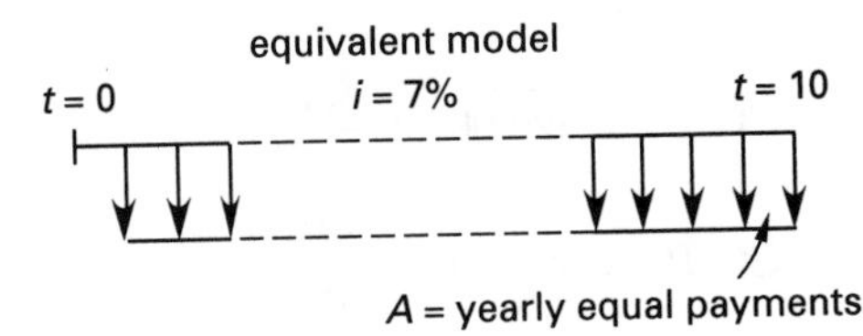

8760 hr/yr
\$8 operational cost/hr
total of (8760)(\$8)= \$70,080 operational cost/yr

$$\begin{aligned} A = \$70{,}080 \\ &+ (\$6000 + \$2000)(A/P, 7\%, 10) \\ &+ (\$7000 + \$2200 - \$2000) \\ &\times (P/F, 7\%, 5) \\ &\times (A/P, 7\%, 10) \\ &- (\$2000)(A/F, 7\%, 10) \end{aligned}$$

$$(A/P, 7\%, 10) = 0.1424$$

$$(A/F, 7\%, 10) = 0.07238$$

$$(P/F, 7\%, 5) = 0.7130$$

$$\begin{aligned} A &= \$70{,}080 + (\$8000)(0.1424) \\ &\quad + (\$7200)(0.7130)(0.1424) \\ &\quad - (\$2000)(0.0724) \\ &= \$71{,}805.42/\text{yr} \end{aligned}$$

Therefore, the uniform annual cost per ton each year will be

$$\frac{\$71{,}805.42}{20 \text{ ton}} = \boxed{\$3590.27}$$

(b)

tons/yr	cost of using A	cost of using B	cheapest
0–20	\$ 94,893 (1x)	\$ 71,805 (1x)	B
20–40	\$ 94,893 (1x)	\$143,610 (2x)	A
40–50	\$ 94,893 (1x)	\$215,415 (3x)	A
50–60	\$189,786 (2x)	\$215,415 (3x)	A
60–80	\$189,786 (2x)	\$287,220 (4x)	A

25.

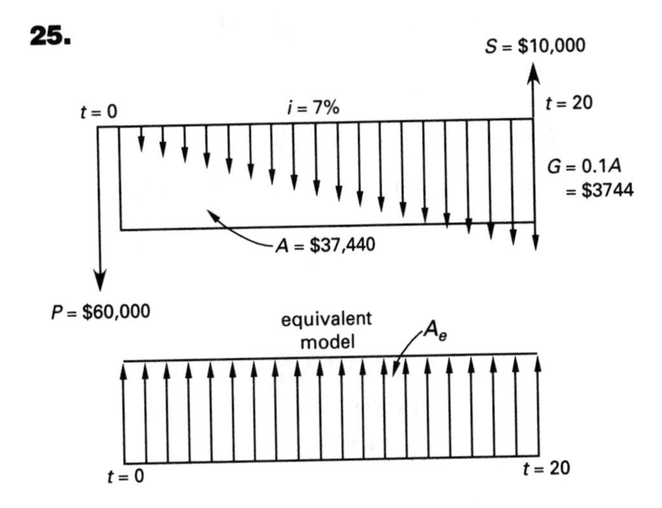

$$\begin{aligned} A_e &= (\$60{,}000)(A/P, 7\%, 20) + A \\ &\quad + G(P/G, 7\%, 20)(A/P, 7\%, 20) \\ &\quad - (\$10{,}000)(A/F, 7\%, 20) \end{aligned}$$

$$(A/P, 7\%, 20) = 0.0944$$

$$A = (37{,}440 \text{ mi})\left(\frac{\$1.0}{\text{mi}}\right) = \$37{,}440$$

$$G = 0.1A = (0.1)(\$37{,}440) = \$3744$$

$$(P/G, 7\%, 20) = 77.5091$$

$$(A/F, 7\%, 20) = 0.0244$$

$$\begin{aligned} A_e &= (\$60{,}000)(0.0944) + \$37{,}440 \\ &\quad + (\$3744)(77.509)(0.0944) \\ &\quad - (\$10{,}000)(0.0244) \\ &= \$70{,}253.78 \end{aligned}$$

(a) With 80,000 passengers per year, the break-even fare per passenger would be

$$\begin{aligned} \text{fare} &= \frac{A_e}{80{,}000} = \frac{\$70{,}253.78}{80{,}000} \\ &= \boxed{\$0.878/\text{passenger}} \end{aligned}$$

(b)

$$\$0.878 = \$0.35 + G(A/G, 7\%, 20)$$

$$\begin{aligned} G &= \frac{\$0.878 - \$0.35}{7.3163} \\ &= \boxed{\$0.072 \text{ increase per year}} \end{aligned}$$

(c) As in part (b), the subsidy should be

$$\text{subsidy} = \text{cost} - \text{revenue}$$

$$\begin{aligned} P &= \$0.878 - (\$0.35 + (\$0.05)(A/G, 7\%, 20)) \\ &= \$0.878 - (\$0.35 + (\$0.05)(7.3163)) \\ &= \boxed{\$0.162} \end{aligned}$$

26.

$$\begin{aligned} P(\text{A}) &= (\$4800)(P/A, 12\%, 25) \\ &= (\$4800)(7.8431) \\ &= \$37{,}646.88 \end{aligned}$$

$$(4 \text{ quarters})(25 \text{ years}) = 100 \text{ compounding periods}$$

$$\begin{aligned} P(\text{B}) &= (\$1200)(P/A, 3\%, 100) \\ &= (\$1200)(31.5989) \\ &= \$37{,}918.68 \end{aligned}$$

$$\boxed{\text{Alternative B is economically superior.}}$$

27.

$$\begin{aligned} \text{EUAC(A)} &= (\$1500)(A/P, 15\%, 5) + \$800 \\ &= (\$1500)(0.2983) + \$800 \\ &= \$1247.45 \end{aligned}$$

$$\begin{aligned} \text{EUAC(B)} &= (\$2500)(A/P, 15\%, 8) + \$650 \\ &= (\$2500)(0.2229) + \$650 \\ &= \$1207.25 \end{aligned}$$

$$\boxed{\text{Alternative B is economically superior.}}$$

28. The data given imply that both investments return 4% or more. However, the increased investment of $30,000 may not be cost effective. Do an incremental analysis.

$$\text{incremental cost} = \$70{,}000 - \$40{,}000 = \$30{,}000$$

$$\text{incremental income} = \$5620 - \$4075 = \$1545$$

$$\begin{aligned} 0 &= -\$30{,}000 \\ &\quad + (\$1545)(P/A, i\%, 20) \\ (P/A, i\%, 20) &= 19.417 \\ i &\approx 0.25\% < 4\% \end{aligned}$$

$$\boxed{\text{Alternative B is economically superior.}}$$

(The same conclusion could be reached by taking the present worths of both alternatives at 4%.)

29. (a)

$$\begin{aligned} P(\text{A}) &= (-\$3000)(P/A, 25\%, 25) - \$20{,}000 \\ &= (-\$3000)(3.9849) - \$20{,}000 \\ &= -\$31{,}954.70 \\ P(\text{B}) &= (-\$2500)(3.9849) - \$25{,}000 \\ &= -\$34{,}962.25 \end{aligned}$$

$$\boxed{\text{Alternative A is better.}}$$

(b)

$$\text{CC(A)} = \$20{,}000 + \frac{\$3000}{0.25} = \$32{,}000$$

$$\text{CC(B)} = \$25{,}000 + \frac{\$2500}{0.25} = \$35{,}000$$

$$\boxed{\text{Alternative A is better.}}$$

(c)

$$\begin{aligned} \text{EUAC(A)} &= (\$20{,}000)(A/P, 25\%, 25) + \$3000 \\ &= (\$20{,}000)(0.2509) + \$3000 \\ &= \$8018.00 \end{aligned}$$

$$\begin{aligned} \text{EUAC(B)} &= (\$25{,}000)(0.2509) + \$2500 \\ &= \$8772.50 \end{aligned}$$

$$\boxed{\text{Alternative A is better.}}$$

30. (a)

$$\begin{aligned} \text{BV} &= \$18{,}000 - (5)\left(\frac{\$18{,}000 - \$2000}{8}\right) \\ &= \boxed{\$8000} \end{aligned}$$

(b) With the sinking fund method, the basis is

$$\begin{aligned} &(\$18{,}000 - \$2000)(A/F, 8\%, 8) \\ &\quad = (\$18{,}000 - \$2000)(0.0940) \\ &\quad = \$1504 \end{aligned}$$

$$\begin{aligned} D_1 &= (\$1504)(1.000) \\ &= \boxed{\$1504} \end{aligned}$$

$$\begin{aligned} D_2 &= (\$1504)(1.0800) \\ &= \boxed{\$1624} \end{aligned}$$

$$\begin{aligned} D_3 &= (\$1504)(1.0800)^2 = (\$1504)(1.1664) \\ &= \boxed{\$1754} \end{aligned}$$

(c) $D_1 = \left(\frac{2}{8}\right)(\$18{,}000)$

$= \boxed{\$4500}$

$D_2 = \left(\frac{2}{8}\right)(\$18{,}000 - \$4500)$

$= \boxed{\$3375}$

$D_3 = \left(\frac{2}{8}\right)(\$18{,}000 - \$4500 - \$3375)$

$= \boxed{\$2531}$

$D_4 = \left(\frac{2}{8}\right)(\$18{,}000 - \$4500 - \$3375 - \$2531)$

$= \boxed{\$1898}$

$$D_5 = \left(\frac{2}{8}\right)(\$18{,}000 - \$4500 - \$3375 - \$2531 - \$1898)$$

$= \boxed{\$1424}$

$$\text{BV} = \$18{,}000 - \$4500 - \$3375 - \$2531 - \$1898 - \$1424$$

$= \boxed{\$4272}$

31. $T = \left(\frac{1}{2}\right)(8)(9) = 36$

$D_1 = \left(\frac{8}{36}\right)(\$14{,}000 - \$1800) = \$2711$

$\Delta D = \left(\frac{1}{36}\right)(\$14{,}000 - \$1800) = \$339$

$D_2 = \$2711 - \$339 = \$2372$

$$\begin{aligned}\text{DR} &= (0.52)(\$2711)(P/A, 15\%, 8) - (0.52)(\$339)(P/G, 15\%, 8) \\ &= (0.52)(\$2711)(4.4873) - (0.52)(\$339)(12.4807) \\ &= \boxed{\$4125.74}\end{aligned}$$

32.

$$\begin{aligned}\text{EUAC}_{5\text{ ft}} &= (\$600{,}000)(A/P, 12\%, 15) + \left(\frac{14}{50}\right)(\$600{,}000) + \left(\frac{8}{50}\right)(\$650{,}000) + \left(\frac{3}{50}\right)(\$700{,}000) + \left(\frac{1}{50}\right)(\$800{,}000) \\ &= \$418{,}080\end{aligned}$$

$$\begin{aligned}\text{EUAC}_{10\text{ ft}} &= (\$710{,}000)(A/P, 12\%, 15) + \left(\frac{8}{50}\right)(\$650{,}000) + \left(\frac{3}{50}\right)(\$700{,}000) + \left(\frac{1}{50}\right)(\$800{,}000) \\ &= \$266{,}228\end{aligned}$$

$$\begin{aligned}\text{EUAC}_{15\text{ ft}} &= (\$900{,}000)(A/P, 12\%, 15) + \left(\frac{3}{50}\right)(\$700{,}000) + \left(\frac{1}{50}\right)(\$800{,}000) \\ &= \$190{,}120\end{aligned}$$

$$\begin{aligned}\text{EUAC}_{20\text{ ft}} &= (\$1{,}000{,}000)(A/P, 12\%, 15) + \left(\frac{1}{50}\right)(\$800{,}000) \\ &= \$162{,}800\end{aligned}$$

$\boxed{\text{Build to 20 ft.}}$

33. Assume replacement after one year.

$$\begin{aligned}\text{EUAC}(1) &= (\$30{,}000)(A/P, 12\%, 1) + \$18{,}700 \\ &= (\$30{,}000)(1.12) + \$18{,}700 \\ &= \$52{,}300\end{aligned}$$

Assume replacement after two years.

$$\begin{aligned}\text{EUAC}(2) &= (\$30{,}000)(A/P, 12\%, 2) + \$18{,}700 + (\$1200)(A/G, 12\%, 2) \\ &= (\$30{,}000)(0.5917) + \$18{,}700 + (\$1200)(0.4717) \\ &= \$37{,}017\end{aligned}$$

Assume replacement after three years.

$$\begin{aligned}\text{EUAC}(3) &= (\$30{,}000)(A/P, 12\%, 3) \\ &\quad + \$18{,}700 + (\$1200)(A/G, 12\%, 3) \\ &= (\$30{,}000)(0.4163) + \$18{,}700 \\ &\quad + (\$1200)(0.9246) \\ &= \$32{,}299\end{aligned}$$

Similarly, calculate to obtain the numbers in the following table.

years in service	EUAC
1	\$52,300
2	\$37,017
3	\$32,299
4	\$30,207
5	\$29,152
6	\$28,602
7	\$28,335
8	\$28,234
9	\$28,240
10	\$28,312

$$\boxed{\text{Replace after 8 yr.}}$$

34. Assume the head and horsepower data are already reflected in the hourly operating costs.

Let N = no. of hours operated each year.

$$\begin{aligned}\text{EUAC(A)} &= (\$3600 + \$3050)(A/P, 12\%, 10) \\ &\quad - (\$200)(A/F, 12\%, 10) + 0.30N \\ &= (\$6650)(0.1770) \\ &\quad - (\$200)(0.0570) + 0.30N \\ &= 1165.65 + 0.30N\end{aligned}$$

$$\begin{aligned}\text{EUAC(B)} &= (\$2800 + \$5010)(A/P, 12\%, 10) \\ &\quad - (\$280)(A/F, 12\%, 10) + 0.10N \\ &= (\$7810)(0.1770) \\ &\quad - (\$280)(0.0570) + 0.10N \\ &= 1366.41 + 0.10N\end{aligned}$$

$$\begin{aligned}\text{EUAC(A)} &= \text{EUAC(B)} \\ 1165.65 + 0.30N &= 1366.41 + 0.10N \\ N &= \boxed{1003.8 \text{ hr}}\end{aligned}$$

35. (a) From Eq. 66.78,

$$\begin{aligned}\frac{T_2}{T_1} &= 0.88 = 2^{-b} \\ \log(0.88) &= -b\log(2) \\ -0.0555 &= -(0.3010)b \\ b &= 0.1843 \\ T_4 &= (6)(4)^{-0.1843} \\ &= \boxed{4.65 \text{ wk}}\end{aligned}$$

(b) From Eq. 66.79,

$$\begin{aligned}T_{6\text{–}14} &= \left(\frac{6}{1 - 0.1843}\right) \\ &\quad \times \left(\left(14 + \tfrac{1}{2}\right)^{1-0.1843} - \left(6 - \tfrac{1}{2}\right)^{1-0.1843}\right) \\ &= \left(\frac{6}{0.8157}\right)(8.857 - 4.017) \\ &= \boxed{35.6 \text{ wk}}\end{aligned}$$

36. First check that both alternatives have an ROR greater than the MARR. Work in thousands of dollars.

Evaluate alternative A.

$$\begin{aligned}P(\text{A}) &= -\$120 + (\$15)(P/F, i\%, 5) \\ &\quad + (\$57)(P/A, i\%, 5)(1 - 0.45) \\ &\quad + \left(\frac{\$120 - \$15}{5}\right)(P/A, i\%, 5)(0.45) \\ &= -\$120 + (\$15)(P/F, i\%, 5) \\ &\quad + (\$40.8)(P/A, i\%, 5)\end{aligned}$$

Try 15%.

$$\begin{aligned}P(\text{A}) &= -\$120 + (\$15)(0.4972) + (\$40.8)(3.3522) \\ &= \$24.23\end{aligned}$$

Try 25%.

$$\begin{aligned}P(\text{A}) &= -\$120 + (\$15)(0.3277) + (\$40.8)(2.6893) \\ &= -\$5.36\end{aligned}$$

Since $P(\text{A})$ goes through zero,

$$(\text{ROR})_\text{A} > \text{MARR} = 15\%$$

Next, evaluate alternative B.

$$\begin{aligned} P(\text{B}) &= -\$170 + (\$20)(P/F, i\%, 5) \\ &\quad + (\$67)(P/A, i\%, 5)(1 - 0.45) \\ &\quad + \left(\frac{\$170 - \$20}{5}\right)(P/A, i\%, 5)(0.45) \\ &= -\$170 + (\$20)(P/F, i\%, 5) \\ &\quad + (\$50.35)(P/A, i\%, 5) \end{aligned}$$

Try 15%.

$$\begin{aligned} P(\text{B}) &= -\$170 + (\$20)(0.4972) + (\$50.35)(3.352) \\ &= \$8.72 \end{aligned}$$

Since $P(\text{B}) > 0$ and will decrease as i increases,

$$(\text{ROR})_\text{B} > 15\%$$

ROR > MARR for both alternatives.

Do an incremental analysis to see if it is worthwhile to invest the extra $170 – $120 = $50.

$$\begin{aligned} P(\text{B} - \text{A}) &= -\$50 + (\$20 - \$15)(P/F, i\%, 5) \\ &\quad + (\$50.35 - \$40.8)(P/A, i\%, 5) \end{aligned}$$

Try 15%.

$$\begin{aligned} P(\text{B} - \text{A}) &= -\$50 + (\$5)(0.4972) \\ &\quad + (\$9.55)(3.3522) \\ &= -\$15.50 \end{aligned}$$

Since $P(\text{B}-\text{A}) < 0$ and would become more negative as i increases, the ROR of the added investment is $< 15\%$.

Alternative A is superior.

37. Use the year-end convention with the tax credit. The purchase is made at $t = 0$. However, the tax credit is received at $t = 1$ and must be multiplied by $(P/F, i\%, 1)$.

(Note that 0.0667 is actually two-thirds of 10%.)

$$\begin{aligned} P &= -\$300{,}000 + (0.0667)(\$300{,}000)(P/F, i\%, 1) \\ &\quad + (\$90{,}000)(P/A, i\%, 5)(1 - 0.48) \\ &\quad + \left(\frac{\$300{,}000 - \$50{,}000}{5}\right)(P/A, i\%, 5)(0.48) \\ &\quad + (\$50{,}000)(P/F, i\%, 5) \\ &= -\$300{,}000 + (\$20{,}000)(P/F, i\%, 1) \\ &\quad + (\$46{,}800)(P/A, i\%, 5) \\ &\quad + (\$24{,}000)(P/A, i\%, 5) \\ &\quad + (\$50{,}000)(P/F, i\%, 5) \end{aligned}$$

By trial and error,

i	P
10%	$17,616
15%	–$20,412
12%	$ 1448
13%	–$ 6142
$12\frac{1}{4}$%	–$ 479

i is between 12 and 12¼%.

38. Assume loan payments are made at the end of each year. Find the annual payment.

$$\begin{aligned} \text{payment} &= (\$2{,}500{,}000)(A/P, 12\%, 25) \\ &= (\$2{,}500{,}000)(0.1275) \\ &= \$318{,}750 \\ \text{distributed profit} &= (0.15)(\$2{,}500{,}000) \\ &= \$375{,}000 \end{aligned}$$

After paying all expenses and distributing the 15% profit, the remainder should be zero.

$$\begin{aligned} 0 = \text{EUAC} &= \$20{,}000 + \$50{,}000 + \$200{,}000 \\ &\quad + \$375{,}000 + \$318{,}750 \\ &\quad - \text{annual receipts} \\ &\quad - (\$500{,}000)(A/F, 15\%, 25) \\ &= \$963{,}750 - \text{annual receipts} \\ &\quad - (\$500{,}000)(0.0047) \end{aligned}$$

This calculation assumes $i = 15\%$, which equals the desired return. However, this assumption only affects the salvage calculation, and since the number is so small, the analysis is not sensitive to the assumption.

$$\text{annual receipts} = \$961{,}400$$

The average daily receipts are

$$\frac{\$961{,}400}{365} = \$2634$$

Use the expected value approach. The average occupancy is

$$\begin{aligned} &(0.40)(0.65) + (0.30)(0.70) + (0.20)(0.75) \\ &\quad + (0.10)(0.80) = 0.70 \end{aligned}$$

The average number of rooms occupied each night is

$$(0.70)(120 \text{ rooms}) = 84 \text{ rooms}$$

The minimum required average daily rate per room is

$$\frac{\$2634}{84} = \boxed{\$31.36}$$

39.

$$\begin{aligned}\text{annual savings} &= \left(\frac{0.69 - 0.47}{1000}\right)(\$3{,}500{,}000) = \$770 \\ P &= -\$7500 + (\$770 - \$200 - \$100) \\ &\quad \times (P/A, i\%, 25) = 0 \\ (P/A, i\%, 25) &= 15.957\end{aligned}$$

Searching the tables and interpolating,

$$i \approx \boxed{3.75\%}$$

40. Work in millions of dollars.

$$\begin{aligned}P(\text{A}) &= -(\$1.3)(1 - 0.48)(P/A, 15\%, 6) \\ &= (\$1.3)(0.52)(3.7845) \\ &= -\$2.56 \quad \text{[millions]}\end{aligned}$$

Since this is an after-tax analysis and since the salvage value is mentioned, assume that the improvements can be depreciated.

Use straight line depreciation.

Evaluate alternative B.

$$\begin{aligned}D_j &= \frac{2}{6} = 0.333 \\ P(\text{B}) &= -\$2 - (0.20)(\$1.3)(1 - 0.48)(P/A, 15\%, 6) \\ &\quad - (\$0.15)(1 - 0.48)(P/A, 15\%, 6) \\ &\quad + (\$0.333)(0.48)(P/A, 15\%, 6) \\ &= -\$2 - (0.20)(\$1.3)(0.52)(3.7845) \\ &\quad - (\$0.15)(0.52)(3.7845) \\ &\quad + (\$0.333)(0.48)(3.7845) \\ &= -\$2.206 \quad \text{[millions]}\end{aligned}$$

Next, evaluate alternative C.

$$\begin{aligned}D_j &= \frac{1.2}{3} = 0.4 \\ P(\text{C}) &= -(\$1.2)\big(1 + (P/F, 15\%, 3)\big) \\ &\quad - (0.20)(\$1.3)(1 - 0.48) \\ &\quad \times \big((P/F, 15\%, 1) + (P/F, 15\%, 4)\big) \\ &\quad - (0.45)(\$1.3)(1 - 0.48) \\ &\quad \times \big((P/F, 15\%, 2) + (P/F, 15\%, 5)\big) \\ &\quad - (0.80)(\$1.3)(1 - 0.48) \\ &\quad \times \big((P/F, 15\%, 3) + (P/F, 15\%, 6)\big) \\ &\quad + (\$0.4)(0.48)(P/A, 15\%, 6) \\ &= -(\$1.2)(1.6575) \\ &\quad - (0.20)(\$1.3)(0.52)(0.8696 + 0.5718) \\ &\quad - (0.45)(\$1.3)(0.52)(0.7561 + 0.4972) \\ &\quad - (0.80)(\$1.3)(0.52)(0.6575 + 0.4323) \\ &\quad + (\$0.4)(0.48)(3.7845) \\ &= -\$2.436 \quad \text{[millions]}\end{aligned}$$

Alternative B is superior.

41. This is a replacement study. Since production capacity and efficiency are not a problem with the defender, the only question is when to bring in the challenger.

Since this is a before-tax problem, depreciation is not a factor, nor is book value.

The cost of keeping the defender one more year is

$$\begin{aligned}\text{EUAC(defender)} &= \$200{,}000 + (0.15)(\$400{,}000) \\ &= \$260{,}000\end{aligned}$$

For the challenger,

$$\begin{aligned}&\text{EUAC(challenger)} \\ &\quad = (\$800{,}000)(A/P, 15\%, 10) + \$40{,}000 \\ &\quad + (\$30{,}000)(A/G, 15\%, 10) \\ &\quad - (\$100{,}000)(A/F, 15\%, 10) \\ &\quad = (\$800{,}000)(0.1993) + \$40{,}000 \\ &\quad + (\$30{,}000)(3.3832) \\ &\quad - (\$100{,}000)(0.0493) \\ &\quad = \$296{,}006\end{aligned}$$

Since the defender is cheaper, keep it. The same analysis next year will give identical answers. Therefore, keep the defender for the next three years, at which time the decision to buy the challenger will be automatic.

Having determined that it is less expensive to keep the defender than to maintain the challenger for ten years, determine whether the challenger is less expensive if retired before ten years.

If retired in nine years,

$$\begin{aligned}\text{EUAC(challenger)} &= (\$800{,}000)(A/P, 15\%, 9) \\ &\quad + \$40{,}000 \\ &\quad + (\$30{,}000)(A/G, 15\%, 9) \\ &\quad - (\$150{,}000)(A/F, 15\%, 9) \\ &= (\$800{,}000)(0.2096) \\ &\quad + \$40{,}000 + (\$30{,}000)(3.0922) \\ &\quad - (\$150{,}000)(0.0596) \\ &= \$291{,}506\end{aligned}$$

Similar calculations yield the following results for all the retirement dates.

n	EUAC
10	$296,000
9	$291,506
8	$287,179
7	$283,214
6	$280,016
5	$278,419
4	$279,909
3	$288,013
2	$313,483
1	$360,000

Since none of these equivalent uniform annual costs is less than that of the defender, it is not economical to buy and keep the challenger for any length of time.

Keep the defender.

67 Engineering Law

PRACTICE PROBLEMS

There are no practice problems corresponding with Ch. 67 of the *Electrical Engineering Reference Manual.*

68 Engineering Ethics

PRACTICE PROBLEMS

There are no practice problems corresponding with Ch. 68 of the *Electrical Engineering Reference Manual.*

69 Electrical Engineering Frontiers

PRACTICE PROBLEMS

There are no practice problems corresponding with Ch. 69 of the *Electrical Engineering Reference Manual*.

70 Engineering Licensing in the United States

PRACTICE PROBLEMS

There are no practice problems corresponding with Ch. 70 of the *Electrical Engineering Reference Manual.*

Visit www.ppi2pass.com for more resources to help you pass the Electrical PE Exam!

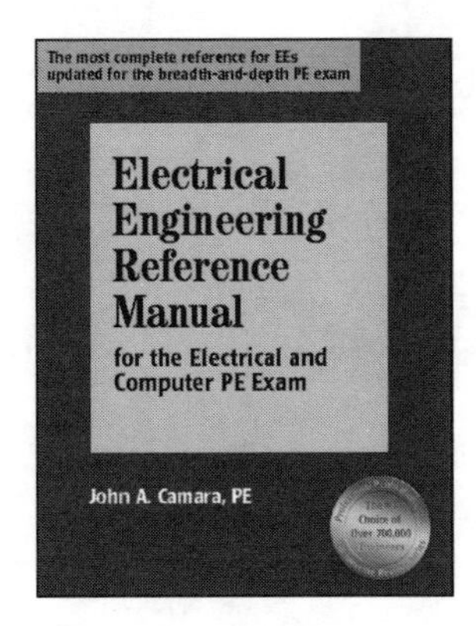

Electrical Engineering Reference Manual for the Electrical and Computer PE Exam
John A. Camara, PE

The **Electrical Engineering Reference Manual** is the most complete study guide available for engineers preparing for the electrical PE exam. It provides a clear, complete review of all exam topics, reinforcing key concepts with more than 370 solved example problems. The text is enhanced by hundreds of illustrations, tables, charts, and formulas, along with a detailed index. After you pass the PE exam, the **Reference Manual** will continue to serve you as a comprehensive desk reference throughout your professional career.

Six-Minute Solutions for Electrical and Computer PE Exam Problems
John A. Camara, PE

Each of the more than 100 practice problems in **Six-Minute Solutions** is designed to be solved in six minutes or less—the average time per problem you'll have during the exam. Working the problems here will give you an edge on exam day.

Electrical and Computer PE Sample Examination
John A. Camara, PE

Nothing can prepare you for the actual experience of the electrical PE exam better than taking a sample test. This book includes four full-length sample exams—one for the morning breadth module and one for each of the three afternoon depth modules. Use these to test your readiness and accustom yourself to the time pressures of this exam.

Quick Reference for the Electrical and Computer Engineering PE Exam
John A. Camara, PE

During the exam, it's vital to have fast access to the information you need to work problems. **Quick Reference** puts the important formulas at your fingertips, conveniently organized by subject.

For everything you need to pass the exams, go to
www.ppi2pass.com
where you'll find the latest exam news, test-taker advice, the Exam Forum, and secure online ordering.

Professional Publications, Inc.
1250 Fifth Avenue • Belmont, CA 94002
(800) 426-1178 • Fax (650) 592-4519

Source Code BOC

☞ Quick — *I need additional PPI study materials!*

Please send me the PPI exam review products checked below. I understand any book may be returned for a full refund within 30 days. I have provided my credit card number and authorize you to charge your current prices, plus shipping, to my card.

For the FE and FLS Exams
- ☐ FE Review Manual: Rapid Preparation for the General FE Exam
- ☐ Civil Discipline-Specific Review for the FE/EIT Exam
- ☐ Mechanical Discipline-Specific Review for the FE/EIT Exam
- ☐ Electrical Discipline-Specific Review for the FE/EIT Exam
- ☐ Engineer-In-Training Reference Manual
  - ☐ Solutions Manual, SI Units
- ☐ Land Surveyor Reference Manual

For the PE, SE, and PLS Exams
- ☐ Civil Engrg Reference Manual ☐ Practice Problems
- ☐ Mechanical Engrg Reference Manual ☐ Practice Problems
- ☐ Electrical Engrg Reference Manual ☐ Practice Problems
- ☐ Environmental Engrg Reference Manual
  - ☐ Practice Problems
- ☐ Chemical Engrg Reference Manual
  - ☐ Practice Problems
- ☐ Structural Engrg Reference Manual
- ☐ PLS Sample Exam

These are just a few of the products we offer for the FE, PE, SE, FLS, and PLS exams. For a full list, plus details on money-saving packages, call Customer Care at 800-426-1178 or visit www.ppi2pass.com.

For fastest service,
Call **800-426-1178** Fax **650-592-4519**
Web **www.ppi2pass.com**

Please allow two weeks for UPS Ground shipping.

NAME/COMPANY ____________________

STREET ____________________ SUITE/APT ________

CITY ____________________ STATE ________ ZIP ________

DAYTIME PH # ____________________ EMAIL ____________________

VISA/MC/DSCVR # ____________________ EXP. DATE ________

CARDHOLDER'S NAME ____________________

SIGNATURE ____________________

Email Updates Keep You on Top of Your Exam

To be fully prepared for your exam, you need the current information. Register for PPI's Email Updates to receive convenient updates relevant to the specific exam you are taking. These will include notices of exam changes, useful exam tips, errata postings, and new product announcements. There is no charge for this service, and you can cancel at any time.

Register at **www.ppi2pass.com/cgi-bin/signup.cgi**

Free Catalog of Tried-and-True Exam Products

Get a free PPI catalog with a comprehensive selection of the best FE, PE, SE, FLS, and PLS exam-review products available, user tested by more than 700,000 engineers and surveyors. Included are books, software, videos, calculator products, and all the NCEES sample-question books—plus money-saving packages.

Request a Catalog at **www.ppi2pass.com/catalogrequest.html**

Reporting an error . . .

BEFORE reporting this error, please take a moment to check the errata listing on the PPI website at **www.ppi2pass.com/Errata/Errata.html**. The error(s) you noticed may have already been identified.

Title of this book: ____________________ Edition: ________ Printing: ________

The error occurs on page ____________. Please describe the error and how you think it should be corrected:

Name ____________________

Address ____________________

City/State/Zip ____________________

Phone ____________________ Email ____________________

PROFESSIONAL PUBLICATIONS, INC.
800-426-1178 • Fax 650-592-4519
errata@ppi2pass.com

NO POSTAGE
NECESSARY
IF MAILED
IN THE
UNITED STATES

BUSINESS REPLY MAIL
FIRST CLASS MAIL PERMIT NO. 33 BELMONT, CA

POSTAGE WILL BE PAID BY ADDRESSEE

PROFESSIONAL PUBLICATIONS INC
1250 FIFTH AVE
BELMONT CA 94002-9979

NO POSTAGE
NECESSARY
IF MAILED
IN THE
UNITED STATES

BUSINESS REPLY MAIL
FIRST CLASS MAIL PERMIT NO. 33 BELMONT, CA

POSTAGE WILL BE PAID BY ADDRESSEE

PROFESSIONAL PUBLICATIONS INC
1250 FIFTH AVE
BELMONT CA 94002-9979